ORGANIC STRUCTURE DETERMINATION

PRENTICE-HALL INTERNATIONAL SERIES IN CHEMISTRY

PRENTICE-HALL, INC.
PRENTICE-HALL INTERNATIONAL, INC., UNITED KINGDOM AND EIRE
PRENTICE-HALL OF CANADA, LTD., CANADA

PRENTICE-HALL, INC., ENGLEWOOD CLIFFS, N.J.

ORGANIC STRUCTURE

DETERMINATION

Daniel J. Pasto

Department of Chemistry
University of Notre Dame

Carl R. Johnson

Department of Chemistry
Wayne State University

PRENTICE-HALL INTERNATIONAL, INC., London
PRENTICE-HALL OF AUSTRALIA, PTY. LTD., Sydney
PRENTICE-HALL OF CANADA, LTD., Toronto
PRENTICE-HALL OF INDIA PRIVATE LTD., New Delhi
PRENTICE-HALL OF JAPAN, INC., Tokyo

Current printing (last digit):

10 9 8 7 6 5 4 3 2 1

13-640854-0

Library of Congress Catalog Card Number 69-15046
Printed in the United States of America

Preface

The rapid development of physical techniques during the last fifteen to twenty years, both in instrumentation and interpretation, has revolutionized the approach to structure determination problems. The importance of this field of study requires a broad understanding of the various techniques available that will facilitate the research in structure.

The philosophy of the authors is that "Organic Structure Determination" should be taught as an introduction to research techniques. In reality, it is probably the first laboratory course in which the student is essentially "on his own." The student does not have a time-tested procedure to follow which will lead him down the correct path to the solution of the problem. After each experiment, the student must evaluate the information he has obtained and then choose what course of action he must follow next.

In an attempt at modernization and unification of the material presented, we have used several new approaches in preparing this text. Extensive discussions of solubility classification and factors affecting acidity and basicity as well as extensive lists of functional group tests and tables of compounds with their physical properties and derivatives have been drastically reduced, or in some cases completely eliminated. We believe that modern instrumental methods have eliminated the requirement for wide use of solubility classification tests. The background chemistry required for determining acidity and basicity effects, and the chemistry involved in many of the functional group classification tests, is assumed to have been acquired in the undergraduate organic chemistry course and will not be discussed in this text except in special cases. To provide tables of compounds and their derivatives that would adequately cover the thousands of compounds available today would be a prohibitive task and would require another separate volume. The student is referred to existing compilations or to the original literature (reviewed in Chapter 13).

An entirely new format has been selected for the presentation of the chemical techniques of structure determination. For each functional group a section is provided combining a statement comparing the utility of spectral versus chemical methods of functional group detection, carefully selected chemical functional group classification tests, selected procedures for the preparation of derivatives, and procedures for conversion of one functional group into a different functional group for further characterization. Special chapters are devoted to outlining general approaches to the obtaining of data and their interpretation and to finding the necessary information on published compounds in the literature. The final section is devoted to the solution of typical structure problems incorporating the techniques outlined in this text. Some of these problems have been kindly provided by other researchers in this area of study in order to provide a greater cross section of types of problems.

Although many schools may not have all the laboratory equipment required to derive the structural information, primarily the spectroscopic, used in this text, it is nevertheless incumbent on the student, and the instructor, to be familiar with the applications of these techniques. The lack of such equipment should not be a deterrent to the understanding of the use of these techniques. Large compilations of spectra have been published and are readily available for reference. The enterprising instructor can locate the necessary spectral information for his students. In their courses the authors have devoted most of the associated lecture time to the discussion of the theory and applications of the various spectroscopic techniques described in this text, culminating in extensive problem-solving sessions.

The authors wish to thank Profs. Orville Chapman and Christopher Foote for their excellent, constructive reviews. Daniel J. Pasto wishes to thank Profs. George Hennion, Gerhard Binsch, and Robert Hayes for reviewing individual sections of the manuscript. Carl Johnson expresses his appreciation to Hope College, Holland, Michigan, for their hospitality during the summer of 1967, during which time much of this work was written.

DANIEL J. PASTO

CARL R. JOHNSON

Contents

Part II   ADSORPTION SPECTROSCOPY

3   Ultraviolet Spectroscopy

4   Infrared Spectroscopy

Introduction

The isolation, purification, and determination of the structures of organic molecules are perhaps the most important tasks facing the organic chemist, whether he is a natural products chemist primarily concerned with structure problems, a synthetic chemist, or a physical organic chemist studying the mechanisms of reactions in which the structure of intermediates and final products must be determined. The rapid development of physical and spectroscopic techniques during the last 15 to 20 years has revolutionized the approach to structure determination problems. Prior to the introduction of these new techniques, the elucidation of the structures of complex molecules was very tedious and time consuming, and generally required considerable quantities of samples that were difficult to obtain. The introduction of these physical techniques to structure determination and the development of empirical spectra-structure correlations have dramatically reduced both the quantity of sample required and the time involved.

A number of steps are involved in the determination of the structure of an unknown substance. The first step involves the isolation, purification, and determination of the purity of the substance. The techniques that may be employed in the isolation and purification procedures depend greatly on the physical and chemical properties of the substance. Initial attempts should be carried out on small portions of the sample to avoid needless and costly waste of the unknown substance. Once a suitable procedure for the separation and purification is devised, the bulk of the sample may then be invested in the procedure. Determination of the purity of the sample is particularly important, since minor amounts of impurities may well lead to data not characteristic of the predominant substance present. The determination of the purity of samples has been greatly facilitated by the recent development of thin-layer and gas-chromatographic techniques.

Once the pure substance is obtained, it is then subjected to a careful examination of its physical and chemical properties. The physical properties, including spectroscopic, are recorded, and an elemental analysis, both qualitative and quantitative, is obtained. The spectral data are interpreted in terms of the types of functional groups present. This is followed by confirmation by chemical tests. On the basis of the physical and chemical properties of the substance, a tentative assignment of the structure, or possible structures, is made. Often the structure of an unknown substance is sufficiently complex that the data obtained on the original substance will not be sufficient to deduce a complete structure. It is then necessary to chemically transform the substance to a different compound which may be characterized, or to a number of compounds of lower molecular weight (degradation) which may be fitted back together much like the pieces of a jigsaw puzzle.

A search of the original literature is then undertaken to see whether the proposed substance has been reported previously in the literature. If the compound has been recorded earlier, a comparison of the properties of the known and unknown is made, including the comparison of the properties of a chemically transformed product (derivative). If the substance has not been reported previously, a full physical and chemical characterization of the substance, and generally one derivative if possible, is made.

In the following chapters of this text, the techniques of isolation, purification, determination of purity, and structure elucidation are discussed. No set procedure may be followed for any particular substance; the investigator must evaluate the data obtained and then decide on the most suitable course of action. Considerable experience is generally required to become proficient in this area.

Part I

PHYSICAL METHODS

OF SEPARATION,

PURIFICATION,

AND CHARACTERIZATION

The Introduction to this text outlined in very general terms the procedure usually followed in the determination of the structure of an unknown substance. Chapter 1 will describe the various techniques which may be applied to the separation, purification, and characterization of an unknown substance.

The procedure to be followed depends greatly on the physical state and general chemical properties of the substance. Regardless of the physical state of the sample, the investigator should immediately record the infrared spectrum of a homogeneous portion of the sample. There are two reasons for this. During the separation and purification procedures, components may be lost, owing to high volatility or high solubility in an aqueous phase with low recovery if extraction procedures are used, or to irreversible adsorption if chromatographic procedures are used. All the absorption peaks in the original sample spectrum must appear in the spectra of the isolated components. Secondly, should additional new peaks appear in the spectra of the isolated components not present in the original sample spectrum, chemical changes have probably occurred during the separation procedures, giving rise to new compounds (artifacts) not contained in the original sample. It should be immediately obvious that pitfalls may await the experimentalist who does not exercise a certain degree of caution.

Another procedure which should be carried out with the unknown substance is an analysis by chromatographic techniques, either thin-layer or gas-chromatographic. This type of analysis will indicate the number of individual components present and also the relative quantities present of these components. Even here, caution must be exercised in the interpretation of the results. Many types of compounds may not be separable with these techniques, or if gas-liquid chromatography is used, they may not be eluted from the column or may undergo decomposition giving rise to one or more peaks in the chromatogram. A comparison should be made of the gas-liquid chromatogram of each individual component isolated from the sample with the chromatogram of the original sample.

Solid samples should be inspected for homogeneity. The presence of two or more differently colored or shaped crystals indicates the probable presence of more than one compound. Occasionally it may be advantageous to separate a few crystals under a low-power magnifying glass until sufficient material, approximately 5 mg, is obtained for a melting-point determination and an infrared spectrum. If a mixture of solids is encountered, a small portion of the material should be subjected to an extraction procedure or to a chromatographic separation in an attempt to separate the mixture into individual fractions. After a successful separation is accomplished on the smaller trial portion, the remainder of the material may then be processed through the separation procedure. One should never invest the entire sample in an untried separation process. Occasionally fractional recrystallization may lead to separation of the components; however, such procedures often prove costly with respect to the amount of material lost in the mother liquors.

Heterogeneous samples containing a liquid and a solid phase can be partially separated by filtration; however, it must be remembered that the liquid phase may still contain considerable amounts of the solid phase, and vice versa. Extractive and chromatographic procedures may then be applied to the liquid and solid phases; but it is usually advisable to subject the entire sample to a single separation scheme.

Liquid samples may be subjected to careful fractional distillation, though certain precautions should be taken. Only a small portion of the sample should be subjected to distillation because some compounds may be thermally unstable and certain mixtures of compounds may undergo chemical interaction at higher temperatures, leading to decomposition or the formation of different compounds. The temperature should not be taken above 150° (all temperatures indicated in this text are in °C) unless trial tests with very small portions of the sample show suitable stability. The distillation may be continued at lower temperatures under reduced pressures (see Sec. 1.2 on distillation). If a fractional distillation does not appear to be successful, extractive or chromatographic procedures must then be employed.

More extensive discussions concerning the separation of mixtures will be presented later (see Secs. 1.5 and 1.6 on chromatography and the separation of mixtures).

Once the chemist has isolated a compound, he is then faced with the problem of ascertaining its purity. It is essential that the compound be of a high state of purity before the chemist proceeds with the physical and chemical characterization of the compound. The presence of impurities, even in relatively minor amounts, may result in misleading data (elemental and functional group tests) which will generally lead to confusion in the overall interpretation of the data.

The melting point, boiling point, density, or refractive index data are generally not good criteria of the purity of a sample; this is particularly true of the latter three when relatively small quantities of materials are involved. The most useful and sensitive method of analysis for both solids and liquids involves the use of chromatographic techniques, either gas-liquid chromatography or thin-layer chromatography. It is highly recommended that such an analysis be carried out before proceeding with chemical and physical characterization of the individual compounds. Once the purity of a component has been ascertained, one may then proceed to the elemental analysis and spectral and chemical functional group analyses.

1

Separation

and Purification

1.1   CRYSTALLIZATION

Crystallization is the deposition of crystals from a solution or melt of a given material. During the process of crystal formation, a molecule will prefer to become attached to a growing crystal composed of the same type of molecules because of a better fit in a crystal lattice for molecules of the same structure than for other molecules. If the crystallization process is allowed to occur under near-equilibrium conditions, the preference of molecules to deposit on surfaces composed of like molecules will lead to an increase in the purity of the crystalline material. Thus, the process of recrystallization is one of the most important methods available to the chemist for the purification of solids. Additional procedures may be incorporated in the recrystallization process to remove impurities. These include filtration to remove undissolved solids and adsorption to remove highly polar impurities.

Recrystallization depends on the differential solubility of a substance in a hot and a cold solvent. It is desirable that the solubility of the substance be high in the hot solvent and low in the cold solvent to facilitate the recovery of the starting material. The proper choice of solvent is critical and may require trial tests with small quantities of the material in a variety of solvents or solvent pairs (combination of two solvents). The general procedure will be reviewed briefly.

1.1.1a Procedure

The solvent, or solvent pair, to be used in the recrystallization of a substance is chosen in the following manner. A few milligrams of the substance are placed in a small test tube and a few drops of solvent are added. In general, one should use the least polar solvent first, for example, hexane or petroleum ether, progressing to the more polar solvents, for example, alcohols, water, and acetic acid. The solvents used in recrystallizations should be relatively low boiling so that the solvent adhering to the crystals can be readily removed by evaporation. Should the sample completely dissolve, chill the solution to see whether crystals will form. If no crystals appear, the material is too soluble in that solvent, and that solvent should not be used for the recrystallization. Other solvents should be tested until a solvent is found in which the sample does not completely dissolve at room temperature but does undergo solution on heating. If crystals reappear on cooling, the solvent is suitable for use. If no single solvent provides suitable results, a mixture of two solvents may be employed, one of the solvents being a good solvent for the sample, and the other being a poor solvent for the sample. The correct proportion of the two solvents must be determined by trial and error.

Once the proper solvent has been chosen, the remainder of the sample is recrystallized.

The material to be recrystallized is placed in a suitable container, a beaker, an Erlenmeyer flask, or a centrifuge tube (depending on the amount of material to be recrystallized), and solvent is added slowly, maintaining a gentle reflux of the solvent in the container, until no more sample dissolves in the solution. Occasionally, highly insoluble materials may be present in the sample, which, regardless of the amount of solvent added, will not undergo solution. At this time, it is best to add a small amount of an adsorbent, usually activated charcoal, to adsorb any highly polar materials present in the sample. The addition of the activated charcoal must be carried out with great care, for if the saturated solution has become superheated, the addition of the charcoal granules will induce violent boiling with the possible loss of material. After the addition of the adsorbent, the mixture is boiled gently for about 30 sec, with rapid stirring to avoid bumping, and then the solids are allowed to settle to the bottom of the container. Any change in the appearance of the supernatant liquid is noted, and an additional amount of the adsorbent is added. The process is repeated until no further change is noted in the appearance of the solution. (Caution must be exercised when dealing with highly polar or colored compounds in order to avoid a great excess of the adsorbent. The use of a great excess of the adsorbent may cause considerable loss of material.) The hot, saturated solution is then filtered through a hot, solvent-saturated filter which is kept warm by refluxing solvent to prevent the premature formation of crystals (see Fig. 1.1). The filter paper is washed with a small portion of hot solvent. The volume of the solvent in the collection flask is then reduced by boiling until a saturated solution is again attained. Occasionally, it is necessary to use a filter aid, for example, Celite, to completely remove the small granules of carbon. Suction filtration techniques, using a Buchner funnel and suction flask, are generally employed when such a filter aid is used.

The rate of crystal growth from the saturated solution is critical. Too rapid a precipitation rate does not allow equilibrium conditions to exist, and the very slow precipitation with the formation of large crystals often leads to extensive solvent inclusion in the crystals. The hot, saturated solution should be allowed to cool at a rate between the two extremes. This can generally be accomplished by allowing the solution to cool on the bench top, followed by chilling in an ice bath. Occasionally, with low boiling solvents such as ether or pentane, it may be desirable to chill the sample in a Dry Ice–acetone bath to induce crystal growth. In such cases the collection funnel should be chilled prior to filtration by pouring a portion of chilled solvent through the filter funnel.

Some materials tend to "oil out" as a liquid instead of forming crystals. This usually occurs when the melting point of the material, or of the liquid phase formed, is below the temperature at which a saturated solution is attained. Additional solvent should be added, maintaining a clear solution, until crystals finally begin to form. If crystals are reluctant to form, scratching the side of the container with a glass stirring

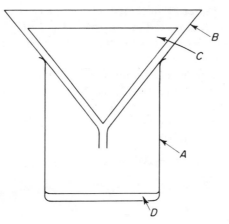

Fig. 1.1. Filtration setup for filtering a hot saturated solution during a recrystallization process: (*A*) beaker, (*B*) glass funnel, (*C*) filter paper, (*D*) refluxing solvent. The entire setup is heated on a steam bath or a hot plate (never over an open flame!).

rod or adding a seed crystal will generally induce crystallization. The crystals are then collected by suction filtration and washed with a small portion of cold solvent to remove any adhering mother liquor. When using filter paper to collect the sample, care should be taken in removing the sample from the paper so as not to dislodge any fibers of the paper which may contaminate the sample. These fibers may interfere with subsequent elemental or spectral analysis.

The recrystallization of milligram quantities of materials may best be carried out in small test tubes or centrifuge tubes. Centrifuge tubes are the most convenient in that the crystalline mass can be forced to the bottom of the tube and the mother liquors removed by a capillary pipette (see Fig. 1.2). The crystalline mass may then be washed with several small quantities of cold solvent and then removed from the tube to be dried.

The recrystallization of highly insoluble compounds may be accomplished by use of a Soxhlet extractor (see Sec. 1.4 on extraction), in which the material is placed in the extraction shell and the purified material is recovered in the solvent distillation flask.

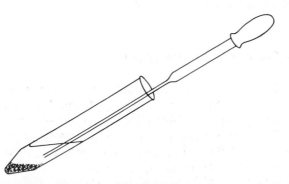

Fig. 1.2. Removing the solvent from a recrystallization with a capillary dropper in a centrifuge tube.

The crystalline material obtained in the procedure outlined above is thoroughly dried to remove adhering solvent. Many compounds can be dried by allowing the sample to set in the open air (care must be exercised to protect the sample from contamination). Hygroscopic compounds or compounds recrystallized from high-boiling solvents must be dried in a vacuum drying apparatus (e.g., a vacuum dessicator). The entire recrystallization procedure should be repeated until a constant melting point of relatively narrow range is obtained (see Sec. 2.2 on melting points for the details of melting-point determinations).

1.2   DISTILLATION

Distillation may be defined as the partial vaporization of a liquid with transport of these vapors and their subsequent condensation in a different portion of the distillation apparatus. Distillation is one of the most useful methods for the separation and purification of liquids. The successful application of distillation techniques depends on several factors. These include the difference in vapor pressure (related to the difference in the boiling points) of the components present, the size of the sample and the distillation apparatus (as well as the type of apparatus employed), the occurrence of codistillation or azeotrope formation, and the care exercised by the experimentalist. Because distillation relies on the fact that the vapor above a liquid mixture is richer in the more volatile component, the composition being controlled by Raoult's and Dalton's Laws, a simple distillation will never lead to the complete separation of two volatile substances. The use of efficient fractionating columns may produce more effective separations when sufficient quantities of the sample are available. When the sample size is too small to carry out an effective distillation, the use of chromatographic techniques is recommended, particularly preparative gas-liquid chromatography. Various types of distillation techniques will be described in the following sections.

1.2.1 Simple (Nonfractional) Distillation

If a sample is known to contain essentially only one volatile component, a simple, nonfractional distillation may be used to effect a suitable purification of the component. A suitable apparatus for such a distillation is illustrated in Fig. 1.3 and employs standard taper 14/20 precision-ground glassware (the correct handling of precision-ground glassware will be discussed in Sec. 1.3.1a). The apparatus shown in Fig. 1.3 may be used for atmospheric or reduced pressure (vacuum) distillations and is suitable for sample sizes down to 0.5 g.

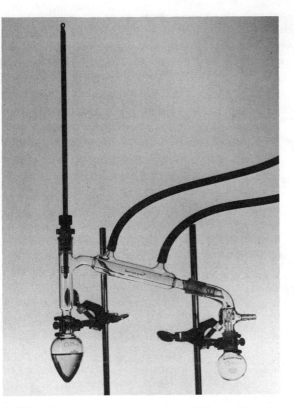

Fig. 1.3. Simple distillation apparatus employing 14/20 precision-ground glassware. (Photograph by courtesy of the Kontes Glass Company, Vineland, N. J.; photograph by Bruce Harlan, Notre Dame, Ind.)

As indicated earlier, a small portion of the sample should be subjected to a micro boiling point determination (see Sec. 2.4) to determine the approximate boiling point of the sample and to determine whether the sample is stable under the conditions of the distillation. In general, if the boiling point of the sample is above 180°, a reduced pressure distillation should be employed. This may be accomplished by the use of a water aspirator capable of producing pressures down to 12 to 15 mm-Hg, or a vacuum pump capable of producing pressures down to 0.01 mm-Hg. Figure 1.4 shows the recommended setup for a vacuum distillation, including a vacuum gauge for determining the pressure at which the distillation is being carried out, a cold trap (to be used only when a vacuum pump is employed) to prevent vapors being sent into the vacuum pump, and an air bleed to introduce air into the system to bring the internal pressure back to atmospheric pressure before disassembling the apparatus. The vacuum should never be broken by the removal of a flask or the thermometer. A safety shield should be routinely used to shield the experimentalist carrying out a vacuum distillation. The

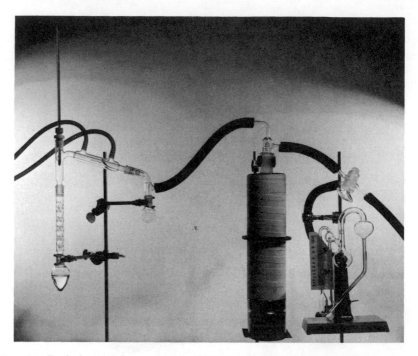

Fig. 1.4. Typical setup for a vacuum distillation including a fractionating column, cold trap, and vacuum gauge. (Photograph by courtesy of the Kontes Glass Company, Vineland, N. J.; photograph by Bruce Harlan, Notre Dame, Ind.)

comparison of boiling points obtained at different pressures is readily accomplished by use of a boiling point nomograph (see Appendix I).

The sample is placed in the distilling flask; a boiling chip or a fine capillary, through which is passed a slow stream of nitrogen or helium, is then added to promote a smooth and continuous boiling action. In many cases, boiling chips or sticks do not provide sufficient agitation of the sample to promote a smooth and continuous boiling action. In such cases it is necessary to provide the necessary agitation by means of a slow passage of nitrogen or helium through a capillary into the liquid. A magnetic stirrer may also be used. The distillation flask should be heated by means of an oil or wax bath, never with a heating mantle or the open flame of a Bunsen burner. Adequate control of the temperature of the distillation flask generally cannot be maintained with the latter two methods of heating.

1.2.2 Fractional Distillation

Fractional distillation is employed when the separation of two or more volatile components is required. The principle of fractional distillation is based on the establishment of a large number of theoretical vaporization-condensation cycles. A fractionating column is used which allows equilibration of the descending condensed liquid with the ascending vapors, thus producing the effect of a multiple vaporization-condensation cycle.

The length and type of fractionating column required depends on the boiling points of the components to be separated. Suitable separations of components differing in boiling points by 15 to 20° can be accomplished by

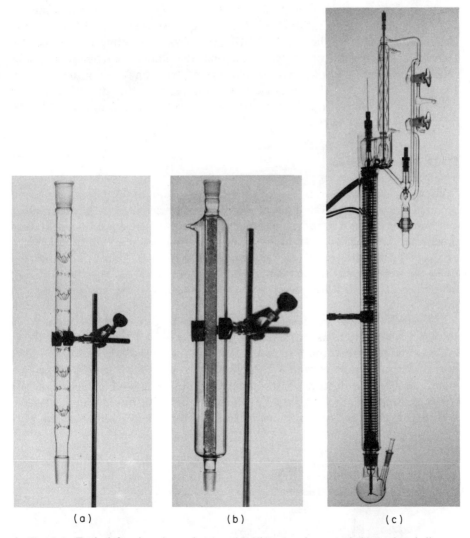

(a)   (b)   (c)

Fig. 1.5. Typical fractionating columns. (a) Vigreux column and (b) a glass helices packed, vacuum-jacketed fractionating column. (Photographs by Bruce Harlan, Notre Dame, Ind.) (c) A spinning band column. (Reproduced by courtesy of the Nester/Faust Manufacturing Corporation, Newark, Del.; photograph by Willard Stewart, Wilmington, Del.)

means of Vigreux columns (see Fig. 1.5a). For separations of components with closer boiling points, packed columns (see Fig. 1.5b) or spinning-band-type columns (see Fig. 1.5c) may be used.

Equilibrium conditions must be maintained in the fractionating column at all times in order to achieve a successful separation. The ratio of distillate to the amount of condensed material returning to the distillation flask (referred to as the reflux ratio) should always be much larger than 1, generally in the region of 5 to 10 for relatively easily separated components. Maintaining the reflux ratio in this region requires a very careful control of the amount of heat applied to the distillation flask. Flooding of the column should be avoided at all times. Fractional distillations carried out under reduced pressures should involve the use of fraction cutting devices, as shown in Fig. 1.5c. These devices do not require the breaking of the vacuum which would in turn destroy the equilibrium conditions established on the fractionating column.

1.2.3 Microdistillation

Frequently the chemist is faced with the problem of distilling milligram quantities of a liquid. Small quantities of liquids recovered from preparative gas-liquid or column chromatographic separations require a final distillation to remove solvents or high-boiling residues (such as the liquid stationary phases used in gas-liquid chromatography which are continually eluted from the column) prior to elemental analysis or recording the physical properties of the sample. Such microdistillations can be accomplished in several ways.

One method may be described as a simple trap-to-trap distillation. A simple double-U tube is prepared as shown in Fig. 1.6a. The sample is introduced into one of the U sections, and that end of the tube is sealed. The portion of the U tube containing the sample is cooled by means of an ice-water or Dry Ice–acetone bath, depending on the boiling point of the sample, and a vacuum is applied at the other end. In this manner, volatile solvents may be effectively removed. The cold trap is removed from the por-

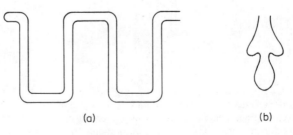

(a)   (b)

Fig. 1.6. Microdistillation tubes.

tion of the double-U tube containing the sample, and the other empty portion of the U tube (the receiver) is placed in the cold bath. Heat is then applied to the portion of the U tube containing the sample, and the sample condenses in the U tube immersed in the cold bath, leaving behind the nonvolatile impurities.

Microdistillations employing small tubes as illustrated in Fig. 1.6b can also be accomplished. The sample is placed in the bottom of the tube, and upon application of heat, the material vaporizes and condenses on the upper portion of the apparatus and drains down into the extended ring receiver portion of the tube. This apparatus suffers from the lack of an efficient condenser, difficulty of operation under reduced pressures, and difficulty of tube preparation by the student.

A more sophisticated apparatus is illustrated in Fig. 1.7. The sample is placed in the inner sample space; it is vaporized, condensed on the cold finger, and collected in the small flask at the bottom of the apparatus. The collection flask may also be immersed in a cold bath. The value of this apparatus lies in its reusability and greater control of the heating source and pressure. Sample sizes from 10 to 200 mg can be conveniently distilled in this apparatus. Distillation in this apparatus is essentially a molecular

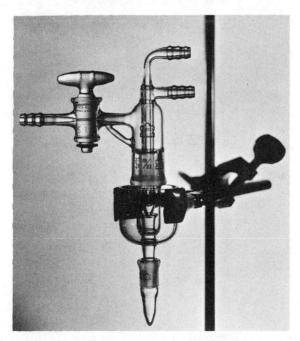

Fig. 1.7. Micromolecular distillation apparatus. (Marketed by, and photograph by courtesy of the Kontes Glass Company, Vineland, N. J.; photograph by Bruce Harlan, Notre Dame, Ind.)

distillation, and very little fractionation can be accomplished, as is true with all microdistillations.

Several other techniques and types of micro stills are available. See Sec. 1.7, References.

1.2.4 Steam Distillation

The attempted distillation of a two-phase immiscible system results in the partial distillation of both phases, the amount of each phase distilled being dependent on their relative volatilities. If water is employed as one of the phases, this type of distillation is called steam distillation. Steam distillations are very useful for separating volatile components from relatively non-volatile components, particularly when the more volatile component possesses a very high boiling point and may be subject to decomposition if a direct distillation is attempted. This technique is also useful if the presence of other lesser volatile impurities causes extensive destruction of the desired fraction under normal distillation conditions. Steam distillation may often

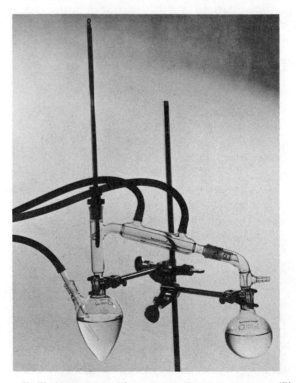

Fig. 1.8. Steam distillation setup with an external steam source. (Photograph by courtesy of the Kontes Glass Company, Vineland, N. J.; photograph by Bruce Harlan, Notre Dame, Ind.)

be used to separate isomeric compounds, particularly when one isomer is capable of extensive intramolecular hydrogen bonding; hence, that isomer is made more volatile, whereas the other isomers may participate only in intermolecular hydrogen bonding.

A typical steam distillation setup is illustrated in Fig. 1.8. Small quantities of materials may be steam distilled by the direct distillation of a mixture of the sample and water, deleting the external steam source and water entrainment trap. The organic material is then recovered from the distillate by extraction with an organic solvent.

1.3 SUBLIMATION

Many materials when heated pass directly into the vapor phase and, on cooling pass directly back to the solid phase. Such a process is termed *sublimation* and, as such, provides a useful method for the purification of materials. Unfortunately however, relatively few organic substances undergo sublimation at atmospheric pressure. Generally, sublimations are carried out under reduced pressures. Types of compounds that readily sublime include many α-aminoacids, ketones, carboxylic acids, and most anhydrides and quinones.

A typical vacuum sublimation apparatus is shown in Fig. 1.9. The finely divided sample is placed in the bottom of the cup, and the cold-finger condenser is then inserted into the container cup. A vacuum is applied, and the temperature of the sample is slowly increased until sublimation occurs; however, the temperature of the sample *must* be kept well below the melting point of the sample. The flow of coolant to the condenser is shut off, the condenser is removed, and the sample is scraped from the cold finger. Repeated sublimations may be required to obtain suitably pure material. If the apparatus such as shown in Fig. 1.9 is not available, the sample may be placed in a side-arm test tube and a small test tube inserted to act as the condenser. The inner tube may be filled with ice or constructed as a condenser.

1.3.1a Handling and care of precision-ground glass apparatus

The glassware illustrated in the figures of this text is precision-ground, standard-taper (**$**) glassware. The use of precision-ground glassware eliminates the necessity of using corks or rubber stoppers for connecting the various pieces of the apparatus. This reduces the possibility of contamination of the sample by contact with the cork or rubber stoppers. It also provides better seals for reduced-pressure distillations. Although the use of ground glassware provides many advantages, it has the disadvantage of higher initial costs. Therefore certain precautions should be exercised in the use and care of the apparatus.

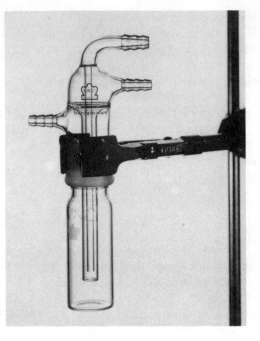

Fig. 1.9. Vacuum sublimation apparatus. (Photograph by courtesy of the Kontes Glass Company, Vineland, N. J.; photograph by Bruce Harlan, Notre Dame, Ind.)

The size of standard-taper joints is indicated by a fraction, for example, 14/20, in which the numerator is the diameter of the widest part of the joint which tapers 1 mm for every 10 mm of length, and the denominator is the length in millimeters of the ground surface. All joints of the same number are interchangeable. Non-$\mathbf{S}$ joints should not be paired with $\mathbf{S}$ joints.

To prevent sticking or freezing of the ground joints, a lubricant is applied before the joints are joined. A small amount of stopcock lubricant is applied around the central part of the inner portion of the joint; the joint is fitted together, and the lubricant is distributed throughout the joint by a gentle twisting action (do not force or apply great pressure when pairing joints). An excess of the lubricant must be avoided because excess lubricant may flow inside the apparatus and contaminate the sample. This is particularly critical in distillation receivers in which the distillate must flow over a lubricated joint.

The choice of the lubricant depends on the several factors: the temperature of operation, the vacuum required, and the chemical reactivity of the system. Three general categories of lubricant available are the hydrocarbon greases, silicone greases, and Kel-F greases. Several types of hydrocarbon greases are generally available. Apiezon greases L, M, and N possess very low vapor pressures and are recommended for high-vacuum systems. However, their maximum operating temperatures are 30°. Apiezon grease T may be used up to 110°. The hydrocarbon greases are easily removed by washing with organic solvents and do not tend to undergo extensive oxidation or hydrolysis reactions as do the silicone greases. Dow Corning silicone grease possesses a low vapor pressure and may be used over a wide temperature range (−40° to 150°). The only disadvantage of using a silicone grease is the problem of removing the silicone grease after use (see the following paragraphs). Kel-F greases are recommended when dealing with very reactive chemical systems, for example, fluorine.

The ground glassware is assembled as desired and supported by clamps from a ring stand or rack. Care must be exercised to ensure that the glassware is not placed under strain when securely clamped in position and that all joints are closed. The apparatus should be disassembled as soon as possible after use and the lubricant removed from the ground-glass surfaces. If difficulty is encountered in the separation of a ground-glass joint, carefully tap the top edge of the outer portion of the joint with a solid glass stopper or the wooden handle of a spatula. If this is not successful, carefully apply heat to the outer portion of the joint with a smoky or yellow flame of a microburner and then gently tap the edge of the joint. The lubricant is readily removed if the ground-glass areas are wiped with a tissue and then washed with ether, benzene, or chloroform (caution must be exercised to avoid excessive contact of the skin with benzene or chloroform). If the lubricant is not immediately removed from the joint or from the inside parts of the apparatus, it may undergo oxidation and hydrolysis, resulting in a cloudy deposit on the surface of the glass. These deposits may be removed if the piece of glassware is allowed to stand for a short period of time in ethanolic potassium hydroxide to remove silicone grease, or in a dichromate-sulfuric acid bath to remove a hydrocarbon grease. The glassware should not be allowed to remain for long periods of time in the ethanolic potassium hydroxide because the strong base will slowly attack the ground-glass surfaces.

All organic residues should be washed from the glassware immediately after use. If warm soapy water does not suffice, allow acetone to stand over the residue. Quite often, it may be necessary to allow the piece of glassware to stand in a dichromate-sulfuric acid cleaning bath for a short period of time (30 min).

1.4   EXTRACTION

The distribution of a substance between two immiscible phases is the basis of extraction. The distribution equilibrium constant K is defined in Eq. (1.1)

$$K = \frac{[C_A]}{[C_B]} \tag{1.1}$$

where $[C_A]$ is the equilibrium concentration of a substance in phase A and $[C_B]$ is the equilibrium concentration in phase B. The distribution co-efficient may be calculated using the limiting equilibrium solubilities of the material in the two phases. In actuality, the equilibrium constant expression may be more complex than is illustrated in Eq. (1.1) in that the substance may exist in dimeric or higher polymeric forms in one or both of the phases, necessitating the introduction of coefficients and exponents in Eq. (1.1) to adequately describe the concentration dependences in the two phases.

In extraction procedures normally employed in organic chemistry, one phase is aqueous and the other phase a suitable organic solvent. A change in solvent will change the solubility in that phase and thus alter the distribution coefficient. In general, the rule of "like dissolves like" is a suitable guideline in choosing the appropriate solvent.

Various materials may be added to either phase, to either increase or decrease the solubility of a substance in that phase. The solubility of a material in one phase may be increased by the addition of a reagent which is capable of forming a stable, soluble complex in that phase. For example, many reactive olefins and polyolefins may be extracted into an aqueous phase in the presence of silver ion, or chelating agents may be extracted either into or from an aqueous phase in the presence of a complexing metal ion. These complexes may be subsequently degraded, and the desired material recovered.

The solubility of a substance may be decreased, particularly in aqueous phases, by the addition of neutral salts which reduce the solubility of the substance in that phase. Various salts may be used, sodium chloride probably being the most widely used. Other salts that may be similarly used are sodium sulfate, potassium carbonate, or calcium chloride.

The distribution coefficient for acidic or basic materials can be greatly altered by changing the pH of one phase. For acidic materials, one may show that the distribution coefficient between two phases, one of which is the aqueous phase with a hydrogen ion concentration of $[H^+]$, may be repre-

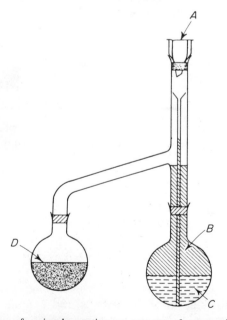

Fig. 1.10a. Diagram of a simple continuous extractor for use with lighter-than-water extracting phase: (*A*) condensor, (*B*) lighter-than-water extracting phase, (*C*) water phase being continuously extracted, and (*D*) distillation reservoir. The solvent is vaporized from *D*, condensed in the condenser *A* and is conducted to the bottom of the extracting flask by the small inner tube. The organic phase percolates up through the water phase and returns to the distillation flask where the extracted material concentrates.

sented by an effective distribution coefficient K_{eff} [see Eq. (1.2)] where K_D is the distribution coefficient between an organic phase and pure water, and K_A is the ionization constant of the acid.

$$K_{eff} = \frac{K_D K_A}{[H^+]} \qquad (1.2)$$

This approach is the basis for the facile separation of strong acids from weak acids by extraction with aqueous bicarbonate. After extraction of the acid or base from an organic phase, the compound may be recovered from the aqueous phase by acidification or basification.

In many situations, however, it is not possible to attain a favorable distribution coefficient by any of the foregoing procedures. When an unfavorable distribution coefficient is encountered, it may be shown by Eq. (1.3) that several extractions with small volumes of the extracting phase will be more efficient than one extraction utilizing all of the extracting phase.

$$[C_A] = [C_A^0]\left(\frac{V_A}{K V_B + V_A}\right)^n \qquad (1.3)$$

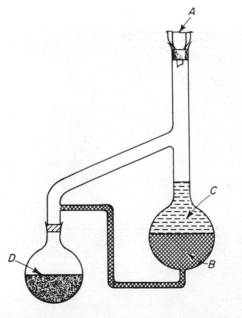

Fig. 1.10b. Diagram of a continuous extractor for use with heavier-than-water extracting phases: (A) condenser, (B) heavier-than-water extracting phase, (C) water phase being extracted, and (D) distillation reservoir. The extracting solvent is vaporized from the distillation flask, condensed in the condenser and drips down through the phase being extracted. The extracting phase returns to the distillation flask by the small return tube and the material extracted concentrates in the distillation flask.

$[C_A]$ and $[C_A^o]$ are the final and original concentrations of the material in phase A, K is the distribution coefficient, V_A is the volume of phase A, V_B is the volume of the extracting phase used in each extraction step, and n is the number of extraction steps.

Continuous liquid-liquid extractions may be required when particularly unfavorable distribution coefficients are encountered. Continuous liquid-liquid extractors suitable for use with lighter- and heavier-than-water organic phases are shown in Figs. 1.10a and 1.10b.

Another type of liquid-liquid extraction is termed *countercurrent distribution*. Countercurrent distribution involves two immiscible phases which are in contact with each other as they flow in opposite directions. The apparatus shown in Fig. 1.11 is typical of the more elaborate countercurrent distribution apparatus. The extracting phase is introduced at one end of the apparatus and encounters the phase to be extracted progressing from the other end. The extracting phase, and similarly the extracted phase, do not collect in a single receiver, and therefore a true fractionation process occurs. In all the continuous liquid-liquid extraction systems mentioned, the fresh extracting phase is continuously in contact with the other phase.

Continuous liquid-solid extractions may be carried out by the use of a

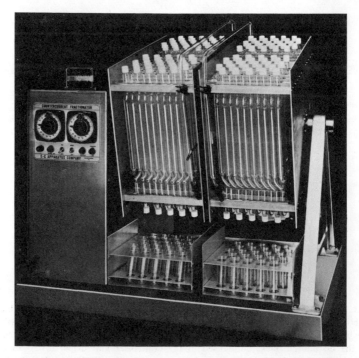

Fig. 1.11. Countercurrent distribution apparatus. (Reproduced by courtesy of the E-C Apparatus Company, Philadelphia, Pennsylvania.)

Soxhlet extractor (see Fig. 1.12). The finely ground solid is placed in the extraction shell, and fresh solvent is allowed to percolate through the sample, the extracted material accumulating in the solvent distillation flask.

In any extraction, both phases become saturated with respect to the other phase. When extracting an aqueous phase with an organic phase, it is necessary to remove the dissolved water before recovering the extracted material by evaporation of the solvent. Generally, most of the dissolved water in the organic phase may be removed by washing the organic phase with a saturated aqueous solution of sodium chloride. Final drying of the organic phase is accomplished by allowing the organic phase to stand for a

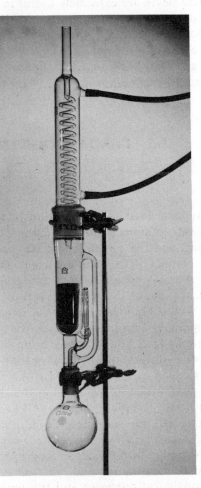

Fig. 1.12. Soxhlet extractor for continuous liquid-solid extractions. (Photograph by courtesy of the Kontes Glass Company, Vineland, N.J.; photograph by Bruce Harlan, Notre Dame, Ind.)

period of time over an anhydrous inorganic salt. Salts commonly used for this purpose include sodium sulfate (neutral), magnesium sulfate (acidic), and potassium carbonate (basic). The choice of the proper drying agent depends on the type of substrate contained in the solution. For example, a basic drying agent should be used for drying amine solutions and vice versa for acids. Magnesium sulfate is the most effective drying agent, but it will adsorb quantities of very polar compounds.

Molecular sieves (see Sec. 1.5.6 on molecular sieves) may also be used as drying agents. Molecular sieves adsorb molecules on the basis of their size. The molecular sieve must be chosen so that only molecules the size of water, or smaller, will be removed and that the material in the solution will not be removed. These sieves are extremely useful in the drying of solvents for reactions demanding anhydrous conditions. Generally, the sieve is placed in a chromatography column, and the solution is allowed to pass slowly through the column. Molecular sieves may occasionally induce chemical reactions, particularly with sensitive substrates which may undergo dehydration or condensation reactions.

1.5   CHROMATOGRAPHIC SEPARATIONS

1.5.1 Introduction

Chromatography may be defined as a separation of molecular mixtures by distribution between two or more phases, one phase being essentially two-dimensional (a surface) and the remaining phase, or phases, being a bulk phase brought into contact in a countercurrent fashion with the two-dimensional phase. Various types of chromatography are possible, depending on the physical states of the phases involved. The use of a solid as the stationary phase with a liquid mobile phase is generally referred to as adsorption chromatography. Employing a gas as the mobile phase is termed gas-solid chromatography (gsc). Separations using adsorption or gas-solid chromatography involve surface adsorption phenomena. The use of a stationary liquid phase, supported on the surface of a solid stationary phase, with a mobile liquid phase which is insoluble in the stationary liquid phase is usually termed partition chromatography. Utilization of a gas as the mobile phase in contact with the stationary liquid phase is called gas-liquid chromatography (glc), and occasionally termed vapor-phase chromatography (vpc). Partition and gas-liquid chromatographic separations involve solubility phenomena. Gas-liquid and gas-solid chromatography are extremely useful for analytical purposes and the tentative identification of compounds. Under appropriate conditions, all forms of chromatography can be used for preparative scale separations.

1.5.2 *Adsorption Chromatography*

The separation of the components of a mixture by adsorption chromatography depends on adsorption-desorption equilibria between compounds adsorbed on the surface of the solid stationary phase and in the moving liquid phase. The extent of adsorption of a single component depends on the polarity of the molecule, the activity of the adsorbent, and the polarity of the mobile liquid phase. The actual separation of the components in a mixture is dependent on the relative values of the adsorption-desorption equilibrium constants for each of the components in the mixture.

In general, the more polar a functional group in the compound is, the more strongly it will be adsorbed on the surface of the solid phase. Table 1.1 lists typical classes of organic compounds in the order of elution (increasing polarity). Minor alterations in this series may occur if a functional group is highly sterically protected by other less polar portions of the molecule.

Table 1.1. Compound Elution Sequence

Hydrocarbons	
Olefins	
Ethers	
Halogen compounds	General order of
Aromatics	elution
Ketones	
Aldehydes	
Esters	
Alcohols, amines, mercaptans	
Acids and strong bases	

The activity of the adsorbent (adsorptive power) depends on the type of material and on the mode of its preparation. Table 1.2 lists several of the more common adsorbents in order of increasing activity.

Table 1.2. Adsorbents for Adsorption Chromatography

Cellulose	
Starch	
Sugars	
Magnesium silicate	General order of
Calcium sulfate	increasing activity
Silicic acid	
Florisil	
Magnesium oxide (magnesia)	
Aluminum oxide (alumina)	
Activated charcoal	

The choice of the proper adsorbent will depend on the types of compounds to be chromatographed. Cellulose, starch, and sugars are primarily used with very labile, polyfunctional plant and animal products. Magnesium silicate has been found to be suitable for the separation of acetylated sugars, steroids, and essential oils. Silicic acid is a relatively mild adsorbent useful for the separation of esters and acids. Florisil is quite similar in properties to silicic acid and may be used for the separation of most functional classes including carboxylic acid derivatives and amines.

Alumina, the most widely used adsorbent, can be obtained in three forms: acidic, basic, and neutral. Acidic alumina is an acid-washed alumina giving a suspension in water with a pH of approximately 4. This alumina is useful for the separation of acidic materials such as acidic amino acids and carboxylic acids. Basic alumina (pH of approximately 10) is useful for the separation of basic materials such as amines. Neutral alumina (pH of approximately 7) is useful for the separation of nonacidic and nonbasic materials.

In addition to varying the acid-base properties of the adsorbent, the activity of alumina can be varied by controlling the moisture content of the sample. For example, alumina can be prepared as activity grades I through V (Brockman scale) in which the activity decreases as the activity grade number increases. In general, the more active grades of alumina should be used when the separation of the components of a mixture is difficult. Detailed procedures are available for the preparation and standardization of the various types and grades of alumina in ref. 8, pp. 24–26, in Sec. 1.7. Acidic, basic, and neutral activity I aluminas are commercially available[1] and may be converted to the other activity grades by the addition of 3% (activity II), 6% (III), 10% (IV), and 15% (V) by weight of water.

The use of the more active adsorbents, particularly alumina, may lead to the destruction of certain types of compounds during chromatography. Owing to the presence of water in the lower-activity grades of alumina, many esters, lactones, and acid halides may undergo hydrolysis, the resulting carboxylic acid being very strongly adsorbed. The low molecular weight aldehydes and ketones undergo extensive aldol and ketol condensations on the surface of the more active adsorbents. The correct choice of adsorbent will be contingent on some knowledge of the types of compounds to be separated. Even then, a small trial chromatographic separation on a small portion of the sample is recommended to elucidate the proper conditions for separation of the remainder of the sample.

In addition to controlling the pH of the adsorbent, the properties of the adsorbent may be altered by coating the adsorbent with various chemicals. For example, silver nitrate coated alumina, prepared by dissolving 20% by weight silver nitrate in aqueous methanol and slurrying with the alumina followed by removal of the solvent, is particularly useful for the separation of olefins. A stationary liquid phase, for example, water, ethylene glycol,

---

[1] M-Woelm-Eschwege, distributed by Alupharm Chemicals, 616 Commercial Place, P.O. Box 30628, New Orleans, Louisiana, and from Bio-Rad Laboratories, 32nd and Griffin Avenue, Richmond, California.

or a low molecular weight carboxylic acid, may be applied to some adsorbents. Separations employing such preparations are termed liquid-liquid partition chromatographic separations.

The recovery of the material from a chromatogram can be accomplished in either of two ways. The individual components of the mixture may be separated, developed, into distinct bands on the column by allowing a solvent, or solvent mixture, of sufficient polarity to proceed down the column affecting the separation. If the individual bands can be discerned by their color, fluorescence under the influence of ultraviolet light, or reaction with a colored indicator, the developed chromatogram can be extruded from the column and the separated components recovered by leaching cut portions of the column with a solvent of high polarity. The more general procedure is, however, to continually flush the column with solvents of increasing polarity to remove each component individually. This type of procedure is generally referred to as elution chromatography.

The solvents generally used as eluents are listed in Table 1.3 in the order of increasing polarity. Since the entire separation is dependent on the establishment and maintenance of equilibrium conditions, the polarity of the solvent system is gradually increased by the slow increase of the concentration of the following, more polar solvent. The rate of change of solvent polarity will depend upon the similarity, or dissimilarity, of the components in the mixture to be separated. If closely related compounds are to be separated, for example, isomeric olefins or alcohols, the change in solvent composition may be limited to a few percent every 100 ml. Occasionally a single solvent system may be used to elute the entire column. The separation of compounds having distinctly different functional groups can often be accomplished by a much more rapid change in solvent composition, for example, 25% every 100 ml.

The choice of eluent solvents will depend on the type of adsorbent used.

Table 1.3. Eluotropic Series

Petroleum ether	
Cyclohexane	
Carbon tetrachloride	
Benzene	Increasing
Methylene chloride	polarity
Chloroform (alcohol free)	
Diethyl ether	
Ethyl acetate	
Pyridine	
Acetone	
n-Propanol	
Ethanol	
Methanol	
Water	
Acetic acid	

The comments made earlier with regard to the types of compounds which can be successfully chromatographed on each adsorbent also pertain to the solvents which can be used. In addition, the solvents must be of high purity. The presence of traces of water, alcohols, or acids in the lesser polar solvents will completely destroy the adsorption activity of the solid stationary phase.

In many respects, the application of chromatographic separation procedures is more an art than a science. Experience gained in the use of these techniques leads to a more efficient use of these techniques in future work. When faced with a separation problem involving a totally unknown mixture, it is usually best to carry out a crude and rapid trial chromatographic separation with a small portion of the mixture, using the information and experience gained in the trial run to carry out a more efficient separation on the remainder of the material. Such a procedure usually results in a saving of time and material. Thin-layer chromatography (Sec. 1.5.3) can often serve as an excellent guide to the conditions for column chromatography. For a discussion of the relationship between thin-layer and column chromatography see Sec. 1.5.3b.

1.5.2a Techniques of column chromatography

The success of an attempted chromatographic separation depends on the care exercised in the preparation of the column and in the elution procedure. Figure 1.13 shows a typical adsorption chromatography column. The con-

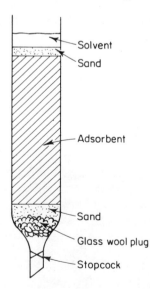

Fig. 1.13. Construction of a chromatography column.

tainer is usually a glass tube provided with some means of regulating the flow of eluent through the column (stopcock preferably made of Teflon). If a ground-glass stopcock is used to control the rate of flow of solvent, all lubricating grease should be removed. Stopcock grease is quite soluble in many of the organic solvents used for elution purposes and may be leached from the stopcock contaminating the material being eluted from the column.

The actual packing of the column is very important. The adsorbent must be uniformly packed, with no entrainment of air pockets. A plug of glass wool or cotton is placed in the bottom of the column, and a layer of sand is added on top of the plug to provide a square base for the adsorbent column. Failure to provide a square base for the column may well lead to elution of more than one fraction at a time if the separation is not very great, as is illustrated in Fig. 1.14.

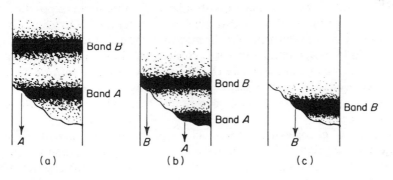

Fig. 1.14. Elution of successive bands from a chromatography column which is not square at the bottom. (a) Elution of A. (b) Elution of A and B. (c) Elution of B.

The packing of the column can be accomplished in several ways. The adsorbent may be dry packed, the solvent being added after the adsorbent. However, this method is generally not suitable because air pockets become lodged within the column and are difficult to remove; in addition, some adsorbents undergo an expansion in volume when wetted with a solvent, thus bursting the column (for example, silica gel and many ion exchange resins). The recommended method involves filling the container with petroleum ether, or the least polar solvent to be used, and then adding the adsorbent in a continuous stream until the desired amount of adsorbent has been added. The tube should be maintained in a vertical position to promote uniform packing. Adsorbents which are not of uniform size and possess a low density tend to produce columns in which layering of large and small particles occurs when filled in this manner. Florisil and silica gel are notorious in this respect. In such cases, the adsorbent is slurried with the solvent, and the slurry is added rapidly to the empty column. This usually prevents segregation of particle sizes in the column. A layer of sand is added to the top of the column to protect the top of the column from being disturbed during the addition of solvent during the elution process.

The quantity of adsorbent and the column size required will depend on the type of separation to be carried out. Generally 20 to 30 grams of adsorbent are required per gram of material to be chromatographed. Ratios as high as 50 or 100 to 1 may be required in cases when the components of a mixture are all quite similar. The height and diameter of the column are also important. Too short a column may not provide a sufficient length of column to effect the separation. In general, a height-to-diameter ratio of 8:1 to 10:1 is recommended.

After the column has been prepared, the solvent level is lowered to the top of the column by draining from the bottom of the column. The material to be separated is dissolved in the least polar solvent that solubility can be attained (10 to 15 ml per gram of material) and carefully added to the top of the column. Elution is begun with a solvent system corresponding to the composition of the solvent used to dissolve the sample. Occasionally, some highly polar components present in mixtures may require a solvent system of such highly polar character for complete solubility that the lesser polar components may be immediately eluted from the column with no separation being achieved. In such cases, the sample may have to be added to the column as a two-phase system. However, this may lead to problems of column congestion.

The rate of change of the composition of the eluting solvent will depend on the type of mixture being separated. Elution with one solvent composition should continue until a decrease in the amount of material being eluted is noted. The fractions should be evaporated as soon as possible, and the weight recovered in each fraction recorded and plotted versus fraction

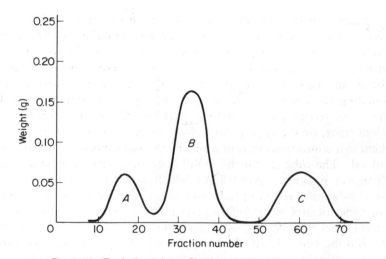

Fig. 1.15. Typical weight vs. fraction number chromatogram.

number. A typical plot of a chromatographic separation appears in Fig. 1.15. A decrease in the weight per fraction indicates the approaching end of a component band. The fractions appearing under one peak may be combined for subsequent purification and identification steps. In general, one should recover 90 to 95% of the material placed on the chromatographic column.

1.5.3 Thin-layer Chromatography

Thin-layer chromatography (TLC) is a special application of adsorption chromatography in which a thin layer of adsorbent supported on a flat surface is utilized instead of a column of adsorbent. Elution or, more properly, development of the chromatogram is accomplished by capillary movement of the solvent up the thin layer of adsorbent. Unfortunately, one is restricted to the use of a single solvent system. However, after use of one solvent system, the chromatogram may be dried and developed further by use of a second solvent system, either in the same direction or at right angles to the direction of the first development.

The most common adsorbents used in TLC in order of importance are silica gel, alumina, keiselguhr, and cellulose. The first two are far more important for general use than the last two. The adsorbents for use in TLC are more finely divided than those for column chromatography and usually contain plaster of Paris as a binding agent. For these reasons only the commercially available adsorbents especially prepared for TLC work should be employed.

The choice of the best eluent for TLC will depend on the type of adsorbent and the components in the mixture to be separated. The same general rules for column adsorption chromatography also apply to TLC. However, since only one solvent system is generally used in TLC, trial chromatographic separations with a variety of solvents should be carried out to determine which solvent leads to the best separation. A series of small individual plates (microscope slides are excellent for this purpose) are prepared, spotted, and developed in the different solvents.

Visualization of the chromatogram to locate the position of each of the components in the separated mixture will depend on the type of molecules present in the original mixture. If all the components are colored, then visual inspection is sufficient to locate the spots. In TLC the most common methods for visualizing compounds are visualization by iodine vapors, the use of ultraviolet light on phosphor-containing layers, and charring with sulfuric acid.

The most commonly used indicator is iodine vapor. The developed and dried plate is placed in a closed container containing a few crystals of iodine. The iodine vapor is adsorbed into the areas of the plate containing organic compounds; brown spots due to iodine charge-transfer complexes appear on a white background. Almost all organic compounds can be detected with this technique. The method is usually nondestructive. The spots may be marked and the plate gently warmed to allow the iodine to sublime out of the layer, leaving the unchanged compounds.

Inert, inorganic phosphors may be added to the slurry prior to the preparation of TLC plates. Such phosphors are frequently supplied with the commercial plaster-of-paris containing adsorbents especially prepared for TLC. These phosphors emit visual light when excited with ultraviolet irradiation. If the compound to be visualized on the thin layer contains a conjugated system, it will quench this emitted light and a dark spot will appear on a bright layer. The method has the advantages of simplicity and high sensitivity as well as being nondestructive.

Many organic compounds char when heated with sulfuric acid. A small amount of sulfuric acid may be incorporated into the slurry prior to the formation of the plates, or the developed chromatogram may be sprayed with sulfuric acid. Charring is then accomplished by heating the developed plate over a hot plate or in a drying oven at 100 to 110°.

Many reagents used for qualitative functional tests may also be used for the visualization of TLC plates. For example, acid-base indicators for acidic or basic compounds, ferric chloride solution for phenols and enols, 2,4-dinitrophenylhydrazine for ketones and aldehydes, dichromate reagents, etc., may be used as reagents. The use of more than one developing reagent may be required to detect all the components present in any given mixture. This can be carried out on a single chromatogram or, if one indicator interferes with subsequent tests, on a different plate.

A value of TLC lies in the relatively small amount of time required per analysis. With thin-layer chromatography on microscope slides the entire operation from the preparation of the layer to the visualization of the spots can be accomplished in about 10 minutes. TLC is very useful for monitoring the progress of reactions, in detecting intermediates in reactions, in analyzing crude products or unknown mixtures to determine the number of components, and for checking the efficiency of purification processes.

TLC can be used for relatively small preparative separations employing thicker layers of adsorbent (2000 μ) and applying the sample as a band instead of spots. Only a narrow section of the chromatogram is developed by use of indicators to locate the bands of material. The bands are then scraped from the plate, and the material is leached from the adsorbent.

TLC may also be used as a tentative means of identification. If plate conditions are maintained constant, a compound in several initial spots will progress up the plate at the same rate relative to an added standard or to the solvent front. The relative displacement is referred to as

$$R_x = \frac{\text{displacement of compound}}{\text{displacement of standard}} \tag{1.4}$$

or

$$R_f = \frac{\text{displacement of compound}}{\text{displacement of solvent front}} \qquad (1.5)$$

For identification purposes, a comparison of a known and unknown should be carried out with a number of different solvent systems on several adsorbents. R_x and R_f values from the literature are not reliable enough for use in final identification; they should be used only as general guides. Whenever possible, direct comparison of a known and unknown should be made. It is wise to spot the known, the unknown, and a mixture of the two on a single chromatogram. Even with direct comparison, a final unambiguous identification is not possible in that many compounds have identical or very similar adsorption properties.

1.5.3a Techniques of thin-layer chromatography

Chromatography on microscope slides

The chemist involved in problems of organic structure determination will find TLC on microscope slides (Fig. 1.16) constantly useful in his work. The remarkable utility and convenience of this method for monitoring reactions and checking composition and purity cannot be overemphasized.

Preparation of the plates. The slides for TLC are most conveniently prepared by dipping the slides into a slurry of commercial TLC adsorbents in organic solvents. For the slurry use 35 g of silica gel G or 60 g of alumina in 100 ml of chloroform-methanol (2:1 v/v). Shake or stir the slurry for about 2 min before use. The slurry may be stored for some time in tightly sealed containers. The plates are prepared as follows:

1. Dip two *clean, dry* microscope slides held back-to-back into the slurry. Remove the slides slowly and allow them to drain on the edge of the container. If the layers are thin and grainy, the slurry should be thickened by the addition of more adsorbent.

2. Separate the two slides and allow them to dry on the desk for 5 min. Remove the excess adsorbent from the edges. Silica gel and alumina plates prepared in this manner are ready for use. If they are not to be used immediately, store them in a dry atmosphere.

Application of the sample. The most convenient applicators are made by pulling the center of an open melting-point capillary to a very fine capillary and breaking at the center. Apply small spots of a dilute solution of the sample in a volatile solvent about 10 mm from the end of the slide. If more than one application is necessary, allow the spot to dry before reapplying more sample. With care, up to three samples can be placed on one slide. Allow the spotting solvent to evaporate before placing the slide in the developing chamber.

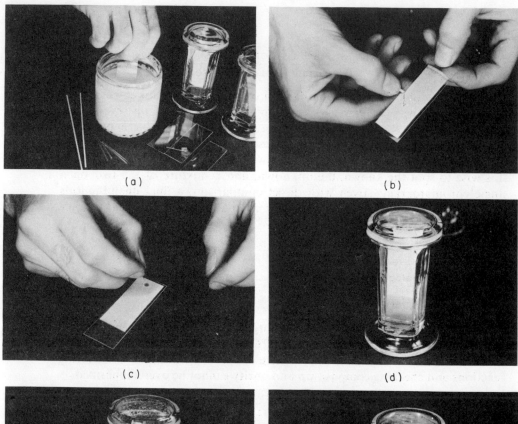

Fig. 1.16. TLC on microscope slides. (a) Two microscope slides, back-to-back, are dipped into the slurry, separated, and allowed to dry. (b) The edges are smoothed by removing a small amount of the adsorbent. (c) The sample is applied with a small capillary. Up to three samples may be applied on a single slide. (d) The chromatogram is developed in a small closed chamber for about 5 min. Jars designed for slide staining make convenient development chambers. (e) The chromatogram is visualized by placing it in a closed chamber containing a few crystals of iodine. (f) Chromatography on shaped layers results in the formation of narrow bands rather than round spots. Only one sample can be placed on a slide, but the method often results in improved separation of the components and facilitates the detection of trace components of low R_f value. (Photographs by L. Pepoy and D. Vegh.)

Development of the plate. Solvent systems are usually chosen by trial and error by using the eluotropic series shown in Table 1.3 as a general guide. The use of mixed solvent systems will often be most valuable.

The spotted plate is placed in a small jar or beaker fitted with a cover (aluminum foil or glass plate). The solvent level in the jar should be below the level of the spots on the plate. Development, such that the solvent runs about three-fourths of the way up the plate, usually requires about 5 min. The plates are then removed from the development chamber and allowed to air dry. The spots are visualized by the methods discussed earlier. If the spots have an R_f of less than 0.4, the chromatogram may be redeveloped in the same or a different solvent system for increased resolution.

Recording of data. R_x or R_f values may be calculated or a picture of the chromatogram may be drawn in the laboratory notebook. One may also use transparent tape to place the layer directly in the notebook. The tape is pressed onto the layer, and the layer will adhere to the tape when it is removed. An additional piece of tape is placed on the back of the layer, completely sealing it between the tapes.

Chromatography on larger plates

The commonly used larger plates are 5 by 20 cm or 20 by 20 cm. Prepared layers are commercially available either on glass plates or plastic sheets. The latter have the advantage that they may be easily cut to any desired size but are flexible and must be supported during the development. The preparation of these larger plates is best accomplished by means of a commerically available apparatus. One can, however, prepare them by spreading a slurry by means of a glass rod. An aqueous slurry is prepared from commercial TLC adsorbent according to the directions on the container. The slurry is spread evenly across the surface of the plate(s) by pulling a glass rod which has been wrapped at both ends with masking tape to the desired thickness of the layer across the surface of the plate(s), forcing the slurry ahead of the glass rod. Alternately an edge of tape of the desired thickness may be placed down each side of the glass plate and the slurry spread between the tapes by means of a glass rod. The plates are air dried and placed in an oven at 105° for 1 hr or more to activate them for use in adsorption chromatography. The spotting and development follows from the directions given for the microscope slide plates.

1.5.3b The relationship between column
and thin-layer chromatography

Thin-layer chromatography may be used as a model for column chromatography. TLC on microscope slides will suggest adsorbent and solvent systems for use in column chromatography and provide details about the complexity of the mixture to be separated. For an effective transition from TLC to column chromatography it is important that the TLC and column adsorbent come from the same manufacturer. The following procedure can be recommended:

1. Find an appropriate solvent system for clean separation on TLC. Such a system should consist of two liquids—one polar and the other less polar.

2. Reduce the polarity of the solvent system by the addition of the less polar solvent until the R_f values of the components are below 0.3 on TLC. If there is only one desired component of the mixture and this component has a high R_f value, reduction of the solvent polarity may not be necessary.

3. Use the less polar solvent system from step 2 to make a slurry and pack the column.

4. Place the sample on the column and develop with the modified solvent system. Analyze the fractions by TLC using microscope slides.

1.5.4 Paper Chromatography

Paper chromatography is somewhat similar to thin-layer chromatography except that high-grade filter paper is used as the adsorbent or solid stationary phase. In actual fact paper chromatography is not strictly adsorption chromatography but a combination of adsorption and partition chromatography. A partitioning of the solute occurs between the water of hydration of the cellulose and the mobile organic phase. Paper chromatography is used primarily when extremely polar or polyfunctional compounds, for example, sugars and amino acids, are to separated. Such materials cannot be chromatographed on more active adsorbents. The method is used as an analytical procedure much like analytical thin-layer chromatography; sample sizes for paper chromatography are usually in the range of 5 to 300 μg.

The selection of solvent systems for paper chromatography is very important. For a discussion of solvent systems, refer to the books listed in Sec. 1.7. In most cases the solvent system should contain some water. The solvent development can be accomplished by suspending the paper strip so that the end of the strip is immersed in a container of solvent (ascending paper chromatography) or by immersing the top of the strip in a small trough of solvent and allowing the solvent movement to occur by a combination of capillary action and gravity (descending paper chromatography). It is very important in all paper chromatography procedures that the development of the chromatogram occurs in an atmosphere of saturated solvent vapors. Otherwise, the solvent will evaporate from the paper faster than it is replaced by capillary action and separation of the components will not be achieved.

After development is complete, the position of the solvent front is noted and the paper strip is allowed to dry. The paper is then sprayed with a visualizing reagent. R_f and R_x values are calculated.

1.5.4a Test-tube technique
for paper chromatography

As in thin-layer chromatography, both elaborate and simple techniques and equipment may be employed in paper chromatography. By far the simplest technique involves the use of an ordinary test tube as the developing chamber; the method can be recommended for most identification work or, at least, for exploratory work in determining conditions for using larger strips or sheets.

A 6 or 8-in. test tube may be used. The paper strip cut from Whatman No. 1 filter paper may be suspended in the test tube by suspending it from a slit cut in the stopper, by putting a fold in the paper, or by cutting the paper with a broad end at the top which will serve to suspend the paper in the tube (Fig. 1.17). Handle the paper only at the extreme top edge since fingerprints will contaminate the chromatogram.

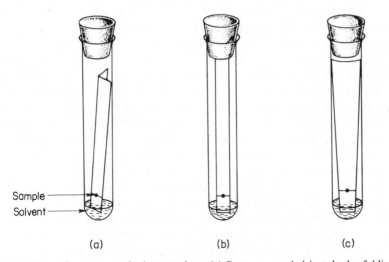

Sample ⎯

Solvent ⎯

(a)   (b)   (c)

Fig. 1.17. Paper chromatography in test tubes. (a) Paper suspended in tube by folding. (b) Paper suspended from slit in rubber stopper. (c) Paper beveled to suspend from sides of tube at the broad end.

Place a light pencil mark about 1 cm from the bottom of the strip. The sample is applied as a small spot with a micropipette (made by pulling a capillary) at the center of the line and allowed to dry. The sample size should be about 10 μg. A small amount of solvent is placed in the bottom of the tube by means of a pipette in such a manner that the sides of the tube above the surface of the solvent remain dry. The spotted and dried paper strip is carefully inserted into the tube so that the end dips into the solvent, but the surface of the solvent is below the pencil line. The tube is stoppered and allowed to stand until the solvent ascends near the top of the strip. This process may take up to 3 hr. The paper is removed with forceps and the solvent front marked. The strip is allowed to dry and the compounds visualized with an appropriate spray.

1.5.5 Gas Chromatography

Gas chromatography utilizes a moving gas phase with a solid stationary phase, termed gas-solid chromatography (gsc), or with a liquid stationary phase, termed gas-liquid chromatography (glc) or occasionally vapor-phase chromatography (vpc). The liquid stationary phase is supported on the surface of a stationary solid support. In gsc the separation of the components of a mixture is dependent on the establishment of adsorption equilibria as in column adsorption chromatography. With glc the separation is dependent on solubility equilibria between the components in the gas phase and liquid phase. Herein lies the great utility of glc. Proper choices of the stationary liquid phases allow the experimentalist to greatly vary the sequence of elution of compounds in a mixture from the column.

Gas chromatography has developed into one of the most powerful analytical tools available to the chemist. The technique allows separation of extremely small quantities of material, of the order of 10^{-4} to 10^{-6} g. The quantitative analysis of mixtures can be readily accomplished (as discussed in greater detail later in this section). The possibility of using much longer columns, producing a greater number of theoretical plates, increases the efficiency of separation beyond that of any other technique available. The technique is applicable over a wide range of temperatures (-70 to $+500°$), making it possible to chromatograph materials covering a wide range of volatilities. Finally, gas chromatographic analysis requires very little time compared with other analytical techniques.

The instrumentation involved may be quite simple or may contain a high degree of sophistication and complexity. The basic components required are illustrated in Fig. 1.18. A relatively high-pressure gas source, 30 to 100 psi., is required to provide the moving gas phase. The gas is introduced into a heated injector block. The sample is introduced into the gas stream in the injector block by means of a microsyringe forced through a syringe septum. Liquids can be injected directly, whereas solids are generally dissolved in a volatile solvent. The injector block is maintained at temperatures ranging from room temperature to approximately 350°. The operating temperature is chosen to ensure complete vaporization of all components present in the mixture. The temperature of the injector, or the column, need not be higher

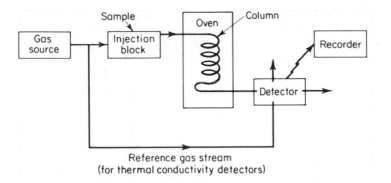

Fig. 1.18. Block schematic of a simple gas chromatograph.

than the boiling point of the least volatile substance. The partial vapor pressure of most substances, in the presence of the high pressure carrier gas, is sufficient to ensure complete vaporization. It must be kept in mind that high injector block temperatures may lead to decomposition of relatively unstable compounds.

The gas stream, after leaving the injector block, is conducted through the chromatographic column, which is housed in an oven. If all the retention times (see later discussion for definition) of the components are quite similar, the column may be maintained at a constant temperature (isothermal operation). If the retention times are greatly different, it may be desirable to slowly increase the temperature of the column during the chromatographic process. Instruments are available for constant-temperature and variable (programmed)-temperature operation. The size of the columns used will depend on the difficulty of separation of the components in a given mixture; this is usually determined by trial runs. Packed columns are usually 0.25 in. in diameter and 2 to 50 ft in length. Normal column lengths are usually 6 to 10 ft; the longer columns are required for the more difficult separations. Larger diameter columns, up to 2 in., are used for preparative purposes (collection of the individual fractions). Capillary columns of up to 300 ft in length are also available.

The effluent gas stream, containing the separated components, is conducted through a detector which measures some physical phenomenon and sends an electronic signal to a strip chart recorder. Many types of detectors are available. These include thermal conductivity devices and ionization devices. The correct interpretation of the data recorded by the strip chart recorder depends on the type of detector used. The types of detectors and the method of handling the data will be discussed in detail in the following paragraphs.

The strip chart recorder provides a chromatogram (not a spectrum!) such as that illustrated in Fig. 1.19. The first question to be asked is: What do the individual peaks represent in the chromatogram? First of all, the

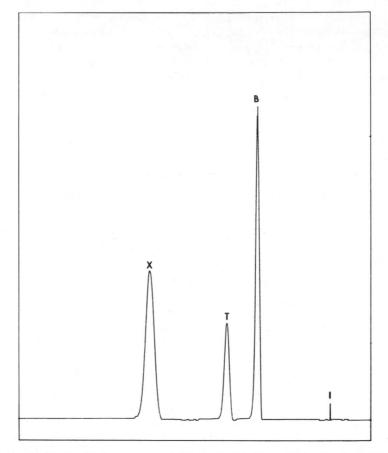

Fig. 1.19. Gas-liquid chromatogram of benzene (B), toluene (T), and xylene (X) on a 5½ ft, 20% Carbowax 20M column at 140°. The small peak labeled I indicates the point of injection of the sample.

peak represents the detection of some material being eluted from the column. The peak can be characterized by a retention volume or retention time. The *retention volume* is the total volume of gas passing through the column to effect elution of the compound responsible for a given peak. The retention volume is a function of the volume of stationary liquid phase, the void volume of the column (volume of gas in the column at a given time), the temperature, and the flow rate. Generally, the experimentalist does not determine retention volumes but, instead, characterizes the peaks with respect to their *retention times*, the elapsed time between injection and elution. In the comparison of retention times, constant column conditions are implied, those conditions being the functions of retention volume cited before.

The identification of an unknown giving rise to a peak in a chromatogram can be tentatively inferred by the comparison of retention times of the unknown with a possible known on a number of columns differing widely in polarity. Additional evidence can be obtained by recording the chromatogram of an admixture of the unknown and known samples. A single peak whose appearance (symmetry) is the same as the unknown and known samples indicates the possible identity of the unknown and known samples. A rigorous identification of an unknown cannot be made by comparison of retention times since it is very conceivable that many other materials may possess similar elution properties.

Rigorous identification of the material giving rise to a peak can be accomplished by collecting the fraction as it emerges from the exit port and then characterizing the material by its physical properties. However, it must be remembered that isomerizations and fragmentations can and do occur in the injector block and column and that the material going through the detector may not be the same as what was injected. Gas chromatographs have been developed to specifically take advantage of the fragmentation of compounds. Injection of highly nonvolatile materials into a pyrolysis injector results in extensive cracking of the material, and the chromatogram obtained is referred to as a *cracking pattern*.

Gas chromatographs may also be used in conjunction with other types of instruments. The emerging gas stream may be directed through an infrared cell of a fast-scan infrared spectrometer, or directly into a mass spectrometer. Such procedures eliminate the necessity of collecting the individual fractions.

The area of the peak of a gas chromatogram represents a quantitative measurement of some physical property of the material going through the detector. For a given material, the response will be linear with concentration. However, the physical constant for a series of different molecules may not have the same absolute value, and the relative response of one compound with respect to another will not be directly proportional to the partial pressures of the materials in the eluent gas stream. The relative response of different compounds depends on the type of detector used, and appropriate corrections in calculations must be made if quantitative data are to be derived.

1.5.5a Detectors

Detectors generally fall into one of two classes: thermal conductivity devices and ionization detectors.

The thermal conductivity device measures changes in the thermal conductivity of the eluent gas stream. The detector usually consists of four heated filaments acting as the four resistances of a Wheatstone bridge. Two of the filaments are immersed in a reference gas stream, a small stream of gas bypassing the injector and column in Fig. 1.16 and entering the detector, to provide a constant reference of resistance. The two remaining filaments are immersed in the eluent gas stream. With a constant flow of pure carrier gas past the filaments, a constant rate of heat transfer from the filaments to the gas stream will occur. As a compound is eluted from the column, the thermal conductivity of the gas stream will change and more or less heat will be transferred from the filament to the gas stream. The change in temperature of the filament results in a change in the resistance of the fila-ment, which is noted by a change in current flow. This signal is relayed to the recorder and plotted on a time scale. Therefore, with the thermal con-ductivity detector, the chromatogram represents the change in the thermal conductivity of the eluent gas stream.

Table 1.4 lists the thermal conductivities of several potential carrier gases and organic compounds in the gas phase at two different temperatures. The

Table 1.4. Thermal Conductivities of Various Substances $\left(\dfrac{\text{g cal}}{\text{sec cm}^2} \text{ per } \dfrac{°C}{\text{cm}}\right)$

Substance	0°C	100°C	Ratio $\dfrac{100°C}{0°C}$
Helium	33.60	39.85	1.18
Hydrogen	39.60	49.94	1.26
Argon	3.88	5.09	1.31
Nitrogen	5.68	7.18	1.26
Carbon dioxide	3.35	4.96	1.48
Methane	7.16	12.3	1.72
Ethane	4.26	7.77	1.83
Pentane	2.94	—	—
Benzene	2.10	4.17	1.99
Ethanol	—	5.03	—

thermal conductivity of a gaseous mixture is equal to the sum of the products of the mole fraction and thermal conductivity. Of the four potential carrier gases listed in Table 1.4, helium and hydrogen provide the greatest sensitivity of measurement, sensitivity meaning the greatest change in the thermal conductivity of the gas stream with respect to change in concentra-tion. Helium is the most widely used carrier gas in thermal conductivity devices, owing to its greater chemical inertness. Argon and nitrogen are not suitable carrier gases for use with thermal conductivity devices, since "nega-tive" peaks may occur with compounds having a thermal conductivity larger than argon or nitrogen.

The conversion of the peak areas of a chromatogram into relative weights, or moles, requires having available the thermal conductivities of the components present in the mixture at the temperature of the detector. Such information is extremely limited, and the measurement of such values would be a prohibitive task reducing the utility of gas chromatography as a quantitative analytical tool. To circumvent this problem, one may determine relative response ratios of the components present in the mixture. This is accomplished by recording the chromatogram of a carefully weighed mixture of the pure components and calculating the (weight ratio)/(area ratio) of each component relative to a single component or to an added internal standard. The area ratios, relative to the same component or internal standard used in the known composition mixture, determined from the unknown mixture chromatogram are converted to relative weight ratios by multiplication of the area ratios by the (weight ratio)/(area ratio) values (illustrated in the sample calculations which follow).

Inspection of the data presented in the last column of Table 1.4 indicates that the thermal conductivities of compounds in the gas phase are not constant with temperature. Secondly, the temperature dependence, which is nonlinear, varies from compound to compound. Therefore, the relative response ratios calculated at a given detector temperature can be used *only* at that temperature. Should an analysis be carried out at a different detector temperature, the relative response ratio must be redetermined at that temperature. It is also apparent from Table 1.4 that the difference in thermal conductivities increases with temperature. To obtain maximum sensitivity with the use of the thermal conductivity device, the temperature of the detector is maintained at a relatively high value, 250 to 350°C.

Quantitative yield determinations of volatile components can also be determined by the use of gas chromatography. This can be accomplished by either of two procedures. An additional weighed amount of one of the components present in the mixture is added to a weighed aliquot of the mixture. The gas chromatograms of the mixture and admixture are recorded. The absolute weights or moles present in the mixture can then be calculated (see the sample problem below). An alternative procedure involves the addition of a weighed amount of an internal standard to an aliquot of the mixture and then recording the gas chromatogram. For this method, one needs the relative response ratio of one of the components with respect to the internal standard. Yields should not be calculated without the use of an internal standard unless *all* products appear as peaks in the chromatogram and there has been *no* mechanical loss of material.

1.5.5b Sample calculations

Reaction of A (2.50 g, 134 mol. wt.) with reagent X gave 2.40 g of a product mixture containing A and two products B (148 mol. wt.) and C (162 mol. wt.). The gas chromatogram of the product mixture showed peaks for A, B, and C, with areas of 77, 229, and 276, respectively. Addition of 36.4 mg of B to a 120.2-mg aliquot of the reaction mixture produced a gas chromatogram with areas of 111, 567, and 396 for A, B, and C. A mixture of 52.7 mg of A, 47.3 mg of B, and 63.2 mg of C produced a gas chromatogram with areas of 287, 275, and 423, respectively. Calculate the yield of B and C based on reacted starting material according to the following equation:

$$A + X \rightarrow B + C$$

Solution:

$$\text{Response ratio } \frac{A}{B} = \frac{\text{wt. A/wt. B}}{\text{area A/area B}} = \frac{52.7/47.3}{287/275} = \frac{1.113}{1.044} = 1.067 \left(\frac{A}{B}\right)$$

$$\text{Response ratio } \frac{C}{B} = \frac{\text{wt. C/wt. B}}{\text{area C/area B}} = \frac{63.2/47.3}{423/275} = \frac{1.337}{1.540} = 0.867 \left(\frac{C}{B}\right)$$

Area of B in admixture
chromatogram due to added B $= \text{area B}_{\text{admix}} - \text{area A}_{\text{admix}} \times \dfrac{\text{area B}_{\text{orig}}}{\text{area A}_{\text{orig}}}$

$$= 567 - 111 \times \frac{229}{77} = 567 - 330 = 237$$

From the proportion

$$\frac{\text{wt. B}_{\text{added}}}{\text{area B}_{\text{added}}} = \frac{\text{wt. B}_{\text{orig}}}{\text{area B}_{\text{orig}}}$$

we have

$$\frac{36.4 \text{ mg B}}{237} = \frac{\text{wt. B}_{\text{orig}}}{330}$$

$$\text{Wt. B}_{\text{orig}} \text{ in aliquot} = \frac{36.4 \times 330}{237} = 50.7 \text{ mg}$$

$$\text{Wt. of B in total sample} = \frac{50.7 \text{ mg}}{0.1202 \text{ g}} \times 2.405 \text{ g} = \underline{1.014 \text{ g of B}}$$

$$\text{Wt. of A in sample} = \frac{\text{area A}}{\text{area B}} \times \text{response ratio} \left(\frac{A}{B}\right) \times \text{wt. B}$$

$$= \frac{77}{229} \times 1.067 \times 1.014 \text{ g A} = \underline{0.364 \text{ g A}}$$

$$\text{Wt. of C in sample} = \frac{\text{area C}}{\text{area B}} \times \text{response ratio} \left(\frac{C}{B}\right) \times \text{wt. B}$$

$$= \frac{276}{229} \times 0.867 \times 1.014 \text{ g C} = \underline{1.060 \text{ g C}}$$

The percent yields are calculated in the usual manner.

In this particular case, the entire crude sample proves to be volatile. This may not be true in all cases.

A similar approach would be used if an internal standard had been added instead of employing additional B.

The second category of detectors may be generally classified as ionization detectors. The eluent stream is subjected to partial ionization; the ions being formed then migrate to a polarized electronic grid, producing a current. The current is amplified and recorded, producing the chromatogram. The peaks of chromatograms derived from ionization detectors represent the number of ions formed. The extent of ionization varies with the type of ionization detector employed and the chemical nature of the eluent.

The first of this type of detector to be discussed is the flame ionization detector. The eluent stream is mixed with a stream of hydrogen and oxygen, and is then subjected to combustion. During this process, the materials eluted from the chromatographic column undergo partial combustion, producing ionic fragments, or they undergo ionization induced by the energy produced from the combustion of the hydrogen. The extent of ionization of individual compounds depends on the number and types of functional groups contained in the molecule. To derive quantitative information from the chromatogram, the approach described earlier for the thermal conductivity detector must be used.

Instead of employing a hydrogen flame to induce ionization, a source of low energy β or γ rays may be used. The mechanism of the ionization process depends on the carrier gas employed. Direct ionization of eluted materials occurs when helium is used as the carrier gas. For each primary electron (β particle) of approximately 1 mev energy, 100 to 200 ionizations per centimeter of path occur, resulting in a total ionization level of $10^{-6}\%$. The relatively high energies involved do not appear to lead to discrimination, and the response is generally linear with respect to the number of molecules of a substance present. Such a detector is referred to as a cross-section ionization detector and employs Ra, ^{90}Sr, ^{90}Y, or ^{3}H as the ionizing particle source.

Substitution of the helium carrier gas by argon results in a change in the ionization mechanism. The primary particle excites an argon atom to a metastable state with a potential of 11.6 ev. The excited argon atom transfers this energy upon collision with the other molecules in the eluent gas stream producing ionization. The response of this type of detector depends on the ionization potential of the materials being chromatographed. Molecules with ionization potentials greater than 11.6 ev will be "transparent." Most organic compounds possess ionization potentials in the 4 to 10 ev region and might be expected to provide a nearly linear response. The level of ionization is approximately 1%, providing an increase in the sensitivity of detection by a factor of 10^6 over that when helium is used as the carrier gas.

The use of nitrogen as a carrier gas in an ionization detector results in the ejection of an electron from a molecule of nitrogen. The secondary electron thus produced attaches itself to a molecule of the material being eluted. The negatively charged sample ions are then counted. The response is dependent on the electron affinities of the compounds being chromatographed. Since electron affinities may vary even more greatly than gas phase thermal conductivities, it is necessary to employ relative response ratios in quantitative determinations.

Several other types of detectors have been described in the literature but do not appear to be as generally applicable as those just described. In general, the use of ionization detectors provides a much more sensitive method of detection than is obtainable with the thermal conductivity device.

1.5.5c Column stationary
solid and liquid phases
for gas-liquid chromatography

The choice of solid phases for gsc is quite limited. Adsorbents which have been used include finely divided alumina, silica gel, charcoal, and molecular sieves. The utility of such solid phases rests in the temperature range available for use, approximately -70 to $+500°C$. The type of samples which can be subjected to gas chromatography using these solid phases is limited. Highly polar compounds generally cannot be successfully chromatographed.

The great utility of glc lies in the great number of combinations of solid supports and stationary liquid phases available. Table 1.5 lists a number of solid supports available with a short description of their properties. The solid supports also can be subjected to specific treatments to improve their utility in certain cases.

These treatments include acid washing to remove traces of metal salts and basic materials which may otherwise interfere in the separations. The solid support is generally treated with concentrated hydrochloric acid for a period of time and is then washed with water until neutral. Treatment with base may be required when acid-sensitive compounds must be chromatographed. Such treatment greatly reduces tailing of highly polar or basic compounds. The support is washed with dilute methanolic potassium hydroxide and is then washed with water until the desired pH is obtained. Occasionally, the support is coated with a layer of potassium hydroxide by evaporation of the methanol used as solvent under reduced pressure. To obtain a nonreactive support, the solid support, which has been acid and

Table 1.5. Solid Supports

Name	Description
Firebrick	Calcined diatomaceous silica, similar in properties to chromosorb P.
Chromosorb P	Highly adsorptive, pink calcined diatomaceous silica, pH 6 to 7.
Chromosorb W	Medium adsorptive, white diatomaceous silica which has been flux-calcined with sodium carbonate, pH 8 to 10.
Chromosorb G	Dense, low adsorptivity, flux-calcined material, pH 8.5 (maximum liquid phase loading, 5%).
Fluoropak 80 Haloport F	Fluorocarbon polymer, maximum usable temperature 260°, reduces tailing of highly polar compounds.
Glass beads	Used only with very low amounts of liquid phase (0.5% and less).

base washed, is stirred for a short time period with a 5% solution of di-chlorodimethylsilane in toluene. The support is washed with toluene and methanol. Such a process is referred to as silanizing.

Table 1.6 lists a number of liquid phases and the classes of compounds which may be separated. The list is by no means exhaustive. It should also be pointed out that the success of a given column substrate depends on the care in preparation, length of use, and on the gross structure of the components being separated. Although Table 1.7 includes recommended liquid phases for a given class of compounds, the experimentalist may find that none of these suggestions is suitable, and he may have to search for a different column. The selection of the proper liquid phase and column conditions, temperature and flow rate, is many times a trial-and-error procedure, but the end results are usually worth the time and effort expended.

Table 1.6 Liquid Substrates

Number	Abbreviation	Trade name or description	Max. temperature, °C
1	Apiezon L	Hydrocarbon grease	300
2	Apiezon M	Hydrocarbon grease	275
3	Carbowaxes (400–6000)	Polyethylene glycols	100–200
4	Carbowax 20M	Polyethylene glycol	250
5	DC-200	Dow Corning methyl silicone fluid	225
6	DC-550	Dow Corning phenyl methyl silicone fluid	225
7	DC-710	Dow Corning phenyl methyl silicone fluid	250
8	DEGS or LAC 728	Diethyleneglycol Succinate	225
9	LAC 446	Diethyleneglycol Adipate	225
10	Flexol Plasticizer 10-10	Didecyl phthalate	175
11	Nujol	Paraffin oil	200
12	QF-1-6500	Dow Corning fluorinated silicone rubber	260
13	SE-30	G.E. methyl silicone rubber	300
14	SE-52	G.E. phenyl silicone rubber	300
15	SF-96	G.E. fluoro silicone fluid	300
16	——	Silicone gum rubber	375
17	——	Silver nitrate-propylene glycol	75
18	TCP	Tricresyl phosphate	125
19	TCEP	1,2,3-Tris(2-cyanoethoxy)propane	180
20	THEED	Tetrakis(2-hydroxyethyl)ethylene-diamine	130
21	Ucon polar	Polyalkylene glycols and derivatives	225
22	Ucon nonpolar	Polyalkylene glycols and derivatives	225
23	——	4,4'-Dimethoxyazobenzene	120–135*

*The stationary phase exists as liquid crystals, referred to as nematic or smectic phases depending on crystal orientations, and the temperature range is thus limited to the regions where these phases exist. Other alkoxyderivatives of azobenzene have also been used [see M. J. S. Dewar and J. P. Schroeder, *J. Am. Chem. Soc.*, **86,** 5235 (1964)].

Table 1.7. Suggested Column Uses

Class of compound	Column number (from Table 1.6)
Acetates	8, 10, 12, 19, 20, 21
Acids	6 (on acid-washed support), 16
Alcohols	3, 4, 5, 6, 16, 19, 20, 21, 22
Aldehydes	3, 4, 10, 11
Amines	3, 4, 6, 20 (on fluoropak or haloport or potassium hydroxide-coated support)
Aromatics	1, 2, 10, 16, 18, 19, 20, 21
Esters	3, 4, 8, 9, 10, 12, 16, 19, 20
Ethers	6, 9, 16, 18, 19
Halides	5, 10, 11, 18, 19
Hydrocarbons	1, 2, 6, 10, 16, 19, 20
Ketones	3, 4, 6, 10, 16, 18, 20
Nitriles	9, 18, 20
Olefins	10, 17, 19, 20
Phenols	4, 12
Isomeric aromatics	23

The column packings are prepared by dissolving the liquid phase in a volatile solvent, usually methylene chloride or methanol, and slurrying with the solid support. The solvent is evaporated under reduced pressure, with continuous agitation of the slurry. The dried packing is poured into the column and firmly packed. Before use, the column must be purged by a stream of gas, usually nitrogen, while maintained at the maximum column operating temperature until equilibrium conditions are reached, as indicated by the production of a steady base line. Additional curing of the column can be accomplished by filling the column with an inert gas under slight pressure, capping both ends of the column, and heating in an oven at 20° above the maximum column operating temperature. After each use, the column should be purged at the maximum temperature to remove any relatively non-volatile materials which may have been left on the column. Capillary columns are coated with the stationary liquid phase by slowly evaporating a solution of the stationary phase inside the column.

1.5.6 Molecular Sieve Chromatography

Molecular sieve chromatography is the separation of molecules primarily on the basis of size, and not polarity or solubility. The two basic types of materials currently available for use as molecular sieves are synthetic zeolites (inorganic) and organic polymers (gels). These materials possess cavities into which molecules may pass and be retained, the larger molecules being eluted very rapidly from the column. The maximum size of a molecule, either in actual size or molecular weight, retained in the column is referred to as the exclusion limit. The operating range is defined as the range of molecular weights which may be separated.

Table 1.8. Properties of Linde Molecular Sieves

Designation	Pore size	Molecules included
Type 3A	3 Å	H_2O, NH_3 (molecules with effective diameter of less than 3 Å).
Type 4A	4 Å	H_2O, NH_3, C_2H_5OH, H_2S, CO_2, SO_2, C_2H_4, C_2H_6, C_3H_6 (not C_3H_8).
Type 5A	5 Å	Molecules with an effective diameter of less than 5 Å, including straight chain alcohols and hydrocarbons to C_{22}, *but* not branched chains and cyclic derivatives.
Type 10X	8 Å	Simple cyclic and substituted aromatics, *but* not polycyclics and highly substituted aromatics.
Type 13X	10 Å	Highly branched hydrocarbons and aromatics (primarily used for gas purification and drying.

The synthetic zeolites (Linde molecular sieves made by the Linde Division of Union Carbide and Carbon Corporation) possess the smallest exclusion limits (see Table 1.8). The synthetic zeolites normally contain water of hydration which is driven off by heating. The cavities thus formed in the crystals are quite uniform in size and are able to accept other molecules into these cavities. More polar molecules will be preferentially adsorbed, but any molecule capable of fitting into the cavity will be included. Linde molecular sieves are particularly useful for removing small amounts of contaminants of small molecular size. The activity of the molecular sieves may be regenerated by heating to drive off the included materials.

Bio-Gel P (Bio-Rad Laboratories) is a series of acrylamide and methylenebisacrylamide copolymers possessing varying exclusion limits. These exclusion limits are quite high (as are those of the Sephadex gels) (see

Table 1.9. Properties of Bio-Gel P Substrates*

Designation	Exclusion limit†	Operating range†
Bio-Gel P-2*	1,600	200– 2,000
P-4	3,600	500– 4,000
P-6	4,600	1,000– 5,000
P-10	10,000	5,000– 17,000
P-20	20,000	10,000– 30,000
P-30	30,000	20,000– 50,000
P-60	60,000	30,000– 70,000
P-100	100,000	40,000– 100,000
P-150	150,000	50,000– 150,000
P-200	200,000	80,000– 300,000
P-300	300,000	100,000–>400,000

*Available in two different particle sizes.
†Molecular weight.

Table 1.9) and find extensive use in biochemistry and in the purification of materials isolated from various natural sources. Molecules possessing molecular weights higher than the exclusion limit are eluted very rapidly from the column with no separation of components, whereas molecules possessing molecular weights below the exclusion limit experience varying degrees of inclusion and will be retarded in the elution sequence, with molecules of lowest molecular weight being eluted last.

Another polymeric organic gel found to be very useful for molecular weight separations is the series of Sephadex gels (Pharmacia Fine Chemicals, Inc.). The properties of the Sephadex gels are given in Table 1.10. The Sephadex gels are quite similar in properties to the Bio-Gel P substrates.

Table 1.10. Properties of Sephadex Gels*

Designation	Exclusion limit†	Operating range†
G-10	700	0– 700
G-25	5,000	100– 5,000
G-50	10,000	500– 10,000
G-75	50,000	1,000– 50,000
G-100	100,000	1,000–100,000
G-200	200,000	1,000–200,000

*Available in several particle sizes.
†Molecular weight.

1.5.7 Ion-exchange Chromatography

Ion-exchange separations are particularly useful for the separation of charged, or potentially charged, materials, such as carboxylic acids, amines, aminoacids, peptides, etc. Ion-exchange separations are carried out by using resin substrates to which are bonded acidic or basic functional groups capable of undergoing exchange. The most commonly used resins are prepared by copolymerization of styrene and divinylbenzene which results in a highly cross-linked polymer. The pore size of the resin (which dictates the size of the counter ion to be accommodated) is dependent on the extent of cross-linking which, in turn, is a function of the styrene-to-divinylbenzene ratio used in the preparation of the resin polymer. The aromatic rings of the polymer are then subjected to substitution reactions, resulting in the introduction of sulfonic acid, carboxyl, amino, quaternary amino, etc., groups which act as the ion-exchange donor groups. Ion-exchange resins are generally classified as cation- or anion-exchange resins, depending on whether a cation, for example, H^+, Na^+, NH_4^+, etc., from acidic groups or their salts, or an anion, for example, Cl^- or OH^- from a quaternary salt group, is exchanged. Ion exchange is very useful in the conversion of carboxylates to acids, ammonium salts to amines, etc., on cation-exchange columns, and the exchange of acetate for chloride, hydroxide for chloride, etc., on anion-exchange columns.

The separation of materials by ion exchange is distribution equilibrium controlled. The position of the equilibrium is controlled by changing the properties of the eluent solvent, either by changing the pH or some ion concentration of the solvent. Ion exchangers may be used to remove acidic or basic materials from a mixture, or ionic complexing agents may be used for the separation of neutral materials. For example, ion exchangers containing bisulfite may be used to remove or separate aldehydes, borates for the separation of polyhydroxy compounds, metal ions for the separation of dicarbonyl compounds, etc.

The amounts of material that can be separated by ion exchange techniques are limited by the capacity, or the number of exchangeable ions, of the column. Exchange capacities are indicated by the number of milli-equivalents (meq) of exchangeable ion per gram of dried resin, and they usually fall in the 2.0 to 10.0 meq/gm range. The capacity of the ion-exchange column should be calculated and care taken not to exceed that capacity when placing the mixture on the column.

The general techniques used in ion-exchange separations are quite similar to the techniques used in column chromatographic separations. The ion-exchange resin is supported in a typical chromatography column. The material to be separated is placed on the column, and the components are eluted by changing the properties of the eluent system. The column may be regenerated after use by contacting the column with a solution containing the replacement ion. Such regeneration procedures allow repeated use of the same ion-exchange resin.

The area of ion exchange is far too large to discuss in greater detail, and the experimentalist is referred to other literature sources listed in Sec. 1.7.

1.6 SEPARATION OF MIXTURES

The separation of a complex mixture poses a challenging problem to the chemist regardless of his area of chemistry. In the previous sections of this chapter, we have discussed several techniques for the purification of individual compounds and some techniques applicable to the separation of complex mixtures (chromatography).

The proper choice of the method of separation depends on the physical and chemical properties of the mixture (number of phases present, the solubility and volatility of the various components, type of functional groups present, quantity of material present, etc.). The authors suggest the use of only three techniques, or the combination of these techniques, in the separation of mixtures. These are fractional distillation, extraction, and chromatography. Specific separation procedures applicable to all cases cannot be prescribed because each mixture presents an individual challenge. The chemist must use his experience and knowledge to successfully achieve the desired separation.

Prior to the start of a separation of a complex mixture, the infrared spectrum and gas-liquid chromatogram or thin-layer chromatogram of a representative sample of the mixture should be recorded.

The physical state of the sample will dictate the individual procedure to be employed in the attempted separation. If the sample is a homogeneous liquid, a fractional distillation may be attempted. A fractional distillation generally requires 2 to 3 ml of sample in a small fractionating column to achieve a suitable separation. For smaller samples, one must employ chromatographic techniques (preparative gas-liquid chromatography). A small portion of the sample is tested for thermal stability up to the maximum temperature to be employed (approximately 180° is recommended). If the sample appears to be stable, i.e., no dramatic color change, evolution of a gas, or the deposition of polymeric material, the entire sample may be subjected to fractional distillation (see Sec. 1.2.2). Care should be exercised not to heat the distillation pot above the temperature tested for the thermal stability of the sample (use an oil bath with thermometer for heating purposes). If the sample does not distill up to 150°, a reduced pressure distillation should be attempted. Care must be exercised to avoid reducing the boiling point of any fraction present to below the temperature of the condenser, in which case that fraction will disappear into the cold trap protecting the vacuum pump (or down the drain with the water when a water aspirator is used!). It is always a good practice to clean the cold traps before and after each distillation so that recovery of such volatile fractions is possible.

The volume collected vs. temperature is recorded so that a distinction between fractions can be noted. The infrared spectrum and gas-liquid chromatogram of each fraction, and the pot residue, if any remains, should be recorded. If the degree of separation in the fractional distillation was not sufficient, each fraction may be individually redistilled, or one must revert to the use of extraction or chromatographic techniques.

Heterogeneous samples, containing two liquid phases or one liquid and a solid phase, may also be subjected to distillation (it is not generally advisable to filter the solid phase since a considerable quantity of the solid phase will undoubtedly be soluble in the liquid phase, and vice versa, only a minor degree of separation being accomplished).

Extraction techniques should be attempted if fractional distillation techniques do not prove satisfactory. Prior to attempting an extractive separation, the solubility of the sample in the two extracting phases should be determined. Water is almost always one of the solvents employed; however, it presents the greatest difficulty of recovery with highly polar compounds that may be contained in the mixture. The solubility of a small weighed portion of the unknown in a small volume of water should be determined before proceeding with the extraction. If the mixture is soluble, the aqueous phase should be extracted several times with small portions of ether or chloroform to determine the amount of material that is recoverable.

If only a small portion of the material is recovered, extraction techniques may not prove satisfactory. (The unknown sample may be recovered by evaporating the water solution.) Compounds which show great water solubility and low organic solvent solubility include polyfunctional compounds, e.g., glycols, aminoalcohols, and aminoacids. The solubility of such compounds in water may be substantially reduced by saturating the water phase with sodium chloride; however, this then forbids recovery of the compounds remaining in the aqueous phase by simple evaporation of the water. The separation of such highly polar or polyfunctional compounds is most readily accomplished by ion-exchange techniques.

Samples which display low water solubilities may be subjected to an extractive separation. A trial run on a small portion of the sample is carried out first; this avoids the possibility of losing the entire sample if the separation is not successful. A small weighed portion of the sample (300 to 500 mg) is dissolved in 25 ml of ether. The ether layer is first extracted with 10-ml portions of 5% sodium bicarbonate (CAUTION: The formation of carbon dioxide may generate considerable pressure in the separatory funnel) until the aqueous phase remains slightly basic (this usually requires only one extraction on this scale). The aqueous phase will contain the salts of the strongly acidic compounds (carboxylic acids, sulfonic acids, etc.) The free acids are regenerated by careful acidification (carbon dioxide evolution!) with concentrated hydrochloric acid. The compounds may be recovered by filtration, if solid, or by extraction with ether. Occasionally the acids may be recovered by precipitation as the calcium or barium salts (particularly with sulfonic acids). The original ether layer is then extracted with a 10-ml portion of 5% sodium hydroxide to remove the weakly acidic compounds (phenols, etc.). The free acids are regenerated by acidification with concentrated hydrochloric acid and recovered by extraction. The aqueous phase remaining after extraction of the weakly acidic fraction is carefully neutralized by the slow addition of 20% sodium hydroxide, carefully observing the solution near the neutralization point. A cloudiness, or a change in color, which develops and disappears on further basification indicates the possible presence of an amphoteric compound, for example, an aminophenol. The solution should be carefully neutralized to the point of maximum cloudiness, or at the point of color change, and repeatedly extracted with ether or an ether-chloroform mixture.

The original ether phase is finally extracted with a 10-ml portion of 5% hydrochloric acid until the aqueous phase remains acidic after extraction. The aqueous phase will contain all the basic components originally in the mixture as their conjugate acids. The acid phase is made strongly basic by the addition of 20% sodium hydroxide, and the free bases are recovered by extraction with ether.

The resulting final ether phase contains the neutral component(s) of the mixture which are recovered by evaporation of the ether.

The weight of each fraction is determined, and the total weight of recovered material is compared with the weight of the initial sample. Total recoveries should range in the 85 to 95% region. The amounts of each of the components in the mixture do not need to be the same; in fact, reaction mixtures are generally composed of one or two major components and one or more minor components.

Each fraction obtained in the separation sequence is analyzed for purity by gas-liquid or thin-layer chromatography. If any fraction still contains more than one component, further separation by chromatographic techniques must be carried out. Chromatographic techniques are outlined in the foregoing sections on chromatography and will not be discussed further here.

It is important for the chemist to realize that the scheme discussed above may not always provide the desired results and that modifications may be necessary. Furthermore, during the extraction sequence, the unknown mixture comes in contact with both acid and base, and the components present in the mixture may undergo chemical reactions under these conditions, for example, the acid-catalyzed hydrolysis of acetals, ketals, esters, or acid chlorides, the base-catalyzed hydrolysis of acid derivatives and the solvolysis of reactive organic derivatives. If all the infrared peaks appearing in the separated fractions are present in the original infrared spectrum of the mixture, and if no additional peaks appear, it may be assumed that it is then safe to invest the remainder of the sample through the separation scheme, being careful to reproduce the exact procedures carried out in the trial run with the smaller portion of the sample.

1.7   REFERENCES

GENERAL

1. Berg, E. W., *Physical and Chemical Methods of Separation*. New York: McGraw-Hill Book Company, 1963.

2. Sixma, F. L. J., and Wynberg, H., *A Manual of Physical Methods in Organic Chemistry*. New York: John Wiley & Sons, Inc., 1964.

3. Wiberg, K. B., *Laboratory Technique in Organic Chemistry*. New York: McGraw-Hill Book Company, 1960.

4. Weissberger, A., *Technique of Organic Chemistry, Vol. III*. New York: Interscience Publishers, Inc., 1956.

ADSORPTION CHROMATOGRAPHY

5. Bobbitt, J. M., Schwarting, A. E., and Gritter, R. J., *Introduction to Chromatography*, Reinhold Book Corp., New York, 1968.

6. Cassidy, H. G., "Adsorption and Chromatography" in *Technique of Organic Chemistry, Vol. V*, A. Weissberger, ed. New York: Interscience Publishers, 1951.

7. Cassidy, H. G., "Fundamentals of Chromatography" in *Technique of Organic Chemistry, Vol. X*, A. Weissberger, ed. New York: Interscience Publishers, 1957.
8. Lederer, E., and Lederer, M., *Chromatography*, 2nd ed. Amsterdam: Elsevier Publishing Co., 1957.
9. Lederer, M., ed., *Chromatographic Reviews*. Amsterdam: Elsevier Publishing Co.
10. *Journal of Chromatography*. Amsterdam: Elsevier Publishing Co.

THIN-LAYER CHROMATOGRAPHY

11. Bobbitt, T. M., *Thin-layer Chromatography*. New York: Reinhold Publishing Corporation, 1963.
12. Randerath, K., *Thin-layer Chromatography*. New York: Academic Press, Inc., 1963.
13. Stahl, E., *Thin-layer Chromatography*. New York: Academic Press, Inc., 1964.
14. Truter, E. V., *Thin-layer Chromatography*. New York: Interscience Publishers, 1963.
15. Wollish, E. G., Schmall, M., and Hawrylyshyn, M., "Recent Developments in Equipment and Applications," *Anal. Chem.*, **33**, 1138 (1961).

PAPER CHROMATOGRAPHY

16. Block, R. J., Durrum, E. L., and Zweig, G., *Paper Chromatography and Paper Electrophoresis*. New York: Academic Press, Inc., 1958.
17. Hais, I.M., and Macek, K., eds., *Paper Chromatography*. New York: Academic Press, Inc., 1963.

GAS CHROMATOGRAPHY

18. Burchfield, H. P., and Storrs, E. E., *Biochemical Applications of Gas Chromatography*. New York: Academic Press, Inc., 1962.
19. Hardy, C. J., and Pollard, E. H., "Review of Gas-Liquid Chromatography" in *Chromatographic Reviews, Vol. I*, M. Lederer, ed. Amsterdam: Elsevier Publishing Co., 1960.
20. Kaiser, R., *Gas Phase Chromatography*. London: Butterworths Scientific Publications, 1963.
21. Keulemans, A. I. M., *Gas-Chromatography*, 2nd ed., C. G. Verver, ed. New York: Reinhold Publishing Corporation, 1959.
22. Littlewood, A. B., *Gas Chromatography, Techniques and Applications*. New York: Academic Press, Inc., 1962.
23. Nogare, S. D., and Juvet, Jr., R. S., *Gas-Liquid Chromatography*. New York: Interscience Publishers, 1962.
24. Purnell, H., *Gas Chromatography*. New York: John Wiley & Sons, Inc., 1962.
25. *Gas Chromatography Abstracts*. London: Butterworths Scientific Publications, 1959–1962.
26. *Journal of Gas Chromatography*, Preston Technical Abstracts Company, Evanston, Ill.

ION EXCHANGE

27. Calmon, C., and Kressman, T. R. E., eds., *Ion Exchangers in Organic and Bio-chemistry.* New York: Interscience Publishers, Inc., 1957.

28. Nachod, F. C., ed., *Ion Exchange Theory and Application.* New York: Academic Press, Inc., 1956.

29. Nachod, F. C., and Shubert, J., eds., *Ion Exchange Terminology.* New York: Academic Press, Inc., 1956.

2

Physical Characterization

2.1   INTRODUCTION

The physical properties of substances, i.e., melting and boiling points, density, refractive index, empirical formula (calculated from its elemental analysis), molecular weight, solubility, optical rotation, and spectral properties (to be considered in a separate section of this text), provide information useful in making structural assignments. The physical properties of materials are dependent on the purity of the sample; therefore, to derive the maximum utility out of the measured data, the samples must be of a high degree of purity. Chapter 1 was devoted to discussions of separation and purification techniques that may be applied to obtain pure samples. This chapter is devoted to discussions concerning the measurement and interpretation of the physical properties of a material.

2.2   MELTING POINTS

The melting point of a solid, or conversely, the freezing point of a liquid, may be defined as the temperature at which the liquid and solid phases are in equilibrium at a given pressure. Implied in this statement is the more quantitative definition of melting point or freezing point as that temperature at which the vapor pressures of the solid and liquid phases are equal.

Although the intersection of the vapor-pressure curves for the liquid and solid phases gives a single point temperature for the melting point of a solid, in an actual experimental determination this is rarely observed owing to the experimental methods employed. In general, very small samples of solids are placed in a capillary tube in an oil bath, or between glass plates on a hot stage, whose temperature is continuously raised by one to two degrees per minute. Since a finite time is required for complete melting to occur, a finite temperature range will be traversed during this process. Therefore, it is necessary to report a range of temperatures for the melting point of any substance, for example, benzoic acid 121.0 to 121.5°. Handbooks and tables of melting points list a single temperature, usually an average of several reported, indicating where a sample of a pure substance should melt. However, in reporting the melting point of a specific sample, a melting-point range should be reported, that is, the temperature at which the first melting is observed and the temperature at which the last crystal disappears. Both the temperature and range of an observed melting point are important quantities. Both reflect the purity of the sample.

The addition of a nonvolatile soluble substance to a liquid produces a decrease in the vapor pressure of the liquid phase (an application of Raoult's Law). From the definition of melting or freezing point, a decrease in the vapor pressure of the liquid phase will result in a lowering of the melting or freezing point (see Fig. 2.1). The vapor pressure of the solid in equilibrium with the liquid will, for all purposes, be unaffected, since the solvent in the solid phase will tend to be present as pure small crystalline units with crystalline solute dispersed throughout. The solute will not, in general, be incorporated into the crystal lattice of the crystalline solvent. Therefore, if the solute remains soluble in the liquid phase, the solute concentration will change as the amount of liquid solvent changes during the melting process. The net result is that the melting point of the mixture will increase as melting proceeds, giving a melting range much greater (several degrees) than that observed for a pure substance. In the determination of molecular weights by melting-point depression techniques (see Sec. 2.8 on molecular weight deter-

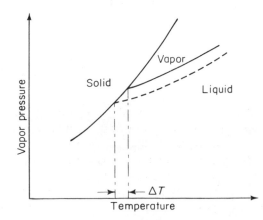

Fig. 2.1. Vapor pressure vs. temperature diagram. Solid lines represent the pure substance A, and the dotted lines represent a solution of a nonvolatile material in A.

mination), the temperature recorded as the melting point of the mixture is the temperature at which the last crystal disappears into the liquid phase.

During the purification of a solid material, the melting range should decrease, and the melting point should increase as the impurities are removed during successive stages of purification. When no further change in the melting range and point is observed on further purification, the material is probably as pure as it can be obtained using that purification technique.

Several methods may be used to determine melting points. The capillary-tube method is one of the most common methods employed. A portion of a finely ground sample is introduced into a fine glass capillary, 1 by 100 mm, sealed at one end. (Glass capillaries are commercially available, or the student may prepare his own by drawing a *clean* piece of 8 to 12 mm soft glass tubing.) Enough sample is placed in the capillary tube and firmly packed until a column of sample 1 to 2 mm is obtained. Occasionally, it is difficult to introduce the sample into the capillary and get the sample firmly packed in the closed end of the capillary. Small quantities of material may be forced into the open end of the capillary and then induced to move down the capillary to the sealed end by vibrations. This may be accomplished by firmly grasping the capillary tube and tapping against the bench or table top, by allowing the capillary to fall down a glass tube and bouncing on the bench top, or by drawing a file across the side of the capillary. After the sample is introduced into the capillary, the capillary may be sealed under vacuum as illustrated in Fig. 2.2. This may be required if the sample appears to undergo oxidative decomposition at elevated temperatures.

The capillary is attached to a thermometer by means of a small rubber band cut from $\frac{1}{4}$-in. rubber tubing, and the thermometer and sample are placed in a heating medium. The sample must be maintained at the same level as the mercury bulb of the thermometer. Mineral oil is usually employed as the heating medium and may be safely used up to 225 to 250°C. For higher temperatures, a hollowed copper block may be used (see Fig. 2.3c).

A variety of liquid heating baths may be employed. Figure 2.3a illustrates the proper use of the Thiele tube. The thermometer and sample are placed in the Thiele tube so that the sample and sensing bulb of the thermometer are slightly below the bottom of the top side-arm neck of the tube. A Bunsen burner is used to gently heat the side arm of the Thiele tube, the circulation of the heating medium occurring by convection. More elaborate setups may be employed which contain internal-resistance wire heaters and mechanical stirrers. A magnifying lens is used to observe the sample during the melting process.

The rate of heating of a sample, in any method of melting-point determination, should be approximately 1°C per minute while traversing the melting-point range. Greater heating rates may be used to raise the temperature of the heating bath to about 10 to 15° below the expected melting point temperature. Many compounds display melting ranges which depend

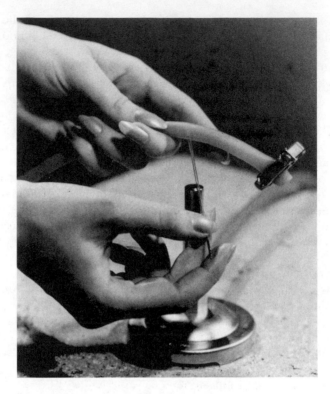

Fig. 2.2. Sealing a melting point capillary under reduced pressure. (Photograph by Bruce Harlan, Notre Dame, Ind.)

on the rate of heating of the sample. Examples of compounds which display such behavior are those which undergo decomposition, chemical isomerization (thermally induced), or changes in crystal form.

Figure 2.3b illustrates the use of a small round-bottom flask as the heating bath vessel; however, maintaining a uniform temperature throughout the heating medium is quite difficult unless some means of stirring is provided, for example, a stream of air bubbles introduced by means of an air tube.

The Fisher-Johns melting-point apparatus (see Fig. 2.4) is a typical hot-stage melting-point apparatus. A small portion of the sample is placed between two 18-mm microscope slide cover glasses, which are then placed in the depression of the electrically heated aluminum block. The rate of temperature rise is regulated by means of a rheostat, and the sample is observed with the aid of a magnifying lens. The Fisher-Johns apparatus is useful for determining melting points between 30 and 300°. This apparatus must be calibrated periodically using a set of melting-point standards.

The Nalge-Axelrod hot-stage melting-point apparatus (Fig. 2.5) may be used to determine melting points while observing the sample with polarized

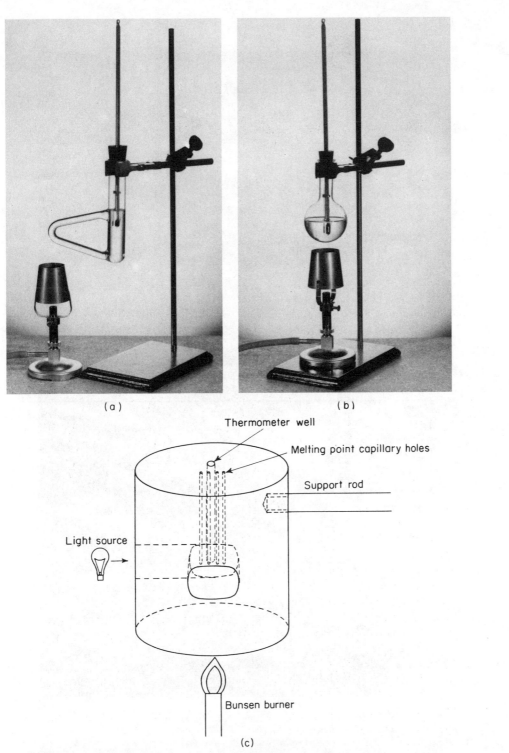

Fig. 2.3. Setups for the determination of melting points. (a) Employing a Thiele tube. (b) Employing a small round bottom flask. (Photographs by Bruce Harlan, Notre Dame, Ind.) (c) Employing a copper block.

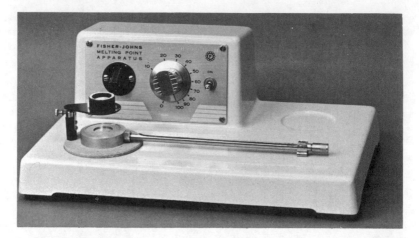

Fig. 2.4. Fisher-Johns hot-stage melting point apparatus. (Reproduced by courtesy of Fisher Scientific Company.)

Fig. 2.5. Nalge-Axelrod hot-stage melting point apparatus. (Reproduced by courtesy of the Nalge Co., Inc., Rochester, N.Y.)

light. The sample is placed between two microscope slide cover glasses and placed on the electrically heated hot stage over a small hole through which the polarized light passes. The microscope tube is adjusted so that polarization of the light by the crystals of the sample occurs, and the temperature of the hot stage is increased until melting occurs. The melting points of anisotropic (exhibiting different properties when tested along axes in different directions, for example, the refraction of light) crystals are easily noted since the polarization colors disappear when melting occurs. The transformation of one crystal form to another form is readily detected. The accessible temperature range of the Nalge-Axelrod melting-point apparatus is 30 to 400°. For materials melting above the ranges of the Fisher-Johns and the Nalge-Axelrod instruments, the capillary tube method utilizing a heated copper block is required.

Mettler Instruments Corporation has developed an automatic melting-point and boiling-point determination apparatus (see Fig. 2.6). For melting points, a temperature digital readout device, capable of handling three samples at the same time, indicates the temperature at which the sample is approximately 80% melted. Melting ranges cannot be determined by use of the digital readout device; however, the apparatus may be connected to a strip chart recorder and the melting range determined from the "melting curve." The advantage of this apparatus is that it removes the human error involved in observing the melting of the compound, and it does not require continuous monitoring by the experimentalist. Micro-boiling points (see Sec. 2.4) are determined quite accurately by this instrument.

Since the temperature-sensing device, usually a thermometer, is not in direct contact with the sample when the Fisher-Johns or the Nalge-Axelrod

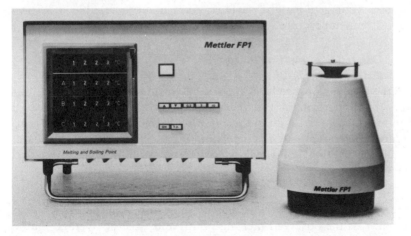

Fig. 2.6. Automatic melting point and boiling point determination apparatus. (Reproduced by courtesy of Mettler Instruments Corporation.)

melting-point apparatus is used, a calibration curve of true temperature vs. the observed temperature should be determined. This is accomplished by recording the melting points of extremely pure standards (a set of melting-point standards is available commercially through the National Bureau of Standards) of known melting points.

The sample should be carefully observed during the melting process for changes in crystal form, decomposition with the formation of gaseous products, or decomposition with general charring. Such observations may be very useful in deriving structural information. Changes in crystal form may be due to a simple change of one allotropic crystal modification to another of the same substance, or they may be due to a thermally induced rearrangement to an entirely new compound, where the final melting point then corresponds to the rearranged material and not to the original sample. The evolution of a gas is generally the result of a thermal decarboxylation of β-carboxy carbonyl compounds, although the loss of small molecules such as water, hydrogen halides, and hydrogen cyanide may occur with some compounds. Should such phenomena occur during the melting point determination, the thermally induced reaction should then be carried out on a larger portion of the sample and the product(s) identified. General decomposition with charring is typical of polyfunctional compounds, such as many sugars, and very high melting compounds.

The melting points of substances that undergo decomposition will vary with the rate of temperature rise of the sample; hence, reliable comparisons of melting or decomposition points are not possible. Determination of the instantaneous melting or decomposition point, accomplished by dropping small quantities of the material on a previously heated hot stage and by increasing the temperature of the hot stage until the sample melts or decomposes upon striking the surface, gives a more reliable melting or decomposition temperature.

2.2.1 *Correction of Melting Points*

In all the foregoing procedures for the determination of melting points, only a very small portion of the thermometer is in contact with the heating medium. Owing to a difference in the coefficient of expansion between mercury and glass, a stem correction must be applied to give a true or "corrected" melting point. This correction is calculated by means of Eq. (2.1)

$$\text{Correction in } ^\circ\text{C} = N(t_1 - t_2)0.000154 \qquad (2.1)$$

where N is the degrees of mercury column above the level of the heating medium, t_1 is the observed melting point, and t_2 is the average temperature of the mercury column. Most research journals recommend the reporting of corrected melting points. If the melting-point apparatus is calibrated with standards, the stem correction is not to be applied.

2.2.2 *Mixture Melting Points*

The phenomenon of melting point depression has many very useful applications in structure determinations. Qualitatively, melting point depressions, or really, the lack of such depression, may be used for the direct comparison of an unknown with possible knowns. For example, chemical and physical properties may limit the number of possibilities for the structure of an unknown to two or three. The melting point of the mixture of the unknown with a known that has the same structure as the unknown does not lead to a depression in the melting point, whereas mixtures of the unknown with knowns that differ in structure from the unknown show depressions in their mixture melting points. The mixtures are prepared by taking approximately equivalent amounts of unknown and known and then melting the mixture until homogeneous. The melt is allowed to crystallize, and the melting point of a small, finely ground portion of the mixture is determined. This procedure has limited utility in that it requires the availability of suitable known compounds. Data derived from such mixture melting point experiments must be used with caution, however, since some mixtures may not display a depression, or only a very small depression, due to compound or eutectic formation. Some mixtures may display an elevation of melting point. This may occur when compound formation occurs or when one has a racemic mixture (see following paragraph).

Mixture melting point data may also be used to distinguish between racemic mixtures (a eutectic of equal amounts of crystals of (−) and (+)-enantiomers), racemic compounds (crystals containing equal quantities of (−) and (+) molecules in a specified arrangement), and racemic solid solutions (crystals containing equal amounts of (−) and (+) molecules arranged in a random manner in the crystal).[1] The addition of a pure enantiomer (1) to a racemic mixture results in an elevation of the melting point, (2) to a racemic compound results in a depression of the melting point, and (3) to a racemic solid solution results in no effective change in the melting point.

The quantitative aspects of melting-point or freezing-point depression are applied in the determination of molecular weights (see Sec. 2.8 on determination of molecular weights).

2.3   FREEZING POINTS

The freezing points of liquids may be used for purposes of characterization; however, freezing points are much more difficult to determine accurately

[1] See E. L. Eliel, *Stereochemistry of Carbon Compounds,* New York: McGraw-Hill Book Company, 1962, pp. 43–46.

and freezing-point data for liquids are much more limited than are melting and boiling point data. In addition, the quantity of sample required to determine a freezing point is much greater than that required for a melting point determination. One to two milliliters of the sample is placed in a test tube and cooled by immersion in a cold bath (the temperature of the cold bath should not be too far below the freezing point of the sample in order to avoid extensive supercooling). The sample is continuously stirred, and the temperature of the sample is recorded periodically. The cooling curve will usually reach a minimum (supercooling) and rise to a constant temperature as crystal growth occurs. This temperature is the freezing point of the sample.

The liquid sample used in a freezing-point determination must be very pure because the presence of even minor amounts of impurities may lead to a substantial depression in the freezing point. This phenomenon is used in determining molecular weights by cryoscopic methods (see Sec. 2.8.2).

2.4   BOILING POINTS

The boiling point of a sample is generally determined during purification of the sample by distillation. Difficulties arise in comparison of boiling points, however, when a boiling point is determined at an applied pressure that is different from the literature reference value. Conversion of a boiling point at one pressure to another pressure can be accomplished by the use of boiling-point nomographs for different classes of compounds (see Appendix I).

The determination of a boiling point by distillation requires the availability of sufficient sample to attain temperature equilibrium in the distilling apparatus. Frequently, insufficient quantities of sample are available to determine boiling points by distillation techniques, and one must use semi-microtechniques. One such technique involves the heating of 0.3 to 0.4 ml of a liquid, with a small boiling chip, in a test tube into which is suspended a thermometer. The sample is gently heated until a continuous reflux of sample from the thermometer and the wall of the test tube is attained. The temperature of the vapors is the boiling point of the sample. Care must be taken not to overheat the sample since superheating of the liquid and vapors will yield erroneous results.

A second microprocedure requires only 2 to 3 drops of sample. The sample is placed in a micro test tube and a very fine capillary, with one end sealed, is placed open-end into the sample. The test tube is suspended in an oil bath and gently heated until a steady stream of bubbles issues from the end of the capillary. The temperature of the oil bath is allowed to slowly decrease until the bubbling stops; this temperature is the boiling point of the sample. The process should be repeated several times until a reproducible boiling point is obtained. One may also use the Mettler apparatus described in Sec. 2.2. This technique is relatively difficult to master, but it should give boiling points within a few degrees of the boiling point that would be obtained during distillation.

2.5 DENSITY MEASUREMENTS

Determination of the density of a substance provides valuable information for the identification of compounds which are not readily converted into derivatives. In addition, the density may be used to calculate the molar refraction (see Sec. 2.6) which may be useful in structure assignments.

Densities are usually determined by a direct comparison of the weights of equal volumes of sample and water at a given temperature, t, and correcting to the density of water at 4° as given in Eq. (2.2).

$$D_t = \frac{\text{weight of sample}}{\text{weight of water at } t} \times D_{\text{H}_2\text{O}} \quad \text{at} \quad t \qquad (2.2)$$

Densities are determined using vessels referred to as pycnometers. Pycnometers with capacities of 1 to 25 ml are commercially available. However, since the structural chemist usually has only limited amounts of sample available, he must occasionally make a pycnometer of suitable size for his use. Suitable micropycnometers can be constructed from 2 to 4 mm soft glass tubing or capillaries, as shown in Fig. 2.7. Pycnometers prepared from 3-mm glass tubing require 0.2 to 0.6 ml of sample and capillaries from 0.02 to 0.05 ml. Density measurements should be carried out using as large a sample size as possible to reduce weighing errors. Densities should be calculated to four significant places, requiring the weighings to be precise to the same degree.

The pycnometer illustrated in Fig. 2.7 is cleaned, completely filled with water, and thermostated at a given temperature. After temperature equilibrium has been attained, the liquid level of the sample in the pycnometer is adjusted to the level of the ring on the right side of the pycnometer by removing liquid from the capillary end with a piece of absorbent paper. The pycnometer is removed from the constant temperature bath, carefully dried,

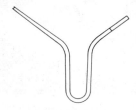

Fig. 2.7. Micropycnometer.

and weighed. The pycnometer is emptied, dried, and filled with the sample, and the process is carried out with the sample. The sample may be recovered uncontaminated for further use.

2.6 REFRACTIVE INDEX

The refractive index, as normally determined, is the ratio of the velocity of light in air to the velocity in the substance being determined. The refractive index can be precisely determined, usually to five significant figures, and is one of the most important physical constants used to describe a compound.

Refractive indices are determined using refractometers, of which there are several types available. Figure 2.8 shows the Abbé refractometer which is the most widely used instrument. The Abbé refractometer consists of a light source, a pair of movable, hinged, water-cooled prisms between which the sample is placed, two movable Amici prisms which compensate for differences in the degree of refraction of light of different wavelengths, a telescope for observing the refraction field, and a scale for reading the refractive

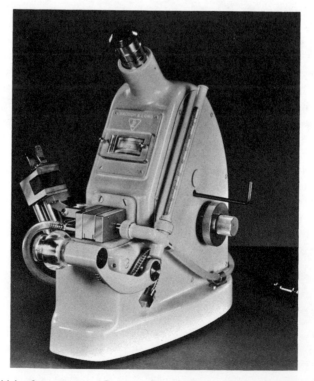

Fig. 2.8. Abbé refractometer. (Courtesy Bausch and Lomb Incorporated, Rochester, N. Y.)

index. If the liquid is quite mobile, a dropping pipette is used to introduce the sample into the small sample depression between the prisms. For viscous samples, the prisms are opened, and one drop of the sample is cautiously applied to the bottom prism; then the prisms are gently closed. Care must be exercised to avoid touching the surface of the prism with the dropper. The light is adjusted to obtain maximum illumination, and the cross hairs in the telescope are brought into focus by adjustment of the eyepiece of the telescope. The Amici prisms are adjusted to remove colors from the field, and the position of the borderline between the black and white fields is adjusted to intersect at the intersection of the cross hairs by adjustment of the course- and fine-adjustment knobs. The refractive index is read from the drum to three decimal places, and the fourth place is estimated. The temperature of the prism assembly is recorded since the refractive index is a function of the temperature. The variation of refractive index with temperature is dependent on the types of functional groups present, but it may be approximated by using a correction of 0.00045 per °C, the refractive index decreasing as the temperature rises. Thus refractive indices determined at one temperature may be reliably compared with values determined at other temperatures. The refractive index n_D^T, for example, for water at 20°, is reported as

$$n_D^{20} \ 1.3333$$

where 20 is the temperature at which the refractive index was measured, and D indicates the wavelength of the light employed as that of the sodium D line.

The simple Fisher refractometer is shown in Fig. 2.9. The sample is placed in a prismatic well formed by a glass slide and the glass eyepiece (see Fig. 2.9b); the prism of liquid refracts the light coming through the sample. If one looks through the eyepiece with no sample present, one sees a scale with an arrow indicating a refractive index of 1.516. When the sample is added to the eyepiece, a secondary image of the arrow will appear either above or below the 1.516 on the scale. The secondary image indicates the refractive index of the sample.

Refractive index data, along with density data, can be used to calculate an observed molar refraction, Mr_D, from the following:

$$Mr_D = \frac{M(n^2 - 1)}{d(n^2 + 2)} \tag{2.3}$$

where M is the molecular weight of the sample, d is the density of the sample, and n is the refractive index. The observed molar refraction may then be compared with the calculated molar refraction obtained by summing group and atomic refractions (see Table 2.1) for the functional groups comprising the molecule. Comparison of molar refractions is particularly useful

(a)

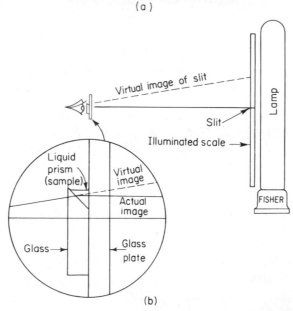

(b)

Fig. 2.9. Fisher refractometer.   (Reproduced by courtesy of the Fisher Scientific Company.)

in the identification of relatively inert compounds which are not readily converted into derivatives, for example, hydrocarbons, halides, and many nitrogen-containing derivatives. However, one cannot distinguish between simple substitution isomers. The following example illustrates the estimation of the molar refraction of 3-hexyne using the atomic refractions contained in Table 2.1.

$$
\begin{array}{lr}
6 \text{ carbon atoms} & 14.508 \\
1 \text{ triple bond} & 2.398 \\
10 \text{ hydrogen atoms} & \underline{11.000} \\
Mr_D & 27.906
\end{array}
$$

Table 2.1 Atomic Refractions*

Atom	Mr_D	Atom	Mr_D
C, singly bound and alone	2.592	N, RNHOH	2.48
C, singly bound	2.418	N, $-NH-NH_2$	2.47
C, double bond†	1.733	N, $RC\equiv N$	3.05
C, triple bond†	2.398	N, $ArC\equiv N$	3.79
C, conjugated	1.27	N, $R_2C=NOH$	3.93
H	1.100	N, $RCO\,NH_2$	2.65
O, hydroxyl	1.525	N, RCONHR	2.27
O, ethereal	1.643	N, $RCONR_2$	2.71
O, ketonic	2.211	NO, as nitrites	5.91
O, ester	1.64	NO, as nitrosoamines	5.37
S, as SH	7.69	NO_2, as alkyl nitrite	7.44
S, as RSR	7.97	NO_2, as alkyl nitrate	7.59
S, in $-CNS$	7.91	NO_2, as nitroparaffin	6.32
S, as RSSR	8.11	NO_2, as nitroaromatic	7.30
N, RNH_2	2.45	NO_2, as nitramine	7.51
N, $ArNH_2$	3.21	F	0.95‡
N, R_2NH	2.65	Cl	5.967
N, Ar_2NH	3.59	Br	8.865
N, R_3N	3.00	I	13.900
N, Ar_3N	4.36		

*Taken from R. R. Dreisback, "Physical Properties of Organic Compounds," *II, Advances in Chemistry Series*, **22**, American Chemical Society, Washington, D.C., 1959.
†Double and triple bond only.
‡For fluorine in polyfluoroderivatives the value is 1.1.

2.7 OPTICAL ROTATION

Conducting a beam of plane polarized light through an optically active medium results in an angular displacement of the plane of the polarized light (see Sec. 7.2 on optical rotatory dispersion for a more detailed discussion of the phenomenon involved). The principal components of a polarimeter (see Fig. 2.10) used to determine optical rotations are a light source, a polarizing Nicol prism, a sample tube compartment, an analyzing Nicol prism, an observation lens, and a scale indicating the optical rotation of the sample.

The purified sample, approximately 200 to 500 mg, is accurately weighed and is dissolved in an appropriate volume of solvent, the volume depending

Fig. 2.10. Polarimeter. (Reproduced by courtesy of O. C. Rudolph and Sons, Caldwell, N. J.)

on the volume of the sample tube to be used. Suitable solvents include water, methanol, ethanol, chloroform, or mixtures of water and an alcohol. If the rotation of the sample is low, higher concentrations may be used. Liquid samples may also be run as neat samples. The sample solution must be free of small particles (dust, lint, etc.) and should also be colorless if possible. The sample is placed in the sample tube, generally a 10-cm tube with glass end plates held firmly in place with rubber washers and a screw cap, by removing the screw cap and end plate from one end of the tube, holding the tube nearly vertical, and adding the sample by means of a capillary dropper. The sample is added until it rises slightly above the end of the sample tube. The glass end plate is carefully placed over the end of the tube so that no air bubbles are trapped in the sample. If bubbles have been trapped, the end plate must be removed and more sample added until the sample level extends above the end of the tube; then the glass end plate is again put in place. When no air bubbles have been trapped, the rubber washer is placed on top of the glass end plate, and the screw cap is firmly secured. One must be careful not to apply too much pressure in screwing the end cap in place since anomalous rotations may be produced. Top-filling polarimeter tubes are also available which provide for a somewhat easier filling operation (see Fig. 2.10).

The *blank* or *zero* reading of the polarimeter is determined before the sample tube is placed in the light beam. The light field is observed through the observation lens. If the orientations of the two Nicol prisms are not the same, the light field will appear as two halves, one dark and one light half, depending on the relative angular displacement from 0°. The observation lens is adjusted until maximum sharpness between the two fields is obtained. The movable prism is then adjusted until no distinction can be made between the two halves of the light field. Finally, the rotation is read from the scale on the movable prism mount. This scale is usually divided into 0.25° divisions with a vernier capable of reading to the nearest 0.01°. Several adjustments of the movable prism are made, and readings are taken and averaged to obtain the rotation. The *blank* or *zero* reading may or may not coincide with the 0° mark on the scale, and an appropriate correction must be applied to the observed rotation of the sample. The sample tube is then placed in the light path and the rotation determined as outlined earlier in determining the instrument *zero* point.

The specific rotation [α] is given by Eq. (2.4)

$$[\alpha]_D^T = \frac{\alpha \times 100}{l \times c} \tag{2.4}$$

where α is the observed rotation of the sample in degrees, l is the length of the sample tube in decimeters, c is the concentration of solute in grams per 100 ml of solution, T is the temperature of the sample at which the rotation was measured, and D specifies the wavelength of the light used as the sodium D line. The solvent used should also be specified as well as the concentration. The rotation of the sample is given, for example, as $[\alpha]_D^{20}$ + 65.2° (c1.0, H_2O). The rotation of a sample should be determined using at least two different nonintegrally related concentrations of solute. Occasionally, materials may possess very large specific rotations, and a single measurement will not allow one to distinguish between, for example, a moderate negative specific rotation and a large positive specific rotation. Specific rotations may be compared only when using the same solvent, since specific rotations may vary with solvent.

2.8 DETERMINATION OF MOLECULAR WEIGHTS

The molecular weight of a substance can be determined in one of several ways. Most methods utilize Raoult's Law, i.e., changes in vapor pressure upon the addition of a nonvolatile solute to a pure solvent. These methods include melting point depression (Rast determination), boiling point elevation, and vapor pressure equalization (isopiestic and osmometric). Molecular weights of volatile compounds can be determined by use of the ideal gas law in which a known weight of material is vaporized at a given temperature and the volume and pressure are measured, or by mass spectrometry. Mass spectrometric methods are the easiest and generally the most reliable methods, although the necessary instrumentation is very expensive.

2.8.1 Rast Method

The addition of a nonvolatile solute to a solid material results in a reduction of the vapor pressure of the liquid phase in equilibrium with the solid phase (see Sec. 2.2 on melting points), and ultimately to a lowering of the melting point of the solid. The magnitude of the melting point depression for a given concentration of the sample is a unique physical constant of the pure solid phase and is referred to as the *molal melting point depression constant*. The molecular weight is calculated from Eq. (2.5)

$$\text{Mol. wt.} = \frac{K \times w \times 1000}{\Delta t \times W} \tag{2.5}$$

where K is the molal melting point depression constant, w is the weight in grams of the solute, Δt is the depression in the melting point, and W is the weight of the solid solvent (or liquid solvent in the case of freezing point depression determinations).

Table 2.2 lists various solid solvents, and their molal freezing point depression constants, which may be used in Rast molecular-weight determinations. The choice of the solid solvent depends on several factors. The solute must be soluble in the liquid phase of the solvent. There must be no chemical or physical (complex formation) reaction of the solute with the solvent which would change the nature of the solvent or the solute. The solute must also be chemically stable at the temperature of the melting point of the mixture. Finally, the solid solvent should be chosen to provide the maximum melting point depression for similar concentrations of various solutions to increase the accuracy of the determination.

Table 2.2. Solid Solvents for Rast Molecular Weight Determinations

	Melting point, °C	Molal depression constant, °C
Benzophenone	48	9.8
Borneol	202	35.8
Bornylamine	164	40.6
Camphene	49	31.08
Camphoquinone	190	45.7
Camphor	178	40.0
Cyclopentadecanone	65.6	21.3
Diphenyl	70	8.0
Naphthalene	80.2	6.9
Perylene	276	25.7

2.8.1a Procedure

An accurately weighed sample of the unknown substance is added to an accurately weighed portion of the solid solvent (the concentration of the solute must be kept below 10 and preferably 5 mole percent). The mixture is melted and mixed until homogeneous.

The melting points of the mixture and pure solid solvent are determined (the melting point being the temperature at which the last crystals disappear into the liquid phase), preferably employing capillary melting point tubes.   The samples are allowed to cool and solidify, and the melting points redetermined.   This process is repeated until consistent results are obtained.

Precautions must be exercised with compounds that undergo decomposition, giving two or more lower molecular weight fragments, for example, chlorohydrins and solvates. In such cases, anomalously low values of molecular weights are derived.

2.8.2 Cryoscopic Method

The determination of molecular weights by freezing point depression measurements is, in theory, the same as melting point depression measurements involving solids. The experimental procedure is much more involved than in the Rast method.   Since the molal depression constants are considerably smaller for liquids (see Table 2.3) compared with the solid solvents, there is a demand for greater accuracy in the measurement of the freezing points, requiring the use of a Beckman thermometer. The apparatus (see Fig. 2.11) is more complex in order to isolate the experiment from disturbing influences of the environment.   An additional complicating factor is that the solubility of many compounds is greatly reduced at the lower temperatures involved in these experiments.

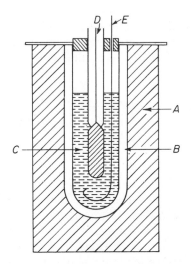
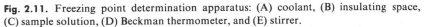

Fig. 2.11. Freezing point determination apparatus: (A) coolant, (B) insulating space, (C) sample solution, (D) Beckman thermometer, and (E) stirrer.

2.8.2a Procedure

An accurately weighed portion of solvent (usually 25 to 50 g) is placed in the inner vessel of the apparatus shown in Fig. 2.11.   The cooling curve is recorded for the pure solvent while a constant stirring rate is maintained.   The freezing point is taken as the

Table 2.3. Solvents for Freezing Point Depression Determinations

	Freezing point, °C	Molal depression constant, °C
Acetic acid	16.7	3.9
Benzene	5.5	5.12
Cyclohexane	6.5	20.0
Sulfuric acid	10.5	6.81
Water	0.0	1.86

constant temperature obtained after supercooling and crystallization begins. The freezing point should be determined at least twice to check the reproducibility of the determination.

An accurately weighed portion of the unknown is then added, and the cooling curve for the solution is recorded. After the supercooling portion, a constant temperature will not be obtained as was found for the pure solvent. The freezing point is determined by extrapolation of the slowly descending cooling curve, after supercooling, to the initial cooling curve; the temperature of the intersection is taken as the freezing point.

2.8.3 Boiling Point Elevation Determination

The addition of a nonvolatile solute to a solvent results in a decrease of the vapor pressure of the solvent and hence an increase in the boiling point of the solution. The elevation of the boiling point can be expressed quantitatively by Eq. (2.6)

$$\text{Mol. wt.} = \frac{K_B \times w \times 1000}{W \times \Delta t} \tag{2.6}$$

where K_B is the molal boiling point elevation constant, w is the weight of the solute, W is the weight of the solvent, and Δt is the difference in the boiling points of the pure solvent and the solution.

Figure 2.12 shows a typical setup for the determination of boiling points. The apparatus must be carefully shielded from drafts. The boiling point of the pure solvent is first determined (again a Beckman thermometer is recommended), whereupon the unknown is added, and the boiling point of the solution is determined. Care must be exercised so as not to cause superheating of the contents of the flask.

Table 2.4 lists several solvents suitable for use in boiling-point elevation determinations.

Fig. 2.12. Boiling point determination apparatus: (A) water-cooled condenser and (B) Beckman thermometer.

Table 2.4. Solvents for Boiling Point Elevation Determinations

	Boiling point, °C	Molal elevation constant, °C
Acetic acid	118	3.07
Acetone	56	1.71
Benzene	80	2.53
Carbon tetrachloride	76	5.03
Chlorobenzene	132	4.15
Chloroform	60	3.63
Cyclohexane	81	2.79
Ethanol	78	1.22
Nitrobenzane	210	5.24
n-Octane	126	4.02
Toluene	110	3.33
Water	100	0.512

2.8.4 Isopiestic Method

In the isopiestic method of molecular weight determination, a solution containing a specific quantity of a standard substance (S) is equilibrated in a closed system, at a constant temperature, with a solution containing a measured quantity of the unknown (X). The vapor pressures of the two solutions will be different at the beginning of the experiment due to the difference in concentrations. At equilibrium, however, the vapor pressures, and thus the mole fractions, must be the same for both solutions. For this equilibrium to be attained, a transfer of solvent from the more dilute to the more concentrated solution must occur by a vaporization-condensation process. The molecular weight (MW_x) of the unknown can be calculated [see Eq. (2.7)] from the final volumes of the two solutions (V_s and V_x), the weights of S and X employed (W_s and W_x), the molecular weight of the standard (MW_s), the density of the solvent (ρ), and the molecular weight of the solvent (MW_{sol}). An approximation is made in the derivation of Eq. (2.7) in the assumption that the final volumes of the two solutions are due only to the solvent. This approximation does not lead to serious deviations as long as the concentrations of S and X are relatively low, of the order of a few weight percent, and of the same magnitude. The apparatus is illustrated in Fig. 2.13.

$$\frac{W_x/MW_x}{W_x/MW_x + V_x\rho/MW_{sol}} = \frac{W_s/MW_s}{W_s/MW_s + V_s\rho/MW_{sol}} \tag{2.7}$$

Side 1   Side 2

Fig. 2.13. Isopiestic tube for molecular weight determinations.

2.8.5 Osmometric Method

When pure solvent and a solution of a relatively nonvolatile solute in that solvent are placed in a closed system, a transfer of solvent from the pure solvent to the solution occurs by a vaporization-condensation process. The adiabatic transfer of solvent produces a temperature differential between the pure solvent and the solution due to the heat of vaporization involved in the evaporation and condensation of the solvent. The temperature differential is proportional to the difference in vapor pressures of the pure solvent and the solution and, hence, is proportional to the solute concentration and the type of solvent used.

Figure 2.14 shows a commercially available, vapor pressure osmometer. Small drops of the pure solvent and the solution are applied to thermistor beads in an environment saturated with pure solvent vapor. The system is allowed to come to equilibrium, and the temperature difference between the thermistor beads is measured as a resistance difference, ΔR. For example, a 0.01 molar solution in benzene produces a temperature differential of about 0.016°C. Since the vapor-pressure osmometer pictured in Fig. 2.14 is capable of distinguishing ± 0.0001°C, excellent accuracy can be obtained at these relatively low concentrations.

The instrument is calibrated with a concentration series (0.01 to 0.1 molar) of a known compound in a given solvent. A ΔR vs. concentration plot is prepared, from which the molar concentration of an unknown solution, in the same solvent, may be determined from observed ΔR values.

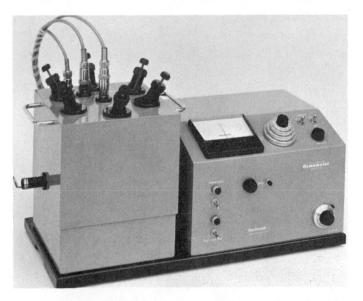

Fig. 2.14. Mechrolab vapor pressure osmometer, Model 302. (Reproduced by courtesy of F & M Scientific Division of Hewlett-Packard, Avondale, Pa.)

It is generally wise to determine the molecular weight of an unknown using at least two different concentrations of the unknown in the specified solvent. Molecular weight measurements require only about half an hour when using the vapor pressure osmometer.

Almost any relatively volatile solvent may be used with the vapor pressure osmometer. In general, one should use polar solvents for polar solutes, and vice versa, in order to reduce association between solute molecules. To obtain an accuracy of about one percent, the vapor pressure of the solute must be below one percent of the vapor pressure of the pure solvent.

2.9   REFERENCES

GENERAL

1. Cheronis, N. D., "Micro and Semimicro Methods" in *Technique of Organic Chemistry*, Vol. VI, Part I, A. Weissberger, ed. New York: Interscience Publishers, 1954.
2. King, J. F., "Applications of Dissociation Constants. Optical Activity, and Other Physical Measurements" in *Technique of Organic Chemistry*, Vol. XI, A. Weissberger, ed. New York: Interscience Publishers, 1963.
3. Overton, K. H., "Isolation, Purification, and Preliminary Observations," in *Technique of Organic Chemistry*, Vol. XI, A. Weissberger, ed. New York: Interscience Publishers, 1963.

Part II

ABSORPTION SPECTROSCOPY

3

Ultraviolet Spectroscopy

3.1 GENERAL INTRODUCTION

The determination of molecular structure has been greatly facilitated by recent advances in spectroscopy, both in the interpretation and correlation of spectral data and in the design and utility of the various spectrometers. Most of these physical techniques are based on the fact that molecules are capable of absorbing radiant energy and undergoing various types of excitation. The modes of excitation available to most molecules include electronic excitation, bond deformations, rotational excitation, and nuclear spin inversions. These modes of excitation require different quantities of energy, and hence the absorptions appear in different regions of the electromagnetic spectrum. Table 3.1 illustrates the regions, wavelengths, energies of the transitions, and the types of transitions occurring in each region of the spectrum. Below the ultraviolet region, the energy available approaches the ionization potential of molecules, leading to ion formation and, at still lower wavelengths, to nuclear transformations. Excitation leading to ionization and subsequent fragmentation is employed in mass spectrometers and yields valuable information with respect to the structure of a molecule.

Certain definitions of terms and equations should be presented before discussing the individual spectral regions. These are presented in Table 3.2. A certain variance in the use of the terms will be noted many times among different scientists. An understanding of the meaning and usage of each of the terms is therefore necessary.

The energies involved in the various transitions are quantized; that is, a given transition in a specific molecule can be effected only with radiant energy corresponding to the energy gap between the two states involved in the transition. This usually leads to a sharp absorption line at a wavelength corresponding to the energy involved. However, in many instances, the

83

Table 3.1. Regions of the Electromagnetic Spectrum

Region	Wavelength	Energy of excitation	Type of excitation
Vacuum ultraviolet	100–200 mμ	286–143 Kcal	electronic
Quartz ultraviolet	200–350 mμ	143– 82 Kcal	electronic
Visible	350–800 mμ	82– 36 Kcal	electronic
Near infrared	0.8–2.0 μ	36– 14.3 Kcal	overtones of bond deformations
Infrared	2–16 μ	14.3– 1.8 Kcal	bond deformations
Far infrared	16–300 μ	1.8– 1 Kcal	bond deformations
Microwave	~ 1 cm	~10^{-4} Kcal	rotational
Radio frequency	~ meters	~10^{-6} Kcal	electron and nuclear spin transitions

Table 3.2. Definition of Terms and Equations

Term	Symbol	Equation	Dimensions
Wavelength	λ	—	Å (Ångström)(10^{-8} cm)
		—	μ (micron)(10^{-4} cm)
		—	mμ(millimicron)(10^{-7} cm)
Frequency	ν	$\dfrac{c}{\lambda}$	Hz (cycles per second)
Wave number	n	$\dfrac{1}{\lambda}$	cm^{-1}
Energy	E	$h\nu, \dfrac{hc}{\lambda}$	depends on the units of h

c = velocity of light (2.9979 × 10^{10} cm-sec^{-1}); h = Planck's constant (6.6237 × 10^{-27} erg-sec)

absorption lines are broadened into bands due to interaction with other types of transitions occurring in the same molecule. These instances will be discussed in the individual sections.

The energy gap between similar transitions in various molecules is a function of the environment of the chromophore undergoing excitation. Correlations of the effects of the molecular environment on the energies of transitions in the various spectral regions have been made and are extremely useful in recognizing certain functional units within a molecule. The following sections will discuss the origin of the transitions, the effects of the molecular environment on the transition energies, and correlations which have been found to be useful in structure determination.

Not all these physical methods will lend themselves to application in every structural problem. Microwave spectroscopy, in general, does not lend itself to applications in structural problems, except with very simple molecules, due to the complexity of the interpretation of the spectral data. In addition, hardly ever will information from one spectral region alone provide sufficient information to derive the complete structure; instead, an integration of spectral data from several spectral regions will be necessary.

Finally, *it cannot be emphasized too strongly that these physical methods by themselves may not lead to the complete and correct structure assignment; an integration of both physical and chemical data is necessary.* Experience has shown that the chemist who tries to rely only on spectral or chemical information wastes a great deal of time and often increases the chances of error in the final assignment of structure. The sections on chemical methods of structure determination will candidly evaluate the utility of physical vs. chemical methods for the detection of individual functional groups.

3.2   INTRODUCTION TO ULTRAVIOLET AND VISIBLE SPECTROSCOPY

The absorption of energy in the ultraviolet and visible regions of the spectrum results in electronic excitation.

The visible region extends from 350 to 800 mμ. The ultraviolet region extending from approximately 100 to 350 mμ is divided into two separate regions: the "vacuum" or "far" ultraviolet region extends from 100 mμ to approximately 200 mμ, and the so-called "quartz" region extends from 200 to 350 mμ. This subdivision is necessary for experimental reasons. Below 200 mμ, air (particularly oxygen and carbon dioxide) and quartz absorb quite strongly, and therefore the sample must be placed in an environment where these materials are absent. The better ultraviolet spectrophotometers are capable of recording down to 180 mμ employing a nitrogen purge system to maintain an atmosphere of nitrogen throughout the instrument. Ultraviolet spectrophotometers capable of recording in the 100 to 180 mμ region are not common laboratory instruments, and the experimental details are sufficiently complex; hence, relatively few studies have been carried out in this region. Fortunately, most compounds capable of absorbing in the general ultraviolet and visible regions display absorption maxima at wavelengths longer than 200 mμ. The "quartz" region is so designated because the sample containers are made of quartz (Pyrex absorbs very strongly below 300 mμ). The light source in this region is usually the hydrogen or deuterium discharge lamp. The visible region allows the use of Pyrex sample containers and the tungsten lamp as the light source.

3.3   PRESENTATION OF ULTRAVIOLET AND VISIBLE ABSORPTION SPECTRAL DATA

Absorption spectral data may be presented as plots of the absorbance or the log of the extinction coefficient vs. wavelength. The absorbance is given by

Eq. (3.1), where I_0 is the intensity of the incident light, I is the intensity of the transmitted light, ϵ is the extinction coefficient, l is the cell path length in centimeters, and c is the concentration in moles per liter.

$$\text{Absorbance} = \log I_0/I = \epsilon c l \tag{3.1}$$

The data usually derived from the ultraviolet spectrometer is a plot of the absorbance vs. wavelength. The extinction coefficients may range from below 10 to 100,000 and thus are generally plotted as the log ϵ vs. the wavelength.

Unless the direct comparison of spectral features is desired, the spectral data can be listed as the wavelength of maximum absorption, $\lambda_{max}^{solvent}$, wavelength of inflections or shoulders, $\lambda_{infl}^{solvent}$, and occasionally the wavelength of minimum absorption, $\lambda_{min}^{solvent}$, followed by the extinction coefficient ϵ, or log ϵ, in parentheses. This method of recording spectral data conserves considerable space.

In all cases, the solvent used must be specified. The reporting of absorbance data is usually reserved for cases in which the molecular weight of the material is not known.

3.4   MODES OF ELECTRONIC EXCITATION

To illustrate the various possible types of electronic excitation, it is convenient to refer to molecular orbital energy diagrams. Let us consider a single bond between two atoms in which all valence shell electrons are involved in single bonds, for example, the C—C bond of ethane or a C—H bond in methane. The molecular orbital energy diagram of this single, isolated bond is shown in Fig. 3.1a, with the bonding electrons occupying the σ bonding level. For convenience and simplicity, in our initial discussion we are

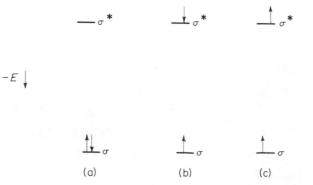

(a)   (b)   (c)

Fig. 3.1. Molecular orbital energy diagrams for the ground state (1a), singlet excited state (1b), and triplet excited state for a single bond system.

considering only a single σ bond in a molecule; however, it must be realized that in a full treatment of any molecule there are several σ bonds. Therefore, the molecular orbital energy diagram will contain several σ and σ^* orbitals. The complexities introduced by the presence of a number of similar orbitals will be discussed later. Electronic excitation may occur with one of the σ bonding electrons being elevated to the σ^* (nonbonding) level (Fig. 3.1b). Such an excitation may be referred to as a $\sigma \longrightarrow \sigma^*$ excitation.[1] The energy required for a $\sigma \longrightarrow \sigma^*$ excitation is relatively large, and absorptions arising from such excitations occur in the far vacuum ultraviolet region. For example, methane absorbs at 125 mμ and ethane at 135 mμ. Cyclopropane absorbs at 190 mμ and is more characteristic of an unsaturated molecule. In general, small-membered rings absorb at slightly longer wavelength than do noncyclic compounds. The excitation of the σ bonding electron to the σ^* bonding level occurs with net retention of electronic spin (as illustrated in Fig. 3.1b). This excited state is referred to as a *singlet* state. The singlet state may then undergo spin inversion (Fig. 3.1c), giving rise to a *triplet* state (both electrons having $\uparrow\uparrow$ or $\downarrow\downarrow$ spins). This state cannot be derived by direct excitation from the ground state. For a more comprehensive discussion of the theory involved in electronic excitation, see Ref. 3 and 8 in Sec. 3.10.

On the basis of the foregoing discussion, one might expect a simple line spectrum in the ultraviolet region. However, this is generally not the case. The total energy of a molecule is composed of the electronic energy term, the vibrational energy term, and a rotational energy term in addition to other energy terms. The overall energy state diagram for a molecule is illustrated in Fig. 3.2. Each electronic state is composed of a series of vibrational states, ν_0, ν_1 etc., and each vibrational state is composed of a series of rotational states. The energies associated with electronic, vibrational, and rotational excitation were presented in Table 3.1.

Electronic excitation occurs from a given rotational level of a given vibrational level of one electronic state to a given rotational level of a vibrational level of a higher electronic state. (Selection rules govern the relationship between the initial and final electronic, vibrational, and rotational levels but will not be discussed here). Electronic excitation generally occurs from the ν_0 level of the ground electronic state because this is the more highly populated state at room temperature.

The possibility of many electronic transitions of slightly varying energies (a few of which are indicated in Fig. 3.2) would be expected to produce a spectrum of closely spaced lines. However, this is usually not the case, except with very small molecules and with instruments of high resolving

---

[1]Several systems of nomenclature for the designation of electronic transitions have been suggested (see ref. 6, 7, and 8 at the end of this chapter). This text will employ the system which relates the type of orbitals involved in the transition since this is the most commonly used system.

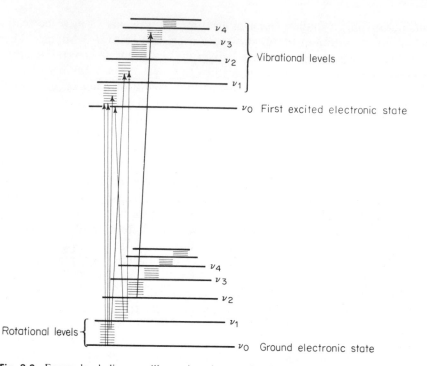

Fig. 3.2. Energy level diagram illustrating the rotational and vibrational levels in the ground electronic and first excited electronic states with a few representative transitions indicated.

power. In complex molecules containing four or more atoms, the number of possible transitions is so great that the observed spectrum is the superposition of all possible lines and appears as a broad band or envelope. The medium in which the absorption spectrum of the molecule is recorded also greatly affects the appearance of the spectrum. For example, in solution, the spectra are usually broadened due to solvent effects.

Highly rigid molecules, such as benzene (see Fig. 3.6 at the end of this chapter) and other similar aromatics, display a series of absorption bands in the ultraviolet region due to vibrational level interactions. As one increases the substitution in such rigid molecules, the fine structure generally disappears, resulting in a broad band spectrum. Spectra of molecules in liquid solution often show less fine structure owing to solvent molecules perturbing the vibrational levels. The spectrum of benzene, Fig. 3.6 was recorded in ethanol solution but still displays fine structure. Gas-phase spectra provide the best conditions to observe the vibrational fine structure. The ultraviolet spectrum of mesityl oxide in ethanol solution is shown in Fig. 3.9 at the end of this chapter. Mesityl oxide, being a much less rigid molecule than benzene, produces only a broad band type spectrum.

An additional mode of excitation becomes available if one of the atoms in a system possesses a pair of nonbonded electrons (see Fig. 3.3). Excitation

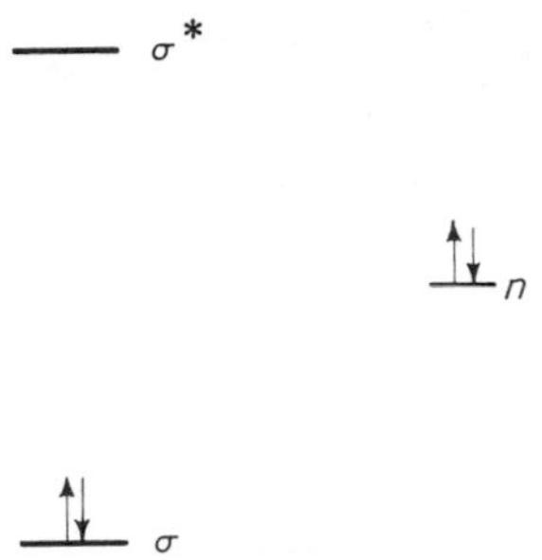

$$\underline{\qquad} \; \sigma^*$$

$$\underline{\uparrow\downarrow} \; n$$

$$\underline{\uparrow\downarrow} \; \sigma$$

Fig. 3.3. Energy level diagram of a system possessing a nonbonded pair of electrons.

of the nonbonded electrons (n) to the σ^* level may occur and is indicated as an $n \longrightarrow \sigma^*$ excitation. The energy required for this type of excitation is less than that required for the $\sigma \longrightarrow \sigma^*$ excitation, and absorption due to $n \longrightarrow \sigma^*$ excitation usually results in end absorption (the maximum occurs below 200 mμ). Table 3.3 lists the absorption maxima of systems containing non-bonded electrons. The positions of maximum absorption of saturated and heteroatom-containing compounds parallel the ionization potentials of the compounds; the lower the ionization potential, the longer the wavelength of absorption. The position of the $n \longrightarrow \sigma^*$ band is quite solvent-sensitive; solvents capable of hydrogen bonding to the nonbonded pair of electrons shift the absorption to shorter wavelengths (*hypsochromic* or *blue shift*).

Table 3.3. Absorption Maxima of Saturated Systems Containing Heteroatoms

Type of compound	λ_{max} ($m\mu$)	Approximate log ϵ
Amines	190–200* also occasionally a longer wavelength shoulder	3.4
Alcohols	180–185*	2.5
Ethers	180–185*	3.5
Epoxides	~170*	3.6
Mercaptans	190–200 and 225–230 (shoulder)	3.2 and 2.2
Sulfides	210–215 and 235–240 (shoulder)	3.1 and 2.0
Cyclic sulfides	slightly longer wavelength than sulfides	
Disulfides	~250*	2.6
Chlorides	170–175*	2.5
Bromides	200–210*	2.6
Iodides	255–260*	2.7

*Long wavelength maxima only.

The molecular orbital energy diagram (again showing only one of the σ and σ^* levels) of a simple olefin, for example, ethylene, is shown in Fig. 3.4. The available modes of excitation include $\pi \longrightarrow \pi^*$ and $\pi \longrightarrow \sigma^*$ transitions. The $\pi \longrightarrow \pi^*$ transition of simple olefins occurs in the 170 to 200 mμ region and, thus, generally below the accessible region of most common laboratory

Fig. 3.4. Molecular energy level diagrams for a simple olefin.

spectrometers resulting only in end absorption. The introduction of alkyl groups in ethylene produces a shift to longer wavelength (*bathochromic* or *red shift*). The bathochromic shift produced is approximately 3 to 5 mμ per alkyl group, but it depends somewhat on the type of alkyl group and the stereochemistry about the double bond. Attempts have been made to correlate the degree of substitution on the double bond with the extinction coefficient of the end absorption in the 200 to 210 mμ region. However, these correlations do not appear to be general. It is easier, and much safer, to determine the substitution pattern on a double bond by nuclear magnetic resonance (see Sec. 5.6).

In contrast to simple olefins, conjugated diolefins absorb in the accessible region of the ultraviolet. The molecular orbital-energy diagram of a simple conjugated diolefin, for example, 1,3-butadiene, is illustrated in Fig. 3.5a. Simple inspection of Fig. 3.5a might lead one to predict that a number of $\pi \longrightarrow \pi^*$ excitations are possible leading to the singly excited states illustrated in Figs. 3.5b–e (we shall ignore excitations involving σ electrons in this and later discussions). Although a number of excitations are potentially possible, not all of these transitions are allowed. The "allowedness" of a particular transition is governed by symmetry considerations. It is beyond

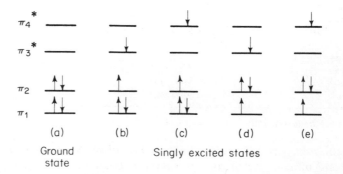

Fig. 3.5. Molecular orbital energy diagrams of the ground state (a) and singly excited states of 1,3-butadiene (b, c, d, and e).

the scope of this text to provide a more detailed discussion of the rules
governing transitions in electronic excitation (see ref. 3 and 8 in Sec. 3.10).

Let us now consider an unsaturated chromophore containing a heteroatom, for example, a simple carbonyl group.   Figure 3.6 displays the
molecular orbital energy level diagram for formaldehyde.   The two nonbonded pairs of electrons on the oxygen are essentially unhybridized *s* and *p*

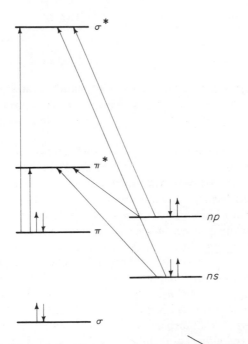

Fig. 3.6. Molecular orbital energy diagram of a $>\!C\!=\!O$ system.

and are indicated in Fig. 3.6 as *ns* and *np* levels. Of the six possible transitions, only the $np \longrightarrow \pi^*$ transition is not allowed, although not all the
transitions have been detected. However, owing to vibrational interaction,
the $np \longrightarrow \pi^*$ excitation does occur to some extent, giving rise to weak absorption in the 270 mμ region. This band is usually referred to as an $n \longrightarrow \pi^*$
transition. The $\pi \longrightarrow \pi^*$ transition occurs in the 180 mμ region and generally
gives rise to strong end-absorption. Table 3.4 lists the absorption maxima
of representative classes of carbonyl containing compounds.

The electronic transitions available with α,β-unsaturated carbonyl systems may be described by an approach similar to that used with 1,3-buta-
diene, except that the *ns* and *np* levels must be added to the molecular
orbital energy diagram. It should be pointed out that the π^* levels are not
localized about the C=C or C=O systems and that the *n* electrons, when
excited, enter the delocalized π^* level embracing the entire α,β-unsaturated
carbonyl system. In α,β-unsaturated ketones, the $\pi \longrightarrow \pi^*$ band occurs in

Table 3.4. Absorption Maxima of Simple Carbonyl Derivatives

Functional group	λ_{max} mμ (log ϵ)
Aldehydes	~190 (1.5–2.0)* and 290–295 (0.9–1.4)†
Ketones	~180 (3–4)* and 275–285 (1.0–2.0)†
Carboxylic acids	~205(1–2)†
Esters	~205(1–2)†
Amides	205–215(1–2)†
Acid chlorides	~235(1–2)†

$\pi \longrightarrow \pi$ transition.
†$n \longrightarrow \pi$* transition.

the 220 to 250 mμ region, and the $n \longrightarrow \pi$* band occurs in the 320 mμ region. Structure-spectra correlations have been developed for α,β-unsaturated ketones, aldehydes, and acids, and will be discussed in the following section.

The position of the $\pi \longrightarrow \pi$* and $n \longrightarrow \pi$* transitions in carbonyl compounds vary with the nature of the solvent used. As the polarity of the solvent increases, from hydrocarbons to water, the $\pi \longrightarrow \pi$* bands undergo a bathochromic (red) shift, whereas the $n \longrightarrow \pi$* bands undergo a gradual, hypsochromic (blue) shift. These shifts are due to solvent stabilization, or destabilization, of the ground or excited electronic states, thus resulting in a change in the energy gap between the levels involved in the transition. The position of the $n \longrightarrow \pi$* band is strongly affected by the presence of hydrogen bonding solvents or protic acids, the band moving to lower wavelengths because the energy of the n-electron levels is significantly lowered by hydrogen bond formation. This phenomenon occurs with all transitions involving n electrons, for example, the $n \longrightarrow \pi$* transitions of aromatic nitrogen heterocyclics (pyridine) and the $n \longrightarrow \sigma$* transitions discussed earlier.

3.5   SPECTRAL CORRELATIONS

Spectral correlations relating the λ_{max} to specific electronic systems have been proposed for olefinic compounds and conjugated ketones. However, these rules must be used with caution. In dealing with relatively simple systems, the rules hold fairly well. Distortion of the planarity of the π-electron system, leading to reduction of p-orbital overlap, or electronic interaction with another functional group within the same molecule, leads to a complete breakdown of the rules and one must revert to the use of model systems.

3.5.1 Olefins

As was pointed out in the previous section, the substitution of a vinyl group for hydrogen in ethylene produces a large bathochromic shift of the $\pi \longrightarrow \pi$*

transition into the accessible ultraviolet region near 220 mμ. Woodward, and Fieser and Fieser, by correlating spectral data from a wide range of compounds, have proposed correlations for the prediction of the wavelength of maximum absorption for substituted dienes. The constants derived for these correlations are given in Table 3.5.

Table 3.5. Constants for Calculation of Absorption Maxima of Substituted Dienes*

Parent diene base absorption†	
heteroannular and acyclic	214 mμ
homoannular	253 mμ
Extended conjugation (per C=C)	+30 mμ
Alkyl substituent (per group)	+ 5 mμ
—O Acyl	+ 0 mμ
—O Alkyl	+ 6 mμ
—S Alkyl	+30 mμ
—Cl, Br	+ 5 mμ
—N Alkyl₂	+60 mμ
Exocyclic double bond	+ 5 mμ

*Taken from Ref. 1 at the end of the chapter.
†Heteroannular and acyclic dienes display extinction coefficients in the 8000–20,000 range, whereas homoannular dienes display extinction coefficients in the 5000–8000 range.

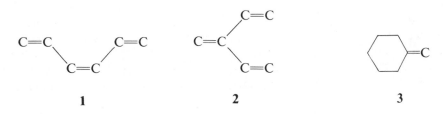

        1   **2**   **3**

A homoannular diene is a conjugated diene in which both double bonds are contained in a single ring, for example, 1,3-cyclohexadiene. The homoannular base is restricted for use only with six-membered rings. A heteroannular diene is one in which the two double bonds are not contained in a single ring. Included in this category are systems in which only one of the double bonds is contained in one ring, for example, 1-vinylcyclohexene, and acyclic dienes. For dienes in which an additional carbon-carbon double bond is in conjugation with the basic diene, an increment of 30 mμ is added. However, the use of this rule must be restricted to molecules displaying "through" conjugation as in example **1**, and not "cross" conjugation as in example **2**. Compounds in the latter class usually do not lend themselves to accurate correlation of predicted with observed absorption maxima. An exocyclic double bond is defined as one in which one of the carbon atoms of the double bond is contained in a ring (**3**). Since a $\pi \longrightarrow \pi^*$ excitation does not lead to great differences of polarity between the ground state and the excited state, changes in λ_{max} with changes in solvent are negligible, as is indicated by a zero solvent correction factor.

A few examples will illustrate the use of these rules. Examples **4** and **5** are quite straightforward.

Parent diene	homoannular	253	mμ
Substituents	4($\times$5)	20	
Exocyclic double bonds	2($\times$5)	10	
Calculated λ_{max}		283	mμ
Observed		282	mμ

Parent diene	heteroannular	214	mμ
Substituents	3($\times$5)	15	
Exocyclic double bonds	1($\times$5)	5	
Calculated λ_{max}		234	mμ
Observed		234	mμ

Parent diene	homoannular	253	mμ
Extended conjugation	1($\times$30)	30	
Substituents	5($\times$5)	25	
Exocyclic double bonds	3($\times$5)	15	
Calculated λ_{max}		323	mμ
Observed		324	mμ

	homoannular	heteroannular
Parent diene	253 mμ	214 mμ
Substituents	25	20
Exocyclic double bonds	15	10 mμ
Calculated λ_{max}	293 mμ	244 mμ
Observed	285 mμ	

Example **6** is a "through" conjugated triene system, and it presents us with a choice of using either the homo- or heteroannular diene base. In general, one will always use the base which requires the least energy for excitation, in other words, the base diene displaying the longer wavelength absorption. In the case of example **6**, the choice is the homoannular diene base. In counting the number of substituents in example **6**, all substituents on the entire triene system are included; it should be pointed out that C-10 of the steroid nucleus is a substituent on two different carbons of the triene system and is counted as two alkyl substituents.

Example **7** presents a "cross"-conjugated triene. Calculations employing both the homo- and the hetereoannular diene bases are shown, neither of which are in acceptable agreement.

The foregoing correlations appear to be applicable to other size ring systems, using the λ_{max} of the cyclic diolefin as the homoannular diene base. Suggested bases are 228 mμ for cyclopentadiene and 241 mμ for cycloheptadiene (the base wavelength includes the two-ring substituent carbons).

It must be pointed out that, in systems where the π-electron system is distorted owing to steric interactions, the preceding rules may not satisfactorily predict the absorption maximum. For example, 1,2-dimethylenecyclohexene absorbs at 220 mμ instead of at the calculated value of 234 mμ. Certain chromophores in highly strained systems can be approximated by the addition of another correction factor to the value calculated from the preceding rules. For example, dienes involving the bicyclo-[3.1.1]-heptane system require the addition of 15 mμ to approximate the observed value.

The use of two-dimensional drawings may not always give a true indication of the distortion in a molecule. Application of the foregoing correlations to questionable cases should be avoided, and spectral comparisons should be made with other known compounds containing the same type of chromophore.

3.5.2 Unsaturated Ketones and Other Carbonyl Derivatives

Table 3.6 contains substituent constants derived by several authors for the prediction of the position of maximum $\pi \longrightarrow \pi^*$ absorption of unsaturated carbonyl derivatives of the general structure **8.**

Table 3.6.

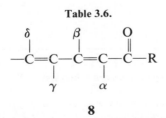

8

The following examples will illustrate the use of these constants:

Parent base	215 mμ
Extended conjugation	30
Substituents	
beta	12
delta	18
Exocyclic double bond	5
Calculated λ_{max}^{ETOH}	280 mμ
Observed	284 mμ

9

Table 3.6. Constants for Calculation of Absorption Maxima of Unsaturated Carbonyl Derivatives*

Parent system‡ (acyclic or six-membered or larger ring ketone)		215 mμ
Five-membered ring ketone		−10
Aldehydes		− 5
Carboxylic acids and esters		−20
Extended conjugation		+30
Homodienic component		+39
Exocyclic double bond		+ 5
Alkyl substituent	α	+10
	β	+12
	γ and higher	+18
Hydroxyl	α	+35
	β	+30
	γ	+50
Alkoxyl	α	+35
	β	+30
	γ	+17
	δ	+31
Acetoxyl	α,β, or δ	+ 6
Dialkylamino	β	+95
Chlorine	α	+15
	β	+12
Thioalkyl	β	+85
Bromine	α	+25
	β	+30
Solvent correction (relative to ethanol)†		
Water		−8 mμ
Methanol		0
Chloroform		+1
Dioxane		+5
Ether		+7
Hexane		+11

*The original correlation is referred to as Woodward's rules (see ref. 9). Additional substituent increments have been taken from ref. 1, 12, and the authors' unpublished observations.

‡The calculated values usually fall within ±3 mμ of the observed values. The extinction coefficients of *cisoid* enones are usually less than 10,000, whereas the extinction coefficients of *transoid* enones are greater than 10,000.

†Suggested by Fieser and Fieser (ref. 1).

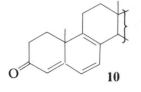

Parent base	215 mμ
Extended conjugation	
(2 × 30)	60
Homodienic component	
(ring **B**)	39
Substituents	
beta	12
3γ or higher	54
Exocyclic double bond	5
Calculated λ_{max}^{ETOH}	$\overline{385}$ mμ
Observed	388 mμ

A homodienic component is included for compound **10** to take into account the occurrence of two double bonds in one ring. Compound **11** illustrates the application to a cyclopentenone derivative. Both the carbonyl group and the double bond must be in the five-membered ring; the five-membered ring correction would not apply to 1-acetylcyclopentene.

Parent base	215 mμ
Substituents	
alpha	10
2 beta	24
Five-membered ring	− 10
Calculated λ_{max}^{ETOH}	$\overline{239}$ mμ
Observed	237.5 mμ

11

$(CH_3)_2C=CHC$ ... **12**

Parent base		215 mμ
Substituents	2 beta	24
Aldehyde		− 5
Calculated λ_{max}^{ETOH}		$\overline{234}$ mμ
Observed		235 mμ

$(CH_3)_2C=CHC$ ... **13**

Parent base		215 mμ
Substituents	2 beta	24
Acid		− 20
Calculated λ_{max}^{ETOH}		$\overline{219}$ mμ
Observed		216 mμ

Compounds **12** and **13** illustrate the application of these rules to aldehydes and acids. The correspondence between calculated and observed positions is much poorer in the application to unsaturated acids (±5 mμ), and certain caution must be exercised. The accuracy of the application of the solvent corrections to carboxylic acids has not been investigated.

Eneone system		$\Delta^{1,2}$		$\Delta^{4,5}$
Parent base		215 mμ		215 mμ
Substituents	beta	12	2 beta	24
Exocyclic double bond		—		5
Calculated λ_{max}^{ETOH}		$\overline{227}$ mμ		$\overline{244}$ mμ
Observed				244 mμ

14

The partial structure **14** illustrates a cross-conjugated system in which the chromophore giving rise to the longest wavelength maximum is to be used.

In all the examples cited, the calculated absorption maximum is for ethanol as the solvent. Suitable corrections must be applied when other solvents are used.

The ultraviolet spectra of derivatives or carbonyl compounds are also quite useful. Oximes absorb at approximately the same wavelength as the original carbonyl compounds, whereas semicarbazones display maxima which are shifted to longer wavelengths by 30 to 40 mμ, with an average increase in the extinction coefficient of 10,000 compared with the original carbonyl compound. 2,4-Dinitrophenylhydrazones of saturated ketones absorb in the 360 to 365 mμ region, whereas similar derivatives of α,β-unsaturated ketones absorb in the 375 to 385 mμ region.

3.5.2a Quinones

Quinones comprise a family of compounds which contain an α,β-unsaturated ketone system; however, because of the nature of the chromophore, they give rise to distinctly different types of absorption spectra.

There are two classifications of quinones: paraquinones and orthoquinones. Distinction between the two categories of quinones is readily accomplished on the basis of their ultraviolet absorption spectra. Paraquinones give rise to three absorption bands in the ultraviolet-visible region. These bands occur in the 240 to 260 mμ (arbitrarily assigned a band I designation), 280 to 340 mμ (band II), and near 430 mμ (band III). For example, p-benzoquinone absorbs at 244 mμ (log ϵ 4.3), 279 mμ (log ϵ 2.8), and 430 mμ (log ϵ 1.3) in ethanol solution. The introduction of alkyl groups produces (1) a bathochromic shift in band I of 4 mμ per alkyl group, (2) substantial bathochromic shifts in band II, (however no consistent trend is noted), and (3) effectively no change in the position of band III. Very minor solvent effects are noted. For example, the ultraviolet spectra of paraquinones recorded in hydrocarbon or chloroform solution display bands in the same region as in ethanol; however, bands I and III generally appear as two and three distinct maxima and/or shoulders in nonpolar solvents.

Hydroxy-*para*-quinones usually display only two absorption bands in ethanol solution. These bands appear in the 250 to 310 mμ (band I) and 350 to 435 mμ (band II) regions. For example, hydroxy-*para*-quinone absorbs at 260 mμ (log ϵ 4.0) and 369 mμ (log ϵ 3.1). The introduction of additional hydroxy groups results in bathochromic shifts of approximately 15 mμ and 20 mμ in bands I and II, respectively. The introduction of alkyl groups in hydroxy-*para*-quinones results in a rather consistent bathochromic shift in band I of 4 mμ per alkyl group, with no apparent effect on the position of band II. Alkoxy-*para*-quinones give rise to two intense bands in the 250 mμ (log ϵ 4) and 360 mμ (log ϵ 3.2) regions; however, no consistent trend with substitution is apparent. A considerable body of ultraviolet data for quinone oximes and their O-methyl derivatives appears in the literature (see ref. 11, Sec. 3.10).

Ortho-quinones display three bands in the ultraviolet and visible region near 260 mμ (log ϵ 3.3), 400 mμ (log ϵ 3.0), and 560 mμ (log ϵ 1.6). The two longer wavelength bands are particularly useful in the characterization of *ortho*-quinones. There appear to be no obvious trends of the wavelength of absorption with substitution.

3.5.3 Aromatic Compounds

Aromatic compounds display several absorption bands, the number and type of which depend on the structure of the molecule. All benzenoid compounds display what are commonly termed E and B bands representing $\pi \longrightarrow \pi^*$ transitions. For example, the E band of benzene occurs in the 180 to 210 mμ region, and the B band occurs in the 250 to 255 mμ region. The band in the 180 to 210 mμ region has been assigned to two overlapping, symmetry-allowed transitions, whereas the band at 250 to 255 mμ represents a symmetry-forbidden transition which has finite, but low, probability due to interaction with a nontotally symmetric vibration.[2]

Aromatic compounds which contain an unsaturated functional group capable of conjugating with the aromatic ring display an additional band which is termed the K band. The molecular orbital energy diagram contains an additional π and π^* orbital, relative to benzene, hence providing for additional possible modes of excitation. Finally, if the functional group attached to the benzene ring possesses a nonbonded pair of electrons, $n \longrightarrow \pi^*$ transitions are possible, giving what are termed R bands. Often the R band is obscured by the more intense B band. The wavelength positions and extinction coefficients for the four bands are given in Table 3.7. Examples of compounds displaying these typical bands are given in Table 3.8. The band positions and shapes may change substantially on changing solvents; the vibronic structure appears in the B band only and frequently disappears in the more polar solvents.

---

[2] Sklar, *J. Chem. Phys.*, **5**, 669 (1937).

Table 3.7. **Positions and Intensities of Aromatic**
Absorption Bands

Band	λ_{max} (mμ)	ϵ
E	180–220	2–6 $\times$ 10^3
K	220–250	1–3 $\times$ 10^4
B	250–290	10^2–10^3
R	275–330	10–100

Table 3.8. Typical Absorption Maxima of Aromatic Compounds

Compound	Band	λ_{max} m$\mu(\epsilon)$			
		E	K	B	R
Benzene		198(8,000)	—	255(230)	—
t-Butylbenzene		208(7,800)	—	257(170)	—
Styrene			244(12,000)	282(450)	—
Acetophenone			240(13,000)	278(1,100)	319(50)

The introduction of substituents onto the aromatic ring produces several changes in the appearance of the absorption spectrum. With substituted aromatics, the *E* band shifts to longer wavelength, usually into the accessible region, regardless of the electronic properties of the substituent. The absorption maximum of the *B* band moves to longer wavelength in cases where the new substituent is electron donating or capable of conjugation. On the other hand, with electron withdrawing substituents, practically no change in the maximum position is observed (see Table 3.9). The intensity of the *B* band generally increases as the electron donating character of the substituent increases. Finally, the fine structure of the benzenoid absorption usually disappears on substitution.

Inspection of the data given in Table 3.9 should point out the fact that no accurate correlation with structure is possible. This is also true of the disubstituted benzenes. The various disubstituted isomers do show slight differences in the positions of the absorption maxima and their intensities; however, the changes are so slight, and the absorption maxima lie in the same general region as monosubstituted benzenes, so that no useful correlation can be made. The careful use of model compounds (see later discussion on model compounds) is recommended in these cases, or, better yet, refer to infrared and nuclear magnetic resonance data.

It is important to note the wavelength shifts involved in going from phenol to phenoxide and from aniline to anilinium ion. The bathochromic shift observed on going from phenol to phenoxide anion is typical of enol-enolate anion systems and is useful as a diagnostic tool to detect enol systems, not only in aromatics but also with nonaromatics. Similarly, the hypsochromic shift observed in the aniline-to-anilinium ion transformation is typical of the spectral changes due to the protonation of basic sites, as indicated earlier.

Table 3.9. Ultraviolet Absorption Maxima of Benzene Derivatives

Compound	E band		B band		Solvent
	$\lambda_{max}^{(m\mu)}$		$\lambda_{max}\ m\mu$		
Benzene	198	8,000	255	230	Cyclohexane
Electron-donating Substituents					
t-Butylbenzene	207.5	7,800	257	170	Ethanol
Phenol	210.5	6,200	270	1,450	Water
Phenoxide anion	235	9,400	287	2,600	Water
Anisole	217	6,400	269	1,480	Water
Thiophenol	236	10,000	269	700	Hexane
Aniline	230	8,600	280	1,430	Water
Electron-withdrawing Substituents					
Fluorobenzene	204	6,200	254	900	Ethanol
Chlorobenzene	210	7,500	257	170	Ethanol
Bromobenzene	210	7,500	257	170	Ethanol
Anilinium ion	203	7,500	254	160	Water
Substituents Capable of Conjugation					
	K band		B band		
Styrene	244	12,000	282	450	Ethanol
Phenylacetylene	236	12,500	278	650	Hexane
Benzaldehyde	244	15,000	280	1,500	Ethanol
Acetophenone	240	13,000	278	1,100	Ethanol
Benzoic acid	230	10,000	270	800	Water
Benzonitrile	224	13,000	271	1,000	Water
Nitrobenzene	252	10,000	270	800	Hexane

The fusion of additional rings to the benzene nucleus, for example, forming naphthalene, anthracene, etc., results in substantial bathochromic shifts. The *B* band of benzene occurs at 255 mμ, whereas the similar bands in naphthalene and anthracene occur at 314 mμ and 380 mμ, respectively.

The introduction of heteroatoms into the aromatic ring, for example, pyridine, quinoline, etc., produces ultraviolet spectra very similar to those obtained from the related benzenoid compounds. The spectra of the hetero-aromatic compounds display less fine structure than the benzenoid aromatics; for example, compare the spectrum of pyridine (Fig. 3.7 at the end of this chapter) with that of benzene (Fig. 3.6). Absorption data for these and other heterocyclic aromatic systems appear in Table 3.10. The position of absorption of heteroaromatic compounds is much more sensitive to substituents than the benzenoid aromatics are. Space does not permit a discussion of substituent effects in these systems, and the reader is referred to more comprehensive texts on ultraviolet spectroscopy or to the original literature.

Table 3.10. Absorption Maxima of Heterocyclic and
Benzenoid Aromatics

		λ_{max}	ϵ
Benzene	(B band)	255 mμ	230
Pyridine	(B band)	252	2,090
Pyridazine	(B band)	246	1,150
Pyrimidine	(B band)	243	2,950
Pyrazine	(B band)	261	6,000
Naphthalene	(B band)	314	250
Quinoline	(B band)	313	2,500
Anthracene	(B band)	380	9,000
Furan		205	6,300
Pyrrole		211	10,000
Thiophene		235	4,500
Imidazole		207	5,000
Pyrazole		210	3,100
Isoxazole		211	4,000
Thiazole		204	4,000

3.5.4 Miscellaneous Chromophores

Table 3.11 contains a list of miscellaneous functional groups capable of absorbing in the ultraviolet region. Conjugated diacetylenes, eneynes, allenes, and cumulenes ($>C{=}C({=}C)_n{=}C<$) display considerable fine structure owing to their comparatively highly rigid structure. Increasing the extent of conjugation with these chromophores causes considerable bathochromic shifts.

3.6 THE USE OF MODEL COMPOUNDS

Often, the partial structure of a complex molecule can be deduced by comparison of the spectral properties of the compound with those of simpler molecules thought to contain the same chromophore. This is particularly true in cases in which the geometry of the molecule causes a distortion of the chromophore, resulting in shifts of the maxima to different wavelengths. Electronic interaction with a neighboring functional group may also lead to anomalous shifts in the band positions. The simpler molecules used for such comparisons are called model compounds. Model compounds are used extensively in natural product chemistry and for highly strained ring compounds.

Although one generally uses the positions of the maxima for correlation purposes, similar changes in the positions of maxima throughout a series of compounds may be very helpful. The chromophores may not be completely

Table 3.11. **Absorption Data for Miscellaneous
Functional Groups***

Functional group	λ_{max} mμ(ϵ)
Acetylenes	170–175 (4,500)
Conjugated diacetylenes	225–235 (200)†
Enynes	220–225 ($\sim$10,000)†
Allenes	175–185 ($\sim$10,000)
Cumulenes (butatriene)‡	241 (20,300)
Nitriles	$\sim$340 (120)
Nitro	$\sim$210($\sim$16,000)($\pi \longrightarrow \pi^*$) and 270–280($\sim$200)($n \longrightarrow \pi^*$)
Nitrate	$\pi \longrightarrow \pi^*$ end absorption and $\sim$260–270(shoulder)(150)
Nitrite	$\sim$350 ($\sim$150)†
Azo	$\pi \longrightarrow \pi^*$ end absorption and $\sim$350 (low)($n \longrightarrow \pi^*$)
Diazo	$\sim$400 ($\sim$3)†
Sulfoxide	210–215 ($\sim$1,600)
Sulfone	no absorption above 208
Vinyl sulfone	$\sim$210 ($\sim$300)

*Functional groups contained in aliphatic systems only.
†Composed of a number of bands for which the approximate
center is given.
‡Cumulenes containing more double bonds absorb at correspondingly longer wavelengths.

identical, but may be similar enough to allow a correlation to be made. The
similarity in the shifts observed in the ultraviolet spectra of the following
steroidal compounds and seven-membered ring compounds indicates a
similarity in the chromophores.

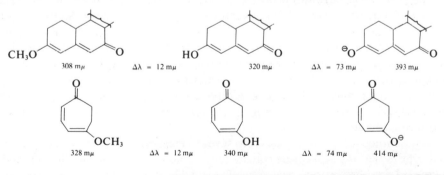

3.7 ADDITIVITY OF CHROMOPHORES

Many molecules may contain more than one absorbing chromophore. In
such cases, the observed ultraviolet absorption spectrum will be the sum of
the absorption bands of the individual chromophores, provided there is no

electronic interaction between the chromophores. If the absorption maxima of the two chromophores lie close to one another, resulting in a great deal of overlapping of the bands, we may subtract an expected absorption of one of the known chromophores to derive the absorption spectrum of the other chromophore. Experimentally, this can be readily accomplished by placing a reference compound containing the desired chromophore into the reference cell of the spectrometer, with the sample in the sample cell. The resulting spectrum will be the difference spectrum between the sample and reference compound. The concentration of the reference compound may have to be adjusted so that the net absorptivity of the reference chromophore equals that of the same chromophore in the sample. A great deal of caution must be exercised in the selection of model compound because the chromophores must be identical.

3.8   PREPARATION OF SAMPLES

The primary consideration in the preparation of samples for recording their ultraviolet spectra is that of concentration. The concentration must be adjusted such that the absorption peak remains on the recording scale of the instrument and, furthermore, in the more accurate region on the recording scale. The region of greatest accuracy is between 0.2 and 0.7 absorbance units. If we specify that the maximum absorbance should be near 0.7 and we have a general idea of what type of chromophore is present and thus a general region for the value of the extinction coefficient, we may use Eq. (1.1) to calculate the concentration required for use in a 1-cm cell. For conjugated diene chromophores with extinction coefficients in the general region of 8,000 to 20,000, the concentration should be near 4×10^{-5} mole per liter, whereas for the $n \longrightarrow \pi^*$ transition of a carbonyl group (ϵ of 10–100), the concentration should be near 10^{-2} mole per liter. Preparation of very dilute samples cannot, in general, be accomplished by a direct weighing technique. The quantity of material would be quite small, in the submilligram range, and large weighing errors would be introduced. In general, such sample concentrations are obtained by successive dilutions of more concentrated solutions.

The choice of solvent rests on several factors. The most important criterion is that the solvent be transparent in the region over which you wish to record the absorption spectrum. Table 3.12 lists several solvents and the cutoff point below which they absorb too strongly to be useful as a solvent. Secondly, the solvent should be a good solvent for dissolving the material. Even though the concentrations are quite small, many compounds may be only partially soluble or may exist as aggregates in solution. Finally, a solvent should be chosen so that it does not react chemically with the sample.

Care should be taken in the purification and drying of the samples. The presence of trace amounts of materials having absorption maxima with large extinction coefficients may give rise to maxima which might be interpreted as belonging to the predominant material. The solvent used should be of high purity, generally referred to as "spectro grade" by distributors. Care should be taken to keep lint and dust from contaminating the final solutions.

Table 3.12. Solvents Used in the Ultraviolet
and Visible Region*

Solvent	Cutoff $(m \mu)^{\dagger}$
Water	205
Methanol	210
Ethanol	210
Diethyl ether	210
Tetrahydrofuran	220
Acetonitrile	210
Cyclohexane	210
Methylcyclohexane	210
2, 2, 4-Trimethylpentane	210
Dichloromethane	235
Chloroform	245
Carbontetrachloride	265
N,N-Dimethylformamide	270
Benzene	280
Acetone	330

*Taken in part from a table of "Solvents for Spectrophotometric Use" appearing in the chemical catalog of Distillation Products Industries, division of Eastman Kodak Company, Rochester, New York.
†The cutoff point is taken as the wavelength at which a 1 cm path length of liquid results in an absorbance of approximately 1.

The sample solution is placed in the sample cell and compartment of the instrument, and pure solvent, or a reference compound if a difference spectrum is to be determined, is placed in the reference cell and compartment. The description of the actual recording of the spectrum will not be described here; the reader is referred to the individual operation manual available with each instrument.

Figures 3.7, 3.8, and 3.9 show the recorded spectra of benzene, pyridine, and mesityl oxide in ethanol solution. The concentration data are indicated beneath each figure.

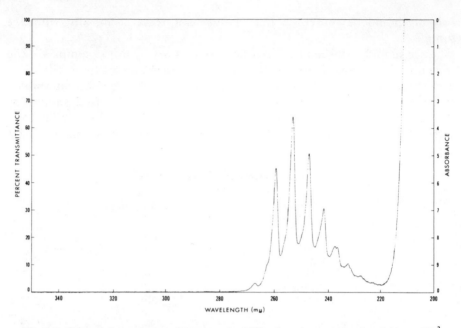

Fig. 3.7. Ultraviolet spectrum of benzene in 95% ethanol: concentration 2.92×10^{-3} mole per liter. (Recorded on a Bausch and Lomb Spectronic 505 ultraviolet and visible spectrophotometer.)

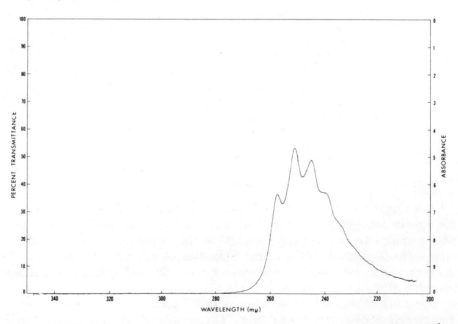

Fig. 3.8. Ultraviolet spectrum of pyridine in 95% ethanol: concentration 1.98×10^{-4} mole per liter. (Recorded on a Bausch and Lomb Spectronic 505 ultraviolet and visible spectrophotometer.)

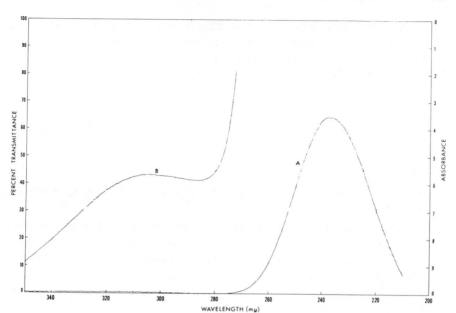

Fig. 3.9. Ultraviolet spectrum of mesityl oxide in 95% ethanol: Trace A displays the $\pi \rightarrow \pi^*$ band, concentration of 6.29×10^{-5} mole per liter, and trace B displays the $n \rightarrow \pi^*$ band, concentration 6.29×10^{-3} mole per liter. (Recorded on a Bausch and Lomb Spectronic 505 ultraviolet and visible spectrophotometer.)

3.9 *LOCATING SPECTRAL DATA RECORDED IN THE LITERATURE*

To compare the spectrum of an unknown with that of another compound, it is necessary to have a sample of that compound available, or to find the required spectral information in the literature. Fortunately, it is not always necessary to search in *Chemical Abstracts*, or other abstracting journals, since collections of ultraviolet spectral data are available (see the references under References in Sec. 3.10). Four volumes entitled *Organic Electronic Spectral Data* (ref. 11), in which compounds are listed by molecular formulas in order of increasing number of carbon atoms, cover the literature from 1946 to 1959. H. M. Hershenson has compiled literature references in two volumes entitled *Ultraviolet and Visible Absorption Spectra* (ref. 10), covering the years 1930 to 1954 and 1955 to 1959, respectively. Reproductions of actual ultraviolet spectra have been compiled by the American Petroleum Institute Research Project 44 (ref. 14), and reproductions of 11,000 ultraviolet spectra have been published by Sadtler Research Laboratories (ref. 13). The *Handbook of Ultraviolet Methods* (ref. 15) is a bibliography of references of ultraviolet analytical procedures dealing with specific compounds or substances. Many other sources, too numerous to mention here, have spectral data for individual, specific classes of compounds.

General references pertaining to the general theory and interpretation of ultraviolet and visible spectra are also given (refs. 1–9).

3.10   REFERENCES

GENERAL

1. Fieser, L. F., and Fieser, M., *Steroids*. New York, N.Y.: Reinhold Publishing, 1959, pp. 15–21.

2. Gillam, A. E., and Stern, E. S., *An Introduction to Electronic Absorption Spectroscopy in Organic Chemistry*. London, England: Edward Arnold Publishers, 1957.

3. Jaffé, H. H., and Orchin, M., *Theory and Applications of Ultraviolet Spectroscopy*. New York: John Wiley & Sons, Inc., 1962.

4. Mason, S. F., "Molecular Electronic Absorption Spectra," *Quart. Revs.*, **15**, 287 (1961).

5. Matsen, F. A., "Applications of the Theory of Electron Spectra" in *Technique of Organic Chemistry*, A. Weissberger, ed. New York: Interscience Publishers, Inc., 1956.

6. Rao, C. N. R., *Ultra-Violet and Visible Spectroscopy*. London, England: Butterworths and Co., Ltd., 1961.

7. "Report on the Notation for the Spectra of Polyatomic Molecules," *J. Chem. Phys.*, **23**, 1997 (1955).

8. Sandorfy, C., *Electronic Spectra and Quantum Chemistry*. Englewood Cliffs, N. J.: Prentice-Hall, Inc., 1964.

9. Woodward, R. B., *J. Am. Chem. Soc.*, **64**, 72 (1952).

SPECTRAL DATA AND REFERENCES

10. Hershenson, H. M., *Ultraviolet and Visible Absorption Spectra*. New York: Academic Press, index for 1930–1954 published in 1956, index for 1955–1959 published in 1961.

11. Kamlet, M. J., *Organic Electronic Spectral Data*. New York: Interscience Publishers, Inc., Volumes I through IV covering the years 1946–1959.

12. Scott, A. I., *Interpretation of the Ultraviolet Spectra of Natural Products*. New York: Pergamon Press, 1964.

13. *Standard Ultraviolet Spectra*, compiled by Sadtler Research Laboratories, 1517 Vine St., Philadelphia 2, Pa.

14. *Ultraviolet Spectral Data*, compiled by the American Petroleum Institute Research Project 44, Chemical Thermodynamic Properties Center, Department of Chemistry, Texas A & M University, College Station, Texas.

15. White, R. G., *Handbook of Ultraviolet Methods*. New York: Plenum Press, 1965.

4

Infrared Spectroscopy

4.1 INTRODUCTION

The infrared region of the spectrum extends from the upper end of the visible region, approximately 0.75 μ, to the microwave region near 400 μ. The portion of this region generally used by the organic chemist for structural work is from 2.5 to 16 μ. This is primarily due to instrument design and cost and to the fact that most of the useful information can be derived in this region. The region from 0.75 to 2.5 μ is referred to as the near-infrared region and contains absorption bands due to harmonic overtones of the fundamental bands and combination bands. The region extending from 16 to 400 μ is referred to as the far-infrared region. The normal and far-infrared regions contain absorptions due to fundamental, harmonic, and combination bands.

The position of absorption in the general infrared region may be expressed in terms of wavelengths (μ) of the absorbed light, or in terms of the wave number (cm^{-1}) of the absorbed light. Both designations currently appear in common use. However, the wave number, or frequency, convention is becoming progressively more popular.

Infrared spectrometers are available which record linear in frequency or linear in wavelength. The use of linear-in-frequency instruments results in a considerable expansion of the high-frequency end of the infrared region, resulting in an increased ability to resolve bands and define their position. The use of linear-in-wavelength instruments results in a considerable compression of the high-frequency region, resulting in a loss sensitivity.

Absorption of radiation in the infrared region results in the excitation of bond deformations, either stretching or bending deformations. Stretching excitation involves changes in the frequency of the vibration of bonded atoms along the bond axis, whereas a bending deformation implies move-

109

ment of the atoms out from the bonding axis. These deformations are designated ν and σ, respectively.

$$A \longleftrightarrow B \qquad A \overset{\curvearrowright}{\underset{\cdot}{}} B$$
$$\text{stretching } (\nu) \qquad \text{bending } (\sigma)$$

The amount of energy required to excite the stretching and bending modes depends on the masses of the atoms or groups A and B, and on the A-B bond order (hybridization of the atoms comprising the bond). In general, the energy required for excitation will decrease with increasing atomic weight of A and B, and will increase with increased bond order (single $\longrightarrow$ double $\longrightarrow$ triple bond) or s character of the orbitals comprising the bond. The effect of mass on the stretching frequency for a simple system in which the mass of A is much smaller than the mass of B, such that there is no substantial movement of the center of mass on stretching excitation, is given by Eq. (4.1). This equation describes the action of a classical harmonic oscillator in which k is the force constant of the bond and is related to the hybridization of the atoms involved in the bond; m is the mass of A. For a system in which the masses of A and B are comparable, Eq. (4.1) becomes Eq. (4.2) (Hooke's law), in which u is the reduced mass of the system.

$$\nu_{osc} = \frac{1}{2\pi} \sqrt{\frac{k}{m}} \tag{4.1}$$

$$\nu_{osc} = \frac{1}{2\pi} \sqrt{\frac{k}{u}} \tag{4.2}$$

For a complex molecule containing many atoms, we might expect to derive a very complex infrared absorption spectrum from which little diagnostic information could be derived. Fortunately, however, such is not always the case. Many functional groups give rise to characteristic absorption bands which vary only slightly in wave number owing to an insensitivity of the large mass of the remainder of the molecule on the localized system (see Eq. 4.1).

A molecule containing n atoms will give rise to $3n - 6$ stretching and bending deformations (normal modes). Linear molecules give rise to $3n - 5$ normal modes. Not all these deformations will, however, result in absorption of energy in the infrared region. As an example, let us consider the linear molecule carbon dioxide (CO_2) which will give rise to $3 \times 3 - 5 = 4$ normal modes, two stretching deformations indicated by ν_1 and ν_2, and two bending deformations indicated by σ_1 and σ_2, as shown in Fig. 4.1 (σ_1 involves a bending deformation in the xy plane, and σ_2 involves a bending deformation in the xz plane). As in electronic excitation, the infrared and Raman (see Sec. 4.6) activity of each normal mode is determined by symmetry considerations which will not be discussed here.[1]

---

[1] See F. A. Cotton, *Chemical Applications of Group Theory*, Interscience Publishers, New York, 1963, chapter 9; and H. Jaffé and M. Orchin, *Symmetry in Chemistry*, John Wiley & Sons, Inc., New York, 1965.

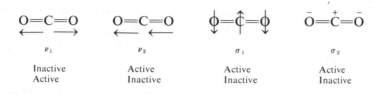

$$\nu_1 \qquad\qquad \nu_2 \qquad\qquad \sigma_1 \qquad\qquad \sigma_2$$

Infrared	Inactive	Active	Active	Active
Raman	Active	Inactive	Inactive	Inactive

Fig. 4.1. Normal modes of carbon dioxide. Arrows indicate the direction of movement of the atoms in the plane of the page; + and − indicate movement forward and backward, respectively, out of the plane of the paper.

In simple terms, the rules governing infrared and Raman activity may be stated as follows:

1. For infrared activity, a change in the dipole moment of the bond, relative to the ground state, is required upon excitation.
2. For Raman activity, a change in the polarizability, relative to the ground state, is required upon excitation.

Of the normal modes available with carbon dioxide, illustrated in Fig. 4.1, only ν_1 is not infrared active, and only ν_1 is Raman active. In general, infrared active normal modes will be Raman active, and vice versa. Very few normal modes are both infrared and Raman inactive, and these occur only with molecules of high symmetry.

Bond deformation excitation occurs from a given rotational level of a vibrational state to a given rotational level of the higher vibrational state. As with electronic excitation, a number of transitions of closely similar energy are available, resulting in a band-type spectrum. Resolution of rotational peaks may be observed with many simple molecules.

The infrared region also contains significant absorption bands which are not due to the fundamental stretching or bending normal modes. These are referred to as overtones, combination, coupled, and Fermi resonance bands.

Overtones are harmonics of the fundamental frequency and occur at near integral multiples of the fundamental absorption frequency. These are referred to as the first, second, or higher overtones of a fundamental, and they decrease in intensity as the order of the overtone increases. First overtones of the most prominent peaks in a given spectrum are quite easy to find. Often, these overtones fall into regions in which the characteristic absorptions of other functional groups occur, leading to a possible misassignment. For example, the first overtone of the C=O stretching frequency falls in the O—H stretch region, and the first overtone of the C—O stretch of ethers, for example, di-*n*-butyl ether, occurs in the carbonyl stretching region.

Combination bands are relatively weak bands appearing at frequencies equal to the sum or difference of two or more fundamental frequencies. Coupling bands occur when two absorption bands of the same portion of the molecule interact, causing the absorption bands to be shifted out of the expected region for each chromophore independently.

Fermi resonance occurs when an overtone or combination band lies close to a fundamental absorption band, resulting in an enhancement of the intensity of the overtone or combination band, or in a splitting of the bands. The coupling between the overtone or combination bands with the fundamental band is called the Fermi resonance interaction.

The intensities of the infrared absorption bands vary from strong to very weak (eventually disappearing if there is little or no change in dipole moment on excitation). The intensity of an infrared absorption band cannot be expressed as a unique constant as was possible in ultraviolet and visible spectroscopy. This is due to the fact that the slit widths (fine slits controlling the amount of infrared radiation passing through the instrument) used in most instruments are of the same order as the typical infrared bandwidth which causes the measured optical density to be a function of the slit width. Despite this problem, infrared spectroscopy can be used as a quantitative analytical tool with certain precautions. Instead of determining the true molar extinction coefficient, one may determine an apparent molar extinction coefficient, ϵ_a, which will be a function of the slit width and even sample concentration. The apparent extinction coefficient is calculated as shown in Equation (4.3).

$$\epsilon_a = \frac{\text{absorbance}}{\text{concentration (moles/liter)} \times \text{cell length (cm)}} \qquad (4.3)$$

A series of apparent extinction coefficients vs. concentration should be determined, and the data plotted for use in final analysis. The apparent extinction coefficients can be used only with the instrument they have been measured on. Analytical limits are approximately $\pm 5\%$ when these methods are used.

The intensities of infrared absorption bands are generally classified as very strong (vs), strong (s), medium (m), weak (w), and very weak (vw). The diagnostic value for structural work of the given bands is indicated as great utility (gu), limited utility (lu) and no practical utility (nu).

4.2   CHARACTERISTIC ABSORPTION BANDS

Not all the absorption bands appearing in an infrared spectrum will be useful in deriving structural information. Certain portions of the infrared region, for example, the region of 1300 cm^{-1} (7.5 μ) to 1000 cm^{-1} (10 μ), is extremely difficult to interpret due to the variety and number of fundamental absorptions occurring in this region. Certain narrow regions of the infrared spectrum provide most of the important information. In the derivation of information from an infrared spectrum, the most prominent bands in these regions are noted and assigned first.

The characteristic bands of various individual functional groups will be discussed in the following sections. These discussions will include the effect of molecular structure and electronic effects on the more prominent absorption bands.

4.2.1 Carbon-Hydrogen Absorption Bands

4.2.1a Alkanes

Carbon-hydrogen absorption occurs in two regions, the C-H stretching region from 3300 to 2500 cm^{-1} (3 to 4 μ) and the bending region of 1550 to 650 cm^{-1} (6.5 to 15.4 μ). The methyl group gives rise to two stretching (ν) bands which occur at 2960 cm^{-1} (3.38 μ) and 2770 cm^{-1} (3.48 μ). The 2960 cm^{-1} (3.38 μ) band is ascribed to an asymmetric stretching deformation involving the entire methyl group, as illustrated in Fig. 4.2a, and the 2770 cm^{-1} (3.48 μ) band is ascribed to the symmetric stretching deformation, as illustrated in Fig. 4.2b. The methyl group also gives rise to two bands in the bending deformation region at 1460 cm^{-1} (6.85 μ) and approximately 1380 cm^{-1} (7.25 μ). The higher-frequency band is due to the asymmetric bending motion illustrated in Fig. 4.2c, and the lower-frequency band is due to the symmetric bending illustrated in Fig. 4.2d. The 1380 cm^{-1} (7.25 μ) region is extremely valuable for the detection of the presence of methyl groups. This region is almost devoid of other types of absorption bands, whereas the other three absorption bands due to the methyl group lie extremely close to similar absorption bands arising from methylene and methine C—H. The symmetrical bending absorption band of the methyl group is quite sharp and of medium intensity (see Figs. 4.13, 4.14, and 4.16). The position of this band is quite sensitive to the type of substituent at-

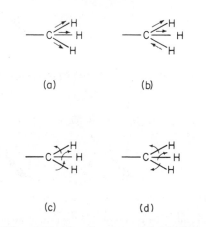

(a)   (b)

(c)   (d)

Fig. 4.2. Symmetric and asymmetric vibration and bending deformations of the methyl group.

tached to the methyl group, and it can be used to detect the possible presence of acetyl and methoxyl groups (see Table 4.1).

The presence of two or more methyl groups bonded to a single carbon atom results in a splitting of the symmetric C—H bending absorption bands. For example, the isopropyl group gives rise to two bands at 1397 cm^{-1} (7.16 μ) and 1370 cm^{-1} (7.30 μ) and the tertiary butyl group gives rise to two bands at 1385 cm^{-1} (7.22 μ) and 1370 cm^{-1} (7.30 μ). Confirmation for the presence of these groups may be obtained by observing the C—C skeletal vibrational bands (see Sec. 4.2.5).

Methylene groups give rise to a number of absorption bands corresponding to the types of deformations illustrated in Fig. 4.3. The symmetric and asymmetric vibration modes, Fig. 4.3a and b, give rise to bands at 2850 cm^{-1} (3.51 μ) and 2930 cm^{-1} (3.42 μ), respectively. Absorption due to the *scissoring* action of the methylene group, Fig. 4.3c, occurs at 1470 cm^{-1} (6.80 μ), very close to the asymmetric bending band of the methyl group. The scissoring of the methylene group varies substantially with changes in the molecular environment. The *rocking* bending deformation, Fig. 4.3d, gives rise to absorption in the 720 cm^{-1} (13.9 μ) region. The *wagging* and *twisting* bending deformations, Fig. 4.3e and f, give rise to absorption in the 1350 cm^{-1} (7.4 μ) to 1180 cm^{-1} (8.5 μ) region. The latter three bending absorptions are not particularly useful in structural work in molecules of any complexity.

The methine group, —$\overset{|}{\underset{|}{C}}$—H, gives rise to single vibrational and bending absorptions at 2890 cm^{-1} (3.46 μ) and 1340 cm^{-1} (7.45 μ) which are very weak and usually are of no practical utility in structural work.

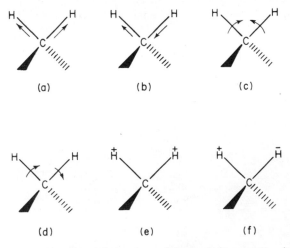

(a)   (b)   (c)

(d)   (e)   (f)

Fig. 4.3. Vibration and bending deformations of the methylene group. Arrows indicate movement in the plane of the page and the + and − signs indicate movement in the forward and rearward directions.

Table 4.1. Selected Carbon-Hydrogen Absorption Bands*

Functional Group	Frequency cm^{-1}	Wavelength μ	Assignment†	Remarks‡
Alkyl				
—CH$_3$	2960	3.38	ν_{as}	s, lu
	2870	3.48	ν_s	m, lu
	1460	6.85	σ_{as}	s, lu
	1380	7.25	σ_s	m, gu
—CH$_2$—	2925	3.42	ν_{as}	s, lu
	2850	3.51	ν_s	m, lu
	1470	6.83	scissoring	s, lu
	1250	~8.00	twisting and wagging	s, nu
—C—H	2890	3.46	ν	w, nu
	1340	7.45	σ	w, nu
—OCOCH$_3$	1380–1365	7.25–7.33	σ_s	s, gu
—COCH$_3$	~1360	~7.35	σ_s	s, gu
—COOCH$_3$	~1440	~6.95	σ_{as}	s, gu
	~1360	~7.35	σ_s	s, gu
Vinyl				
=CH$_2$	3080	3.24	ν_{as}	m, lu
	2975	3.36	ν_s	m, lu
	~1420	~7.0–7.1	σ (in-plane)	m, lu
	~900	~11	σ (out-of-plane)	s, gu
C=C—H	3020	3.31	ν	m, lu
mono-substituted	990	10.1	σ (out-of-plane)	s, gu
	900	11.0	σ (out-of-plane)	s, gu
cis-disubstituted	703–675	13.7–14.7	σ (out-of-plane)	s, gu
trans-disubstituted	965	10.4	σ (out-of-plane)	s, gu
trisubstituted	840–800	11.9–12.4	σ (out-of-plane)	m–s, gu
Aromatic				
C—H	3070	3.30	ν	w, lu
5-adjacent H	770–730	13.0–13.7	σ (out-of-plane)	s, gu
	710–690	14.1–14.5	σ (out-of-plane)	s, gu
4-adjacent H	770–735	13.0–13.6	σ (out-of-plane)	s, gu
3-adjacent H	810–750	12.3–13.3	σ (out-of-plane)	s, gu
2-adjacent H	860–800	11.6–12.5	σ (out-of-plane)	s, gu
1-adjacent H	900–860	11.1–11.6	σ (out-of-plane)	m, gu
1, 2-, 1,4- and	1275–1175	7.85–8.5	σ (in-plane)	w, lu
	1175–1125	8.5–8.9	(only with 1,2,4-)	

Table 4.1. (Continued)

Functional Group	Frequency cm^{-1}	Wavelength μ	Assignment†	Remarks‡
1,2,4-substituted	1070–1000	9.35–10.0	(two bands)	w, lu
1-, 1,3-, 1,2,3-, and	1175–1125	8.5–8.9	σ (in-plane)	
1,3,5-substituted	1100–1070	9.1–9.35	(absent with 1,3,5-)	
	1070–1000	9.35–10.0		
1,2-, 1,2,3- and 1,2,4-substitution	1000–960	10.0–10.4	σ (in-plane)	w, lu
Acetylene				
$\equiv$C—H	3300	3.0	ν	m–s, gu
Aldehyde	2820	3.55	ν	m, lu
—C$\begin{smallmatrix}\nearrow O \\ \searrow H\end{smallmatrix}$	2720	~3.7	overtone or combination band	m, gu

*Values selected from tables presented in Bellamy (ref. 3) and Nakanishi (ref. 11).
†Assignments are designated as follows: ν, vibrational; as, asymmetric; s, symmetric; and σ, bending deformation.
‡Intensities are designated as: s, strong; m, medium; w, weak. Band utilities for structural assignments are indicated as: gu, great utility; lu, limited utility (depends on the complexity of the structure); nu, no practical utility.

The wavelengths of absorption of the vibrational and bending deformation modes for methyl, methylene, and methine all lie quite close together, with the exception of the symmetrical bending mode of the methyl group. These change position only slightly with change in the molecular environment. The differences in position caused by changes in substitution are not sufficiently great to allow resolution by most of the common infrared spectrophotometers available for general use; hence, the amount of structural information that one can derive is somewhat limited. The use of higher-resolution infrared spectrometers, or the harmonics of the normal modes lying in the near-infrared region, may provide more useful structural information.

4.2.1b Olefinic and aromatic compounds

The absorption peaks due to stretching modes of excitation of olefinic and aromatic carbon-hydrogen bonds occur in the 3180 to 2980 cm^{-1} (3.15 to 3.36 μ) region. The terminal methylene group, $\begin{smallmatrix}\diagdown \\ \diagup\end{smallmatrix}$C=C$\begin{smallmatrix}\diagup H \\ \diagdown H\end{smallmatrix}$, displays two stretching bands, one occurring near 3090 cm^{-1} (3.24 μ) and the other near

2980 cm^{-1} (3.36 μ), corresponding to the asymmetric and symmetric stretching modes. The stretching bands of other types of olefinic carbon-hydrogen bonds occur between 3080 and 3020 cm^{-1} (3.24 and 3.31 μ), while aromatic carbon-hydrogen absorption occurs near 3030 cm^{-1} (3.30 μ). These absorption bands are usually quite weak, and the positions vary only slightly with change in substitution. Owing to the possible congestion in this very narrow region, caution must be exercised in using information derived from this region for distinguishing aromatic and olefinic compounds. One must rely on the carbon-carbon double bond stretching region between 1700 and 1450 cm^{-1} (5.9 and 6.9 μ) and the fingerprint region from 1000 to 650 cm^{-1} (10 to 15.5 μ).

The absorption bands due to the bending deformations of olefinic and aromatic compounds are particularly useful in that one may determine the extent and position of substitution, and also the stereochemistry (*cis* or *trans*) of disubstituted olefins.

There are two possible bending deformations: the in-plane bending deformation, in which the hydrogen remains in the nodal plane of the olefinic or aromatic system, and the out-of-plane deformation, in which the hydrogen bends out of the nodal plane of the molecule. The out-of-plane bending deformations are the most valuable for deriving structural information. These peaks occur in the so-called "fingerprint" region of the spectrum, 1000 to 650 cm^{-1} (10 to 15 μ), in which very few other absorptions occur, whereas the in-plane deformations occur in the 1400 to 1000 cm^{-1} (7 to 10 μ) region. The in-plane bending region is useful only in relatively simple molecules due to the great number of bands that appear in this region.

Vinyl groups (—CH=CH$_2$) give rise to two out-of-plane bending deformation bands, one band appearing near 900 cm^{-1} (11 μ) with the second band near 990 (10.1 μ) (see Fig. 4.14). The terminal methylene of 1,1-disubstituted olefins also appears near near 900 cm^{-1} (11 μ). This band is quite intense and usually appears as one of the more prominent bands in a spectrum. *Cis*- and *trans*-disubstituted olefins absorb near 685 cm^{-1} (14.2 μ) and 965 cm^{-1} (10.4 μ), respectively (see Fig. 4.11). The out-of-plane bending deformation in a trisubstituted olefin appears near 820 cm^{-1} (12.1 μ). The position of this band is quite different compared with the corresponding band of the terminal vinyl group which appears at 990 cm^{-1} (10.1 μ).

A monosubstituted benzene displays two out-of-plane bending deformation bands near 750 and 710 cm^{-1} (13.4 and 14.3 μ). *Ortho*-disubstituted benzenes display a single band near 750 cm^{-1} (13.4 μ), characteristic of a four-adjacent-hydrogen system. Although the mono- and ortho-disubstituted benzenes give rise to a band near 750 cm^{-1} (13.4 μ), they are readily distinguished by the presence of, or lack of, the 710 cm^{-1} (14.3 μ) band. *Para*-disubstituted benzenes display a single band near 830 cm^{-1} (12.1 μ), which is characteristic of a two-adjacent-hydrogen system. *Meta*-disubstituted benzenes display two bands: one band is near 780 cm^{-1} (12.8 μ), characteristic of a three-adjacent hydrogen system, and the other band is near 880 cm^{-1} (11.4 μ), characteristic of a one-adjacent hydrogen system. Several of these types of systems are illustrated in Figs. 4.9, 4.10, 4.11, 4.12 and 4.13 at the end of this chapter.

These characteristic absorption patterns for the various adjacent hydrogen systems are also observed with substituted pyridines and polycyclic benzenoid aromatics, for example, substituted naphthalenes, anthracenes, and phenanthrenes.

The in-plane and out-of-plane deformation band positions are summarized in Table 4.1. These bands are quite intense and are subject to considerable change in frequency upon change of substitution. The presence of electron-donating substituents on an olefin or aromatic nucleus results in shifts to lower frequencies (longer wavelengths), and vice versa for electron-withdrawing substituents. Occasionally, a band may shift into the region normally assigned to a different type or degree of substitution. Chemical or other physical data are required in such cases to determine the type of substituent that is present before an accurate assignment of these bands can be made.

The 2000 to 1670 cm^{-1} (5 to 6 μ) region of spectra of aromatic compounds displays several weak bands which are overtone and combination bands of the in-plane and out-of-plane bending deformations. These bands are usually quite weak, and a fairly concentrated solution of the sample must be used to record these bands (see Fig. 4.9). These bands are also useful in deriving structural information; however, the positions of these bands are sensitive to the types of substituents present in the system, as was noted for the out-of-plane deformations discussed in preceding paragraphs. Figure 4.4 displays typical absorption patterns of substituted aromatics in the 2000 to 1670 cm^{-1} (5 to 6 μ) region.

4.2.1c Acetylenic compounds

Acetylenic carbon-hydrogen absorbs very close to 3330 cm^{-1} (3.0 μ). The band is quite sharp and intense.

4.2.1d Aldehyde carbon-hydrogen

Two bands appear below 2860 cm^{-1} (3.5 μ) at approximately 2820 cm^{-1} (3.55 μ) and 2700 cm^{-1} (3.7 μ). The lower-frequency band is quite useful in detecting the presence of an aldehyde group.

4.2.2 Oxygen-Hydrogen Absorption Bands

The oxygen-hydrogen stretching deformation occurs in the 3700 to 2500 cm^{-1} (2.7 to 4.0 μ) region. The position and the shape of the absorption

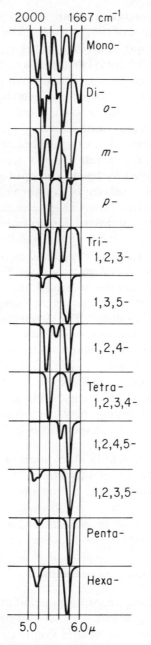

2000　　　1667 cm^{-1}

Mono-

Di-
　　o-

　　m-

　　p-

Tri-
1,2,3-

1,3,5-

1,2,4-

Tetra-
1,2,3,4-

1,2,4,5-

1,2,3,5-

Penta-

Hexa-

5.0　　　6.0 μ

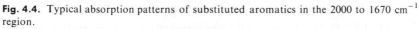

Fig. 4.4. Typical absorption patterns of substituted aromatics in the 2000 to 1670 cm^{-1} region.

bands vary greatly with structure and can be very useful in deriving information concerning the structure of the molecule. The position and shape of the absorption bands are also quite sensitive to the type and extent of hydrogen bonding, either to another hydroxyl group or another acceptor group.

The nonbonded hydroxyl gives rise to a sharp absorption whose position varies slightly from primary to secondary to tertiary hydroxyl. The difference in band position varies only about 10 cm^{-1} (0.01 μ) between primary and secondary, and secondary and tertiary. This difference in position borders on the limits of resolution of spectrometers commonly available in laboratories, and it should not be relied on to decide the environment of the hydroxyl group. Nonbonded hydroxyl absorption can be observed only in dilute solutions in nonpolar solvents. At normal concentrations, approximately 5 to 10% by weight, extensive intermolecular hydrogen bonding occurs, causing an additional broad absorption at lower frequency (see Fig. 4.9). The position and shape of the band depends on the extent of hydrogen bonding, becoming more broad and moving to lower frequency as the strength of the hydrogen bond increases (see Table 4.2).

Occasionally, another functional group may be present in the same molecule as the hydroxyl group to act as an acceptor functional group for intramolecular hydrogen bonding. The absorption shape and intensity is not a function of the concentration of the sample and can be differentiated from intermolecular hydrogen bonding by dilution studies.

The oxygen-hydrogen stretching absorption of carboxylic acids appears as a very broad band with maximum absorption at approximately 2940 cm^{-1} (3.4 μ); it extends to nearly 2500 cm^{-1} (4.0 μ) owing to the very strong intermolecular hydrogen bonding between carboxyl groups. Since acids exist as dimers the position and shape of this band is little affected by concentration.

The oxygen-hydrogen stretching absorption of phenols occurs at slightly lower frequency than the oxygen-hydrogen absorption of alcohols; but usually it is not sufficiently different to allow distinction between phenols and alcohols with most spectrometers.

The oxygen-hydrogen bending absorption occurs in the 1500 to 1300 cm^{-1} (6.7 to 7.7 μ) region and is of no practical value for analysis.

4.2.3 Nitrogen-Hydrogen Absorption Bands

Nitrogen-hydrogen stretching absorptions occur at slightly lower frequencies than the O—H stretching absorptions do. Primary amines display two bands corresponding to asymmetric and symmetric stretching deformations, whereas secondary amines display a single peak (see Table 4.2). Imines, $C{=}NH$, display a single peak in the same region. In general, it is easy to distinguish primary from secondary amines, but distinction between various types of N—H will be difficult.

Table 4.2. Selected X-H Absorption Bands

Functional group	Frequency cm^{-1}	Wavelength μ	Remarks
Alcohols			
(nonbonded) primary	3640	2.72	m, gu, usually determined in dilute solution in nonpolar solvents.
secondary	3630	2.73	
tertiary	3620	2.74	
phenols	3610	2.75	
intermolecularly H-bonded	3600–3500 3400–3200	2.78–2.86 2.94–3.1	m, dimeric, rather sharp s, polymeric, usually quite broad
intramolecularly H-bonded	3600–3500	2.78–2.86	m-s, much sharper than intermolecular hydrogen bonded OH; is not concentration dependent.
Amines			
RNH$_2$	~3500 ~3400 1640–1560	~2.86 ~2.94 6.1–6.4	m, gu, ν_{as} m, gu, ν_s m-s, gu, corresponds to scissoring deformation.
R$_2$NH	3500–3450	2.86–2.90	w-m, gu, ν
ArNHR	3450	2.90	w-m, gu
Pyrroles, indoles	3490	2.86	w-m, gu
Ammonium salts			
NH$_4^+$	3300–3030 1430–1390	3.0–3.3 7.0–7.2	s, gu s, gu
—N$^+$H$_3$	3000 1600–1575 1490	~3.0 6.25–6.35 ~6.7	s, gu, usually quite broad s, gu, σ_{as} s, gu, σ_s
$\geq$N$^+$H$_2$	2700–2250	3.7–4.4	s, gu, ν_{as} and ν_s, usually broad or a group of bands
$\geq$N$^+$H	1600–1575 2700–2250	6.25–6.35 3.7–4.4	m, gu, σ s, gu, ν, σ_{NH^+} band is weak and of no practical utility.
Mercaptans			
—SH	2600–2550	3.85–3.92	s, gu band is often very weak and can be missed if care is not exercised.

Amines display N—H bending deformation absorption bands similar to those of —CH$_2$— and —C̲—H, except that they occur at slightly higher frequency. Protonation of amines, to give the corresponding amine salts, results in the formation of bands similar to those of —CH$_3$, —CH$_2$—, and —C̲—H, again at higher frequencies.

Primary amides display N—H stretching bands at slightly higher frequencies than do amines. The N—H bending deformations of amides occur in the 1610 to 1490 cm^{-1} (6.2 to 6.7 μ) region and are generally referred to as "amide II" bands. These bands are quite intense.

Hydrogen bonding involving N—H gives rise to similar effects noted with hydroxyl absorption bands.

4.2.4 Sulfur-Hydrogen Absorption Bands

The sulfur-hydrogen stretching absorption occurs near 2630 to 2560 cm^{-1} (3.8 to 3.9 μ), is usually quite weak, and is not subject to large shifts due to hydrogen bonding.

4.2.5 Carbon-Carbon Absorption Bands

Carbon-carbon single-bond stretching bands are extremely variable in position and are usually quite weak in intensity. These absorption bands are of little practical use in structure determination.

Specific alkyl group vibrational deformation bands have been characterized. In addition to the characteristic C—H vibrational frequencies produced (see Sec. 4.2.1a), the tertiary butyl group gives rise to two bands at 1250 cm^{-1} (8.00 μ) and near 1208 cm^{-1} (8.28 μ) and the isopropyl group gives rise to a band at 1170 cm^{-1} (8.55 μ) with a weaker band at 1145 cm^{-1} (8.73 μ). A geminal dimethyl group produces two bands at 1125 cm^{-1} (8.23 μ) and 1195 cm^{-1} (8.37 μ). Groups of two, three, and four methylene groups absorb at 790 to 770 cm^{-1} (12.7 to 13.0 μ), 743 to 734 cm^{-1} (13.5 to 13.7 μ), and 725 to 720 cm^{-1} (13.8 to 13.9 μ) respectively. Very few other characteristic group absorptions have been defined.

Carbon-carbon double-bond stretching absorption occurs in the 2000 to 1430 cm^{-1} (5 to 7 μ) region. The position and intensity of these bands are very sensitive to substitution and, in cyclic olefins, the ring size. Alkyl substitution results in a shift of the absorption band to higher frequencies. The shift to higher frequencies with increasing substitution is least with acyclic olefins and increases dramatically with decreasing ring size in cyclic olefins (see Table 4.3). The intensity of the band is a function of the sym-

Table 4.3. Selected C—C and C—N Absorption Bands

Functional group	Frequency cm^{-1}	Wavelength μ	Remarks
Alicyclic C=C($\nu_{C=C}$)			
monosubstituted	1645	6.08	m, these bands are of little
1,1-disubstituted	1655	6.04	m, utility in assigning sub-
cis-1,2,-disubst.	1660	6.02	m, stitution and stereo-
trans-1,2-disubst.	1675	5.97	w, chemistry; the C—H
trisubstituted	1670	5.99	w, out-of-plane bending
tetrasubstituted	1670	5.99	w, bands in the fingerprint region are recommended for this purpose.
Conjugated C=C($\nu_{C=C}$)			
diene	1650 and 1600	6.06 and 6.25	s, gu,
with aromatic	1625	6.16	s, gu,
with C—O	1600	6.25	s, gu,
Cyclic C=C($\nu_{C=C}$)			
6-membered ring and larger	1660–1650	6.03–6.06	m, of limited utility due to closeness to the alicyclic region.
monosubstituted	1680–1665	5.95–6.00	m,
5-m. unsubstituted	1615	6.19	m, can be of great utility
monosubstituted	1660	6.02	m, although the ring size
disubstituted	1690	5.92	w, must be assigned first
4-m. unsubstituted	1660	6.02	m, for those compounds
3-m. unsubstituted	1640	6.10	m, whose absorption falls
monosubstituted	1770	5.65	m, into regions consistent
disubstituted	1880	5.32	m, with other types of absorption bands (nuclear magnetic resonance spectra can be very useful in this respect).
Aromatic C=C	1600	6.24	w-s, in-plane skeletal vibra-
	1580	6.34	s, tions, the intensities of
	1500	6.67	w-s, the 1600 and 1500 cm^{-1}
	1450	6.9	s, may be rather weak.
C—N			
aliphatic amine	1220–1020	8.2–9.8	m-w, lu, $\nu_{C—N}$
aromatic amine	1370–1250	7.3–8.0	s, gu-lu, $\nu_{C—N}$
C=N	1700–1615	5.9–6.2	s, gu, $\nu_{C=N}$

metry of the alkyl substituted olefin; mono- and tri-substituted olefins give rise to more intense bands than the *cis-* or *trans-*disubstituted olefins, with the tetraalkyl olefin giving rise to a very low intensity band.

Decreasing the ring size of cyclic olefins from six- to four-membered results in shifts to lower frequencies; however, a further decrease in ring size to cyclopropene results in a dramatic shift to higher frequencies.[2] These shifts are probably due to changes in hybridization involving the carbon atoms which comprise the carbon-carbon double bond.

Substitution of hydrogen by a nitrogen or oxygen functional group greatly increases the intensity of the C=C absorption band. The effect of conjugation with an aromatic nucleus results in a slight shift to lower frequency, and with another C=C or C=O, a shift to lower frequency of approximately 40 to 60 cm^{-1} (0.15 to 0.20 μ) with a substantial increase in intensity.

The skeletal C=C vibrations of aromatics give rise to a series of four bands in the 1660 to 1430 cm^{-1} (6 to 7 μ) region. These bands occur very close to 1600 cm^{-1} (6.24 μ), 1575 cm^{-1} (6.34 μ), 1500 cm^{-1} (6.67 μ), and 1450 cm^{-1} (6.9 μ). The first and third bands vary greatly in intensity, generally becoming more intense in the presence of electron-donating groups. Table 4.3 lists the positions of absorption of various C=C systems.

Bending deformation absorptions of C—C and C=C bonds occur beyond the limits of the general infrared spectrometers, at wavelengths longer than 15 μ, and they are of little utility in structure work at the present time.

4.2.6 Carbon-Nitrogen Absorption Bands

Aliphatic amines display C—N stretching absorption in the 1220 to 1020 cm^{-1} (8.2 to 9.8 μ) region. These bands are of medium to weak intensity. Owing to the intensity and position of these bands, they are of little practical value for structural work, except in relatively simple molecules. Aromatic amines absorb in the 1370 to 1250 cm^{-1} (7.3 to 8.0 μ) region and give quite intense bands. Carbon-nitrogen double bond absorption occurs in the 1700 to 1615 cm^{-1} (5.9 to 6.2 μ) region and is subject to similar environmental effects as the C=O absorption bands (see following section).

4.2.7 Carbon-Oxygen Absorption Bands

Carbon-oxygen single bond absorption occurs in the 1250 to 1000 cm^{-1} (8 to 10 μ) region. Although both alcohols and ethers absorb in this region, distinction between these classes can be made from information gained from the hydroxyl region.

The position of the carbon-oxygen stretching absorption depends on the

---

[2] K. Wiberg and B. Nist, *J. Am. Chem. Soc.*, **83**, 1226 (1961)).

extent and type of substitution on the carbinol carbon atom. Nakaniski (ref. 11) has developed the following correlation to predict the frequency of absorption of the carbon-oxygen stretching deformation, based on standard frequencies of absorption for primary, secondary, and tertiary alcohols (structure **1**) of 1050, 1100, and 1150 cm^{-1}, respectively.

	α-branching:	$-15\ cm^{-1}$ (per branch in α' and α'')
	α-unsaturation:	$-30\ cm^{-1}$
	α,α'-ring formation	$-50\ cm^{-1}$
	α-unsaturation + α'-branching	$-90\ cm^{-1}$
	α- and α'-unsaturation	$-90\ cm^{-1}$
1	α-, α'-, and α''-unsaturation	$-140\ cm^{-1}$

$$C \overset{\beta}{-} C^{\alpha} \diagdown$$
$$C - C^{\alpha'} - C - OH$$
$$C - C^{\alpha''} \diagup$$

The effect of ring size on the position of absorption in ethers is indicated in Table 4.4. These absorption bands in the cyclic ethers are due to the anti-symmetric C—O—C vibrational modes.

The carbon-oxygen double bond absorbs in the 2000 to 1540 cm^{-1} (5 to 6.5 μ) region, except for ketenes which absorb near 2200 cm^{-1} (4.65 μ). This is probably the most useful portion of the infrared region because the position of the carbonyl group is quite sensitive to substituent effects and the geometry of the molecule. The band positions are also solvent sensitive owing to the high polarity of the C=O bond.

An empirical correlation has been developed which can be used to adequately predict the positions of most carbonyl bands. Table 4.4 lists the band positions for a number of different types of compounds that contain the C=O group. For ketones, aldehydes, acids, anhydrides, acid halides, and esters, the frequencies (wavelengths) cited are for the normal, unstrained (acyclic or contained in a six-membered ring) *parent* compound in carbon tetrachloride solution. The introduction of a double bond or aryl group in conjugation with the C=O results in a reasonably consistent 30 cm^{-1} (0.1 μ) shift to lower frequency. Introduction of a second double bond results in an additional shift of approximately 15 cm^{-1} (0.05 μ) in the same direction.

The position of the C=O band is very sensitive to changes in the C(CO)C bond angle. A decrease in ring size, resulting in a decrease in the C(CO)C bond angle, of ketones and esters results in a reasonably consistent shift of 30 cm^{-1} (0.1 μ) to higher frequency per each decrease in ring size from the six-membered ring. In cyclic structures containing more than six atoms, slight shifts to lower frequencies are observed. For example, a shift of $-10\ cm^{-1}$ (+0.03 μ) for seven-membered ring and an approximate -5 cm^{-1} (0.01 to 0.02 μ) shift for eight- and nine-membered ring systems are observed. Further increases in ring size result in a moderate increase in frequency back to the parent compound position. Highly strained bridge carbonyl compounds, for example, 7-ketobicyclo-[2.2.1]-heptanes and

Table 4.4. Selected Carbon-Oxygen Absorption Bands

Functional group	Frequency cm^{-1}	Wavelength μ	Remarks
C—O Single Bonds			
Primary C—O—H	1050	9.52	s, gu, ν_{C-O} (See discussion for substituent effects.)
Secondary C—O—H	1100	9.08	s, gu, ν_{C-O}
Tertiary C—O—H	1150	8.68	s, gu, ν_{C-O}
Aromatic C—O—H	1200	8.33	s, gu, ν_{C-O}
Ethers-acyclic	1150–1070	8.7–9.35	s, gu, antisymmetric ν_{C-O-C}
C=C—O—C	1270–1200 and	7.9–8.3 and	s, gu, antisymmetric ν_{C-O-C}
	1070–1020	9.3–9.8	s, gu, symmetric ν_{C-O-C}
Cyclic ethers			
6-m. and larger	1140–1070	8.77–9.35	s, gu,
5-m.	1100–1075	9.1–9.3	s, gu,
4-m.	980–970	10.2–10.3	s, gu,
Epoxides	1250	8.0	s, gu,
Cis-disubstituted	890	11.25	s, gu,
Trans-disubstituted	830	12.05	s, gu,
C—O Double Bonds			
Ketones	1715	5.83	s, gu, $\nu_{C=O}$, unstrained C=O group in acyclic and 6 m. ring compounds in carbon tetrachloride solution. (See discussion for effects of conjugation and ring size.)
α, β-unsaturated	1685	5.93	s, gu, $\nu_{C=O}$. For the s-*cis* configuration (C〈O, C=C) the $\nu_{C=C}$ may appear above 1600 cm^{-1} (below 6.26 μ) with an intensity approximately that of the $\nu_{C=O}$. The *trans*-configuration does not show this enhanced intensity of absorption of the C=C.
α- and β-diketones	1720	5.81	s, gu, $\nu_{C=O}$, two bands at higher frequency when in the s-*cis* configuration
	1650	6.06	s, gu, $\nu_{C=C}$ if enolic (C=C)
Quinones	1675	5.97	s, gu, $\nu_{C=O}$

Table 4.4 (Continued)

Functional group	Frequency cm^{-1}	Wavelength μ	Remarks
Tropolones	1650	6.06	s, gu, $\nu_{C=O}$
	1600	6.26	s, gu, $\nu_{C=O}$, if intramoleculary hydrogen bonded as in α-tropolones.
Aldehydes —C(=O)(H)	1725	5.80	s, gu, $\nu_{C=O}$ (See discussion for effects of conjugation.)

Carboxylic acids and derivatives

Functional group	Frequency cm^{-1}	Wavelength μ	Remarks
—C(=O)—OH	1710	5.84	s, gu, $\nu_{C=O}$, usually as the dimer in nonpolar solvents, monomer absorbs at 1730 cm^{-1} (5.78 μ) and may appear as a shoulder in the spectrum of a carboxylic acid; for conjugation effects see the discussion.
Esters	1735	5.76	s, gu, $\nu_{C=O}$, acyclic and 6-m. lactones; for conjugation and ring size effects see the discussion.
	1300–1050	7.7–9.5	s, lu, symmetric and antisymmetric ν_{C-O-C} giving 2 bands, indicative of type of ester, for example, formates: 1178 cm^{-1} (8.5 μ) acetates: 1242 cm^{-1} (8.05 μ) methyl esters: 1164 cm^{-1} (8.6 μ) others: 1192 cm^{-1} (8.4 μ) but the distinction generally is not great enough to be of diagnostic value.
Anhydrides	1820 and 1760	5.48 and 5.68	s, gu, $\nu_{C=O}$, the intensity and separation of the bands may be quite variable. (See discussion for general effects of conjugation and ring size.)
Acid halide	1800	5.56	s, gu, $\nu_{C=O}$, acid chlorides and fluorides absorb at slightly higher frequency while the bromides and iodides absorb at slightly lower frequency. (See discussion for effects of conjugation.)
Amides	1650	6.06	s, gu, $\nu_{C=O}$, "Amide I" band, this frequency is for the associated amide (see COOH), free amide at 1686 cm^{-1} (5.93 μ) in dilute solution, cyclic amides shift to higher frequency as ring size decreases.
	1300	7.7	s, gu, ν_{C-N}, "Amide III" band, free amide at slightly higher frequency.
α, β-unsaturated	1665	6.01	s, gu, $\nu_{C=O}$
Carboxylate (—C(=O)(O$^-$))	1610–1550 and 1400	6.2–6.45 and 7.15	s, gu, antisymmetric and symmetric stretching of —C(=O)(O$^-$)

6-ketobicyclo-[2.1.1]-hexanes absorb at relatively high frequencies, in the general region of 1800 cm^{-1} (5.5 μ).

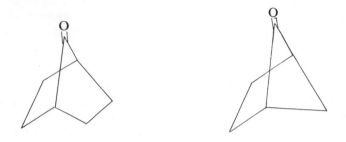

The effects of conjugation and ring size are cumulative and may be used to adequately predict the position of absorption in more complex molecules. For example, the calculated position of absorption in compounds 2 and 3 in carbon tetrachloride solution are illustrated below and are compared with the observed absorptions.

          2   3

	Normal ester (from Table 4.4)	1735 cm^{-1}	5.76 μ
	Ring size correction	+30	−0.10
2	Unsaturation	−30	+0.10
	Calculated $\nu_{C=O}$	1735 cm^{-1}	5.76 μ
	Observed $\nu_{C=O}$	1740 cm^{-1}	5.74 μ
	Normal ketone	1715 cm^{-1}	5.83 μ
	Ring size correction	−10	+0.03
3	Unsaturation (first C=C)	−30	+0.10
	(second C=C)	−15	+0.05
	Calculated $\nu_{C=O}$	1660 cm^{-1}	6.01 μ
	Observed $\nu_{C=O}$	1666 cm^{-1}	6.00 μ

The C=O stretching absorption band of amides displays a different behavior when conjugated with an unsaturated system than do other types of C=O containing functional groups. Instead of moving to lower frequencies when α,β-unsaturated, the band actually moves to higher frequencies owing to the negative inductive effect of the olefinic linkage. The amide group is considered to be highly resonance stabilized and does not interact strongly with the additional C=C.

Although anhydrides were considered earlier, this functional group is quite different from other acid derivatives in that two bands appear in the 1850 to 1725 cm^{-1} (5.4 to 5.8 μ) region owing to vibration coupling of the carbonyl groups. The relative intensities of these bands vary greatly, but both are usually easily discernible.

The introduction of electron withdrawing groups on the α-carbon atom leads to a displacement of the absorption band to higher frequencies (shorter wavelengths), the magnitude of the shift being a function of the electronegativity, or electron withdrawing ability, of the group and the torsional angle of the X—C—C=O group. An example of the latter effect is demonstrated by the α-substituted cyclohexanones in which the axial isomers (**4**) absorb at lower frequency (longer wavelength) than the equatorial isomers (**5**) do.

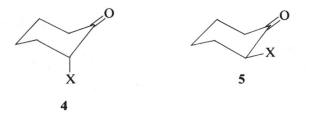

Many of the listed carbonyl containing compounds give rise to absorption bands in other portions of the infrared region which are equally important in assigning partial structures, for example, the O—H and N—H stretching absorptions of carboxylic acids and amides which have been discussed in earlier sections. The identification of these bands, integrated with the results of functional classification tests, should allow the facile assignment of the type of carbonyl group which is present in any given molecule.

4.2.8 Carbon-Carbon Triple Bond Absorption

Carbon-carbon triple bond absorption in terminal acetylenes occurs in the 2140 to 2080 cm^{-1} (4.6 to 4.8 μ) region. The absorption band is relatively weak but quite sharp. A band in this region, along with the C—H stretching absorption band near 3300 cm^{-1} (3.0 μ) are indicative of the presence of a terminal acetylene.

Unsymmetrically disubstituted acetylenes absorb in the 2260 to 2190 cm^{-1} (4.42 to 4.57 μ) region. The intensity of this band is a function of the symmetry of the molecule; the more symmetrical the substitution, the less intense is the absorption band.

Conjugation with an olefinic linkage causes an increase in the band intensity and a shift to slightly lower frequencies. The observed shifts are usually less than those observed with olefins. Conjugation with carbonyl groups does not appreciably alter the position of absorption.

4.2.9 Carbon-Nitrogen Triple Bond Absorption

The carbon-nitrogen triple bond of nitriles absorbs in the 2260 to 2210 cm^{-1} (4.42 to 4.52 μ) region. Saturated nitriles absorb in the higher frequency portion of the region, while unsaturated nitriles absorb in the lower-frequency end. The C≡N absorption band is of much greater intensity than the C≡C absorption bands.

4.2.10 Absorption of X=Y=Z Functional Groups

Functional groups having the general structure X=Y=Z, for example allenes, ketenes, ketenimines, isocyanates, carbodiimides, and azides, absorb in the 2500 to 2000 cm^{-1} (4 to 5 μ) region (see Table 4.5). The bands are very intense and are of great diagnostic value. Distinction between the possible functional groups falling in this region may require functional classification tests or conversion to suitable derivatives.

Table 4.5. Absorption Bands of Y≡Z and X=Y=Z Type Groups

Functional group	Frequency cm^{-1}	Wavelength μ	Remarks
Acetylenes			
terminal	2140–2100	4.67–4.76	w, gu, $\nu_{C\equiv C}$, to be used in conjunction with the ν_{C-H} band.
disubstituted	2260–2190	4.42–4.57	v.w, lu, $\nu_{C\equiv C}$, may be completely absent in symmetrical acetylenes.
Nitriles			
alkyl	2260–2240	4.42–4.47	m-s, gu, $\nu_{C\equiv N}$
aryl	2240–2220	4.46–4.51	s, gu, $\nu_{C\equiv N}$
α-,β-unsaturated	2235–2215	4.47–4.52	s, gu, $\nu_{C\equiv N}$
Allenes	1950	5.11	m, gu, $\nu_{C=C=C}$, terminal allene displays two bands and displays ν_{C-H} at 850 cm^{-1} (11.76 μ)
Ketenes	2150	4.65	s, gu
Ketenimines	2000	5.0	s, gu
Cyanates	2275–2250	4.39–4.45	v.s, gu, extremely intense
Carbodiimides	2145–2130	4.66–4.69	v.s, gu, extremely intense
Azides	2160–2120	4.63–4.72	v.s, gu

4.2.11 Nitrogen-Oxygen Absorption Bands

Nitro groups display two intense absorption bands near 1520 cm^{-1} (6.6 μ) and 1350 cm^{-1} (7.4 μ). Conjugation with an aromatic ring causes a shift to lower frequencies. Nitrates, RONO$_2$, absorb in several regions, including near 1640 cm^{-1} (6.1 μ) and 1280 cm^{-1} (7.8 μ) (see Table 4.6).

Table 4.6. Selected N—O Absorption Bands

Functional Group	Frequency cm^{-1}	Wavelength μ	Remarks
Nitro (R—NO$_2$)	1570–1500	6.37–6.67	s, gu, antisymmetric ν_{NO_2}
	1370–1300	7.30–7.69	s, gu, symmetric ν_{NO_2}, conjugated nitro absorbs in the lower-frequency portions of the cited regions.
Nitrate (RO—NO$_2$)	1650–1600	6.06–6.25	s, gu, antisymmetric ν_{NO_2}
	1300–1250	7.69–8.00	s, gu, symmetric ν_{NO_2}, additional bands at 870–855 cm^{-1} (11.5–11.7 μ) (ν_{O-N}), 763 cm^{-1} (13.1 μ) (out-of-plane bending) and 703 cm^{-1} (14.2 μ) (NO$_2$ bending).
Nitramine (N—NO$_2$)	1630–1550	6.13–6.45	s, gu, antisymmetric ν_{NO_2}
	1300–1250	7.69–8.00	s, gu, symmetric ν_{NO_2}
Nitroso (R—N=O)	1600–1500	6.25–6.66	s, gu, $\nu_{N=O}$, conjugation shifts to lower frequencies in the cited region.
Nitrite (R—ONO)	1680–1610	5.95–6.21	s, gu, $\nu_{N=O}$, two bands usually present due to *cis* and *trans* forms.
Amine Oxide $\left(\overset{+}{\underset{}{\ge N}} - \overset{-}{O} \right)$			
aliphatic	970–950	10.3–10.5	v.s, gu, ν_{N-O}
aromatic	1300–1200	7.7–8.3	v.s, gu, ν_{N-O}
Azoxy $\left(\diagdown N = \overset{+}{N} \diagup \underset{O^-}{} \right)$	1310–1250	7.63–8.00	v.s, gu, ν_{N-O}

Nitroso groups give rise to a single absorption peak in the 1600 to 1500 cm^{-1} (6.2 to 6.7 μ) region. Nitrites, RONO, give rise to two bands, ascribed to the *cis* and *trans* forms, in the 1680 to 1610 cm^{-1} (5.9 to 6.2 μ) region.

Nitrogen-oxygen single bond absorption occurs in the 1320 to 910 cm^{-1} (7.6 to 11 μ) region and may require additional spectral and chemical data for identification purposes (see Table 4.6).

4.2.12 Sulfur-Oxygen Absorption Bands

Functional groups containing a sulfur-oxygen single bond display absorption bands in the 910 to 710 cm^{-1} (11 to 14 μ) region. Sulfur-oxygen bonds of the type appearing in sulfoxides and sulfones absorb at considerably higher frequencies. Sulfoxides display a single absorption band near 1050 cm^{-1} (9.5 μ), whereas sulfones give rise to two bands near 1330 cm^{-1} (7.5 μ) and 1140 cm^{-1} (8.7 μ). These absorption bands are little affected by conjugation or ring strain (see Table 4.7). Functional group absorptions not included in Table 4.7 can be predicted by combinations of the values given in Table 4.7.

A general spectra-structure correlation table of group frequencies in the infrared region is presented in Figs. 4.5, 4.6, 4.7, and 4.8. These tables summarize in more general terms the group frequencies discussed in greater detail in the earlier sections of this chapter.

4.3   ISOTOPE EFFECTS

Substitution of an atom by a heavier isotope, for example, deuterium for hydrogen, results in a shift of the absorption band to lower frequencies. The

Table 4.7. Selected S—O Absorption Bands

Functional group	Frequency cm^{-1}	Wavelength μ	Remarks
—S—O	900–700	11.1–14.2	s, gu, ν_{S-O}
Sulfoxide $\left(\!>\!S\!=\!O\right)$	1090–1020	9.43–9.62	s, gu, $\nu_{S=O}$
Sulfone $\left(\!>\!S\!<^O_O\right)$	1350–1310	7.42–7.63	s, gu, antisymmetric ν_{SO_2}
	1160–1120	8.62–8.93	s, gu, symmetric ν_{SO_2}
Sulfonic acid $\left(-SO_2OH\right)$	1260–1150	7.93–8.70	s, gu, antisymmetric ν_{SO_2}
	1080–1010	9.26–9.90	s, gu, symmetric ν_{SO_2}
	700–600	14.2–16.6	s, gu, ν_{S-O}, may appear outside the general infrared region; in addition, O—H absorption appears.
Sulfonate $\left(-SO_2OR\right)$	1420–1330	7.04–7.52	s, gu, antisymmetric ν_{SO_2}
	1200–1145	8.33–8.73	s, gu, symmetric ν_{SO_2}
Sulfonamide $\left(-SO_2NR_2\right)$	1370–1330	7.30–7.52	s, gu, antisymmetric ν_{SO_2}
	1180–1160	8.47–8.62	s, gu, symmetric ν_{SO_2}

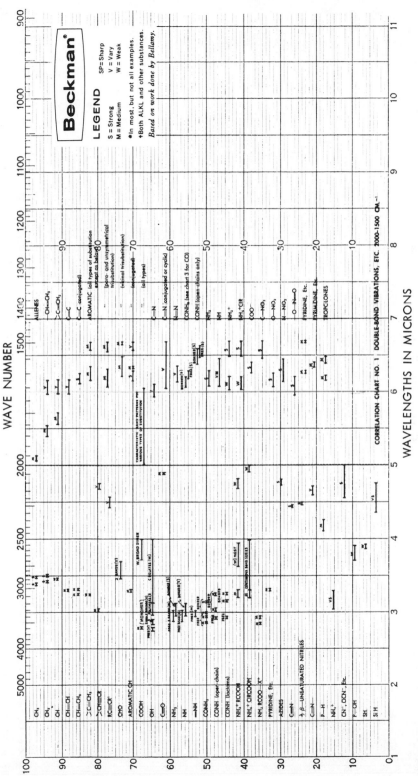

Fig. 4.5. Correlation chart 1. Hydrogen stretching, and double- and triple-bond stretching deformation frequencies (3750 to 1500 cm^{-1}). (Reproduced by permission of Beckman Instruments, Inc., Fullerton, Calif.)

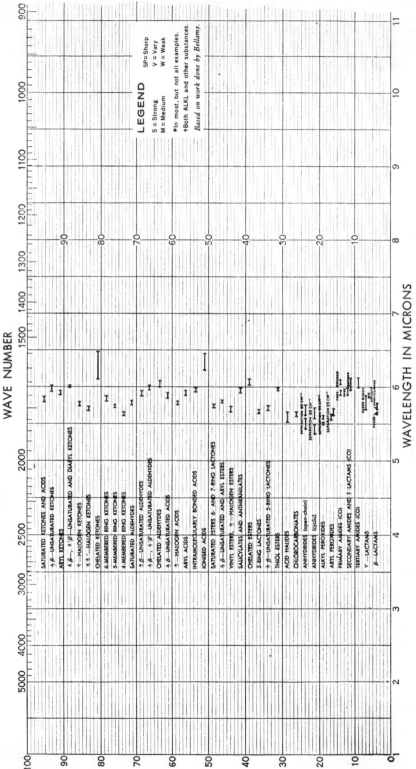

Fig. 4.6. Correlation chart 2. Carbonyl stretching deformation frequencies (1900 to 1500 cm^{-1}). (Reproduced by permission of Beckman Instruments, Inc., Fullerton, Calif.)

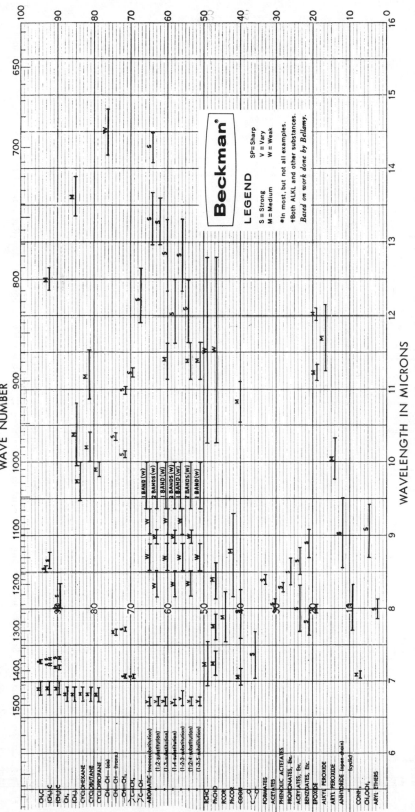

Fig. 4.7. Correlation chart 3. Single-bond stretching and bending deformation frequencies (1500 to 650 cm^{-1}). (Reproduced by permission of Beckman Instruments, Inc., Fullerton, Calif.)

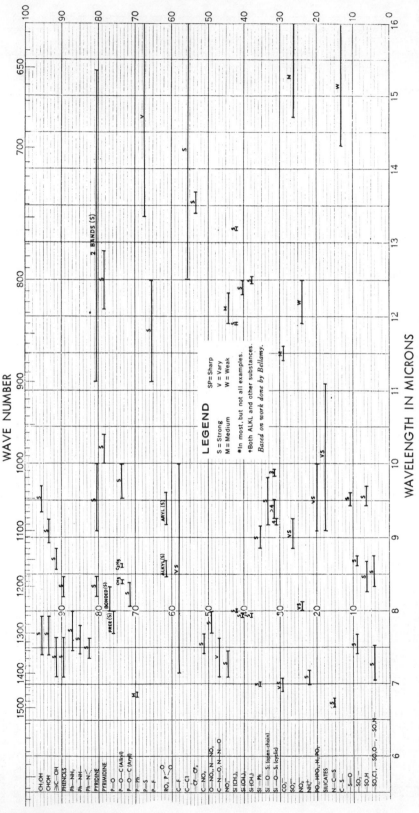

Fig. 4.8. Correlation chart 4. Single-bond stretching and bending deformation frequencies (1500 to 650 cm⁻¹). (Reproduced by permission of Beckman Instruments, Inc., Fullerton, Calif.)

position of absorption of the new bonded system can be approximated fairly closely by Eq. (4.4)

$$\nu_{A-C} \approx \nu_{A-B} \sqrt{\frac{m_B}{m_C}} \qquad (4.4)$$

where C is the isotope of B in the bond A—C, and m_B and m_C are the masses of the two isotopes. Equation 4.4 is derived from Eq. 4.1, assuming that the force constant k is the same for both systems.

This same technique can be applied reasonably well to bonding systems in which one atom is exchanged for another atom of the same family, for example, O—H to S—H and C=O to C=S. This approach is useful when insufficient information is available in the literature on the position of absorption of a particular functional group.

4.4   INTERPRETATION OF REPRESENTATIVE SPECTRA

The infrared spectrum of 1-phenylethanol is shown in Fig. 4.9. The complete trace was taken of a neat liquid film between sodium chloride plates (see discussion on the preparation of samples, Sec. 4.5). The intense, broad band near 3360 cm^{-1} (3 μ) represents the hydrogen bonded O—H stretch. The nonbonded absorption peak is barely perceptible as a shoulder near 3550 cm^{-1} (2.81 μ). Trace B is of the hydroxyl region of a carbon tetra-

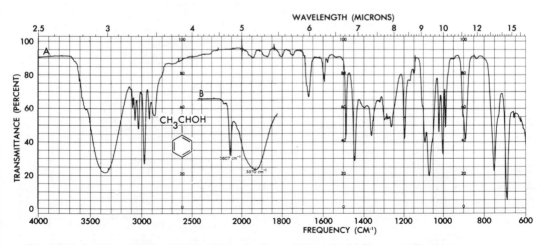

Fig. 4.9. Infrared spectrum of 1-phenylethanol. Trace A, neat liquid film; trace B, 5% in carbon tetrachloride. Trace B is of the 3 μ region and has been displaced on the chart paper. (The infrared spectra reproduced in this book were recorded on a Perkin-Elmer 621 spectrophotometer.)

chloride solution of 1-phenylethanol, which has been displaced on the chart paper for greater clarity, showing the sharp nonbonded band and the broad bonded band. The C—H stretching region shows the aromatic C—H stretch as three bands just above 3000 cm^{-1} (3.33 μ). The band at 2970 cm^{-1} (3.36 μ) represents the asymmetric stretch of the methyl.

The bending deformation bands of the methyl group appear at 1450 cm^{-1} (6.9 μ) and 1365 cm^{-1} (7.3 μ), the former overlapping one of the aromatic skeletal vibration bands. The out-of-plane C—H bending deformation bands occur at 755 cm^{-1} (13.1 μ) and 685 cm^{-1} (14.8 μ); characteristics of a monosubstituted aromatic. The in-plane C—H bending deformation bands appear in the 1100 to 1000 cm^{-1} (9 to 10 μ) region, along with the C—O stretch and other possible skeletal deformation bands, thus making definite assignments rather tenuous. The overtone and combination bands of the out-of-plane and in-plane C—H deformations appear in the 2000 to 1670 cm^{-1} (5 to 6 μ) region. These bands are perceptible in trace A. Note the similarity of these band shapes and positions with those predicted in Fig. 4.4.

Only two of the aromatic skeletal stretching bands are readily visible in the 1650 to 1430 cm^{-1} (6 to 7 μ) region, those appearing at 1590 cm^{-1} (6.3 μ) and 1485 cm^{-1} (6.7 μ).

The immediate information that a reader should derive from this spectrum, if given to him as an unknown, is the presence of O—H, a monosubstituted benzene ring, a methyl group, and probably very little other aliphatic C—H.

Figure 4.10 displays the spectrum of phenylacetylene. The acetylenic C—H band appears at 3290 cm^{-1} (3.03 μ) as a very sharp, intense band. The C≡C stretch on the other hand, appears as the relatively weak band at

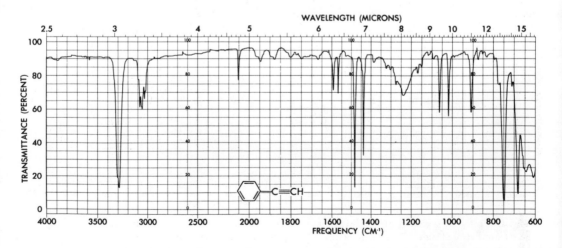

Fig. 4.10. Infrared spectrum of phenylacetylene as a neat liquid film.

2100 cm^{-1} (4.76 μ). The typical out-of-plane C—H bending bands appear at 745 cm^{-1} (13.4 μ) and 680 cm^{-1} (14.6 μ), with the in-plane C—H bending bands appearing in the 1100 to 900 cm^{-1} (9 to 11 μ) region. The typical pattern of a monosubstituted benzene ring is clearly seen in the 2000–1650 cm^{-1} (5 to 6 μ) region. The 1590 cm^{-1} (6.28 μ) and 1570 cm^{-1} (6.37 μ) bands due to the aromatic system are clearly more distinct than in Fig. 4.9, along with the 1480 cm^{-1} (6.74 μ) and 1440 cm^{-1} (6.94 μ) bands.

Figure 4.11 shows the infrared spectrum of *trans-β*-bromostyrene. The olefinic C—H out-of-plane deformation appears at 930 cm^{-1} (10.7 μ); corresponding to a *trans*-disubstituted olefin. The C=C double-bond stretch overlaps the 1600 cm^{-1} (6.2 μ) aromatic band. The interpretation of the remainder of the spectrum is left to the reader.

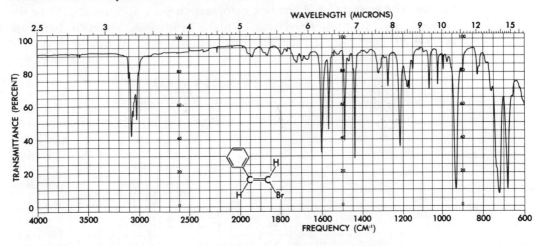

Fig. 4.11. Infrared spectrum of *trans-β*-bromostyrene as a neat liquid film.

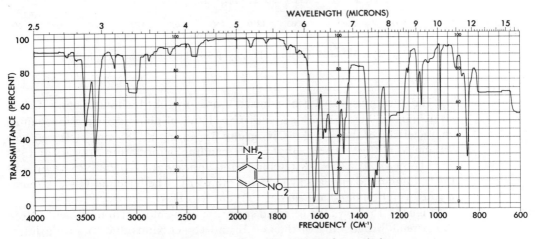

Fig. 4.12. Infrared spectrum of *m*-nitroaniline in chloroform solution.

Figure 4.12 shows the infrared spectrum of *m*-nitroaniline in chloroform solution. The N—H absorption of the primary amine appears as two bands at 3500 cm^{-1} (2.86 μ) and 3400 cm^{-1} (2.93 μ). The aromatic C—H stretch is barely resolved near 3030 cm^{-1} (3.3 μ). The aromatic C—H out-of-plane deformations occur at 860 cm^{-1} (11.7 μ), characteristic of a single isolated C—H, and at 820 cm^{-1} (12.2 μ). This latter band appears only as a slight shoulder where the solvent (chloroform) absorbs very strongly. The characteristically very intense nitro group bands appear at 1515 cm^{-1} (6.6 μ) and 1340 cm^{-1} (7.4 μ). Note the intense band at 1625 cm^{-1} (6.16 μ) due to in-plane bending of the —NH$_2$.

The infrared spectrum of methyl tolyl ketone is shown in Fig. 4.13. The C═O stretch appears at 1675 cm^{-1} (5.95 μ), characteristic of a conjugated ketone. The methyl C—H bending deformation peak appears at 1350 cm^{-1} (7.41 μ). The aromatic C—H out-of-plane bending deformation band appears at 805 cm^{-1} (12.2 μ), characteristic of a two adjacent hydrogen system.

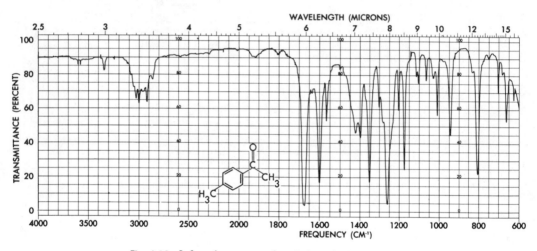

Fig. 4.13. Infrared spectrum of methyl *p*-tolyl ketone as a neat liquid film.

The infrared spectrum of 5-hexen-2-one, Fig. 4.14, shows olefinic C—H stretch at 3080 cm^{-1} (3.24 μ), with C—H out-of-plane deformation bands, characteristic of the —CH═CH$_2$ group appearing at 985 cm^{-1} (10.1 μ) and 905 cm^{-1} (11.0 μ). The methyl C—H bending deformation appears at 1355 cm^{-1} (7.39 μ). The C═O stretching band appears at 1720 cm^{-1} (5.83 μ), and the bending band appears at 1155 cm^{-1} (8.65 μ). The terminal C═C stretching deformation band appears at 1635 cm^{-1} (6.11 μ).

Figure 4.15 shows the infrared spectrum of α-chloropropionic acid. Note the extreme broadening of the O—H and C═O bands due to strong intermolecular hydrogen bonding of the O—H to the C═O. The C═O absorbs

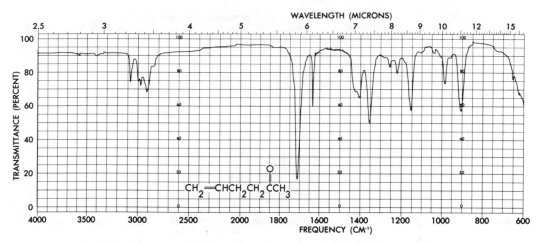

Fig. 4.14. Infrared spectrum of 5-hexen-2-one as a neat liquid film.

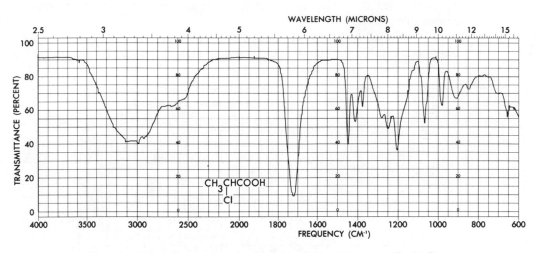

Fig. 4.15. Infrared spectrum of α-chloropropionic acid as a neat liquid film.

at a slightly higher frequency due to the presence of the electronegative chlorine on the α-carbon.

The infrared spectrum of acetamide in dilute chloroform solution is shown in Fig. 4.16. The N—H stretching absorptions appear as two bands at 3530 cm^{-1} (2.83 μ) and 3410 cm^{-1} (2.93 μ). The "Amide I" band, the C=O stretch, appears at 1670 cm^{-1} (5.97 μ), while the "Amide II" bands, the N—H bending deformations, appear at 1585 cm^{-1} (6.29 μ) and near 1400 cm^{-1} (7.2 μ); the "Amide III" band appears at 1330 cm^{-1} (7.52 μ). The band at 1370 cm^{-1} (7.29 μ) represents the C—H bending deformation of the methyl group. The C=O bending band appears close to 1240 cm^{-1} (8.1 μ)

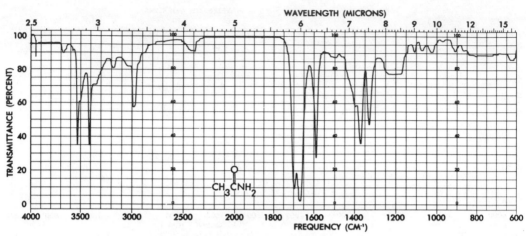

Fig. 4.16. Infrared spectrum of acetamide in chloroform solution.

but is broadened, and partly obscured, by an absorption band of the solvent chloroform.

The infrared spectrum of succinic anhydride appears in Fig. 4.17. Two bands occur in the carbonyl region which are displaced to a higher frequency due to the five-membered ring.

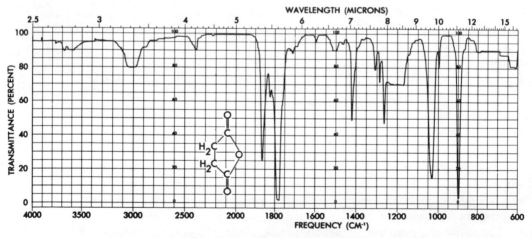

Fig. 4.17. Infrared spectrum of succinic anhydride in chloroform solution.

4.5 PREPARATION OF THE SAMPLE

The method of handling the sample for recording the infrared spectrum is generally dictated by the physical state of the sample and the region one wishes to record.

Gas samples are placed in gas cells whose internal path length can be greatly multiplied by internal mirrors. The effective length of gas cells may extend from a few centimeters up to several meters (generally in multiples of meters). The handling of gases usually requires a vacuum line for storage and transfer to the cell.

Liquid samples may be run either as a neat (pure liquid) sample or in solution. Neat samples can be prepared by placing one or two small drops of the sample between two highly polished pieces of cell material (see Fig. 4.18). The thickness of the capillary film is very difficult to control and

(a)   (b)

Fig. 4.18. (a) Preparation of a neat sample for recording the infrared spectrum of a liquid. A drop of the liquid, or mull, is placed between the sodium chloride plates and the plates are secured between the metal retainers of the cell (bottom of the photograph). (b) The filling of a solution cell. (Courtesy of the Perkin-Elmer Corporation, Norwalk, Conn.)

reproduce, giving spectra with varying absorption intensities. Thin metal foil spaces with thicknesses accurately measured down to 0.001 mm may be used to control the thickness of the sample. A few small drops of the sample are placed in the open area of the spacer on one plate, and the second plate is firmly pressed to the spacer in a cell holder (see Fig. 4.18). Solution spectra are obtained by dissolving the sample in an appropriate solvent (see subsequent paragraphs) and placing in a cell. The concentration range normally employed is 2 to 10% by weight. A variety of cells are available for holding the sample solution. In addition to the sample cell, a reference cell of the same thickness as the sample cell is filled with pure solvent and placed in the reference beam of the instrument. The weak absorption bands occurring in the reference beam offset similar weak absorption bands of the solvent occurring in the sample beam; thus interfering extraneous absorption bands are removed from the spectrum of the compound. Major absorption bands of the solvent absorb so strongly that no effective infrared energy passes through the cell, and no differential absorption due to the sample can be detected.

Solid samples may be run as solutions, mulls, or as a solid dispersion in

potassium bromide. The solutions are prepared as described before. Mulls
are prepared by suspending finely ground sample particles in Nujol (paraffin
or mineral oil) and then recording the spectrum of the mull as a neat sample.
Fluorolube may also be used in the preparation of mulls. Not all solids can
be mulled successfully. In addition, the Nujol displays intense C—H absorption and renders these regions useless for identification purposes. In general,
mulls should be used only if no other method is available.

Solid dispersions of samples, approximately 1% by weight in potassium
bromide, are prepared by carefully grinding a mixture of the sample and
potassium bromide until composed of very finely ground particles. The
finely ground mixture is then pressed into a transparent disc under several
tons of pressure. The disc is mounted on a holder, and the spectrum is then
recorded. Since potassium bromide is transparent in the infrared region,
only bands corresponding to the sample appear in the spectrum. The spectrum obtained thus, as well as with Nujol mull dispersions, will be of the
material in the solid phase and may differ from solution spectra owing to
restrictions of molecular configurations (in microcrystals) or increased
functional group interactions. Broad hydroxyl absorption near 3300 cm^{-1}
(3 μ) is usually present, owing to moisture absorbed by the potassium
bromide, unless one is careful when preparing the disc.

The sample container is placed in the sample beam of the instrument (see
Fig. 4.19) along with the appropriate reference beam blank (pure solvent),
and the spectrum is recorded. Instructions for operating the instrument are
provided in the manufacturer's instruction manual and should be thoroughly
understood before individual operation.

(a)   (b)

Fig. 4.19. (a) The cell compartment of a typical infrared spectrophotometer. The left
cell is the sample cell and the right cell is a variable thickness cell filled with pure solvent
in the reference beam of the instrument. The thickness of the variable thickness cell is
adjustable to match exactly the thickness of the sample cell to balance the solvent
absorption peaks. (b) A typical infrared spectrophotometer. (Courtesy of the Perkin-
Elmer Corporation, Norwalk, Conn.)

Recent advances in infrared spectroscopy involve the use of reflected light to record the infrared spectrum of surfaces of materials (attenuated total reflectance, ATR). This has proved particularly useful with samples which cannot be prepared in any of the foregoing manners (painted surfaces, etc.).

4.5.1 Solvents

The choice of solvent depends on the solubility of the sample and the absorption characteristics of the solvent. Since almost all solvents employed in infrared spectroscopy are organic molecules themselves, they will give absorption bands characteristic of the types of bonds present. It is always desirable to use a solvent having the least amount of absorption in the infrared region. For example, carbon tetrachloride absorbs strongly only in the 830 to 670 cm^{-1} (12 to 15 μ) region, whereas chloroform absorbs strongly near 3030, 1220, and 830 to 670 cm^{-1} (3.3, 8.2, and 12 to 15 μ). Thus carbon tetrachloride is more useful than chloroform, provided suitable concentrations can be achieved in both solvents. Frequently it may be necessary to record the spectrum of a sample in two different solvents to derive all of the available spectral information. For example, the use of carbon tetrachloride as the solvent for olefins and aromatics allows one to observe all absorption bands out to 900 cm^{-1} (11 μ), but the fingerprint region will be obscured. Recording the spectra of these compounds in carbon disulfide, or acetonitrile, allows one to record the absorption bands appearing in the fingerprint region. Figure 4.20 lists several solvents suitable in the infrared region. The darkened areas indicate regions in which the solvent absorbs most, or all, of the available energy. Slight solvent shifts are noted with some functional groups, particularly those capable of entering into hydrogen bonding either as acceptors or donors.

4.5.2 Cell Materials

Cell window materials must be transparent in the desired portion of the infrared region, must not be soluble in the solvent system, and must not react chemically with the solvent or sample. Materials suitable for use as cell windows are tabulated in Table 4.8. Sodium chloride is the most commonly used cell material owing to its relatively low cost and ease of handling; however, Irtran-2, despite being expensive, is finding extensive use because of its chemical and physical durability.

Certain care must be exercised in the use of these cell materials. The reader must be familiar with the general properties of the cell window materials being used, and he must take proper precautions to protect the windows. Particular care must be exercised in preparing the samples and solvents to be used.

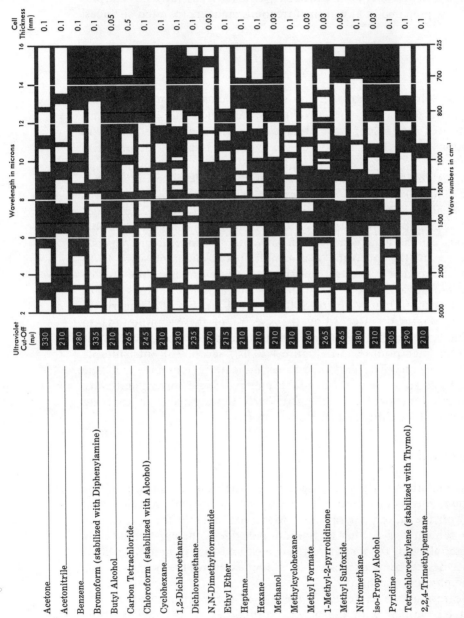

Fig. 4.20. Chart of solvents for spectrophotometric use. The ultraviolet cut-off given is the wavelength at which the absorbance of a 1-cm pathlength of the solvent is about unity. The white windows in the chart show the infrared regions for which these solvents have been found useful in the cell thicknesses indicated. (Courtesy of Distillation Products Industries, Eastman Kodak Company, Rochester, N.Y.)

Table 4.8. Cell Materials

Material	Useful range (μ)	General properties
NaCl	0.25–16	Soluble in water; slightly soluble in low molecular-weight alcohols, glycols, amines, carboxylic acids; low cost; easy to polish.
KBr	0.25–30	Soluble in water, low molecular weight alcohols, glycols, acids; slightly soluble in amines, ethers; low cost; easy to polish; slightly hygroscopic.
CsBr	1–40	Soluble in water, low molecular weight alcohols, glycols, acids; quite expensive; soft; very hygroscopic; used primarily for long wavelength transmission.
CaF$_2$	0.2–10	Insoluble in water; soluble in ammonia and salt solutions; relatively expensive; hard material; useful for high-pressure cells.
BaF$_2$	0.2–12	Insoluble in water; soluble in acids; hard material; sensitive to mechanical and thermal shock.
AgCl	1–22	Insoluble in water; soluble in ammonia and strong bases; should not be extensively exposed to light of wavelength less than 0.6 μ; possesses high reflectivity giving rise to interference fringes (may appear as anomalous peaks).
KRS-5	1–40	Double salt containing TlBr (48%) and TlI (52%); excellent transmission properties, chemical inertness; sparingly soluble in water, soluble in bases, insoluble in acids; high reflectivity; expensive.
KRS-6	1–22	Double salt containing TlBr (40%) and TlCl (60%) (TlCl leaches out in water and bases); physical and chemical properties similar to KRS-5; high reflectivity; expensive.
Irtran-2	2–13	Zinc sulfide; extremely durable; insoluble in virtually all organic and inorganic solvents (except with strong oxidizing agents); useful temperature range -200 to $+800°C$; high reflectivity; expensive.
Glass	0.35–2	Useful in the near-infrared region.
Quartz	0.2–4	Useful in the near-infrared region.

Cells and plates should always be carefully cleaned after each use by thorough washing with purified chloroform, or the solvent used in preparing the sample solution, and drying with a stream of dry nitrogen.

4.6 NEAR-INFRARED SPECTROSCOPY

The near-infrared region, 0.75 to 3 μ, contains the overtones of the fundamental stretching frequencies, and various combination bands, appearing in the infrared region. The band positions of the overtone bands appear approximately at $\nu_0/2$, $\nu_0/3$, $\nu_0/4$, etc., where ν_0 is the wavelength of the fundamental. The intensities of the overtones decrease quite rapidly as the order of the overtone increases (see the extinction coefficients contained in Fig. 4.21), and thus only the first and second overtones are readily visible unless long path-length (up to 10 cm of pure liquid) cells are used. This severely limits the types of functional groups that will give rise to overtone peaks in the near-infrared region to X—H type groups, for example, C—H, N—H,

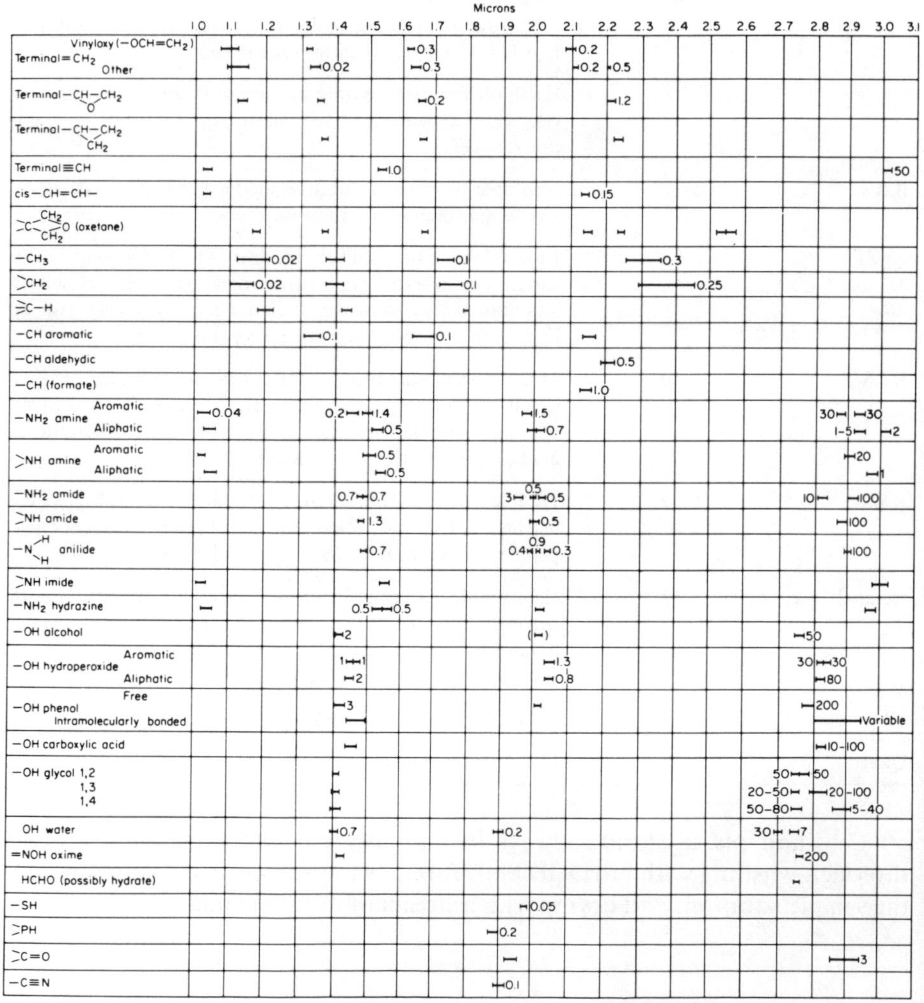

Microns

Group	1.0	1.1	1.2	1.3	1.4	1.5	1.6	1.7	1.8	1.9	2.0	2.1	2.2	2.3	2.4	2.5	2.6	2.7	2.8	2.9	3.0	3.1
Vinyloxy (—OCH=CH₂) Terminal=CH₂								0.3				0.2										
Other					0.02			0.3				0.2 0.5										
Terminal—CH—CH₂ (epoxide O)								0.2					1.2									
Terminal—CH—CH₂ (cyclopropane)																						
Terminal ≡ CH						1.0													50			
cis—CH=CH—												0.15										
oxetane																						
—CH₃			0.02					0.1						0.3								
>CH₂			0.02					0.1						0.25								
≡C—H																						
—CH aromatic					0.1			0.1														
—CH aldehydic												0.5										
—CH (formate)											1.0											
—NH₂ amine Aromatic			0.04			0.2 1.4					1.5								30 30			
Aliphatic						0.5					0.7								1-5 2			
>NH amine Aromatic						0.5													20			
Aliphatic						0.5													1			
—NH₂ amide						0.7 0.7					3— 0.5/0.5								10 100			
>NH amide						1.3					0.5								100			
—N(H)(H) anilide						0.7					0.4 0.9/0.3								100			
>NH imide																						
—NH₂ hydrazine						0.5 0.5																
—OH alcohol					2						(—)								50			
—OH hydroperoxide Aromatic					1 1					1.3								30 30			
Aliphatic					2					0.8								80			
—OH phenol Free					3														200			
Intramolecularly bonded																					Variable	
—OH carboxylic acid																			10-100			
—OH glycol 1,2																50 50						
1,3																20-50 20-100						
1,4																50-80 5-40						
OH water					0.7					0.2						30 7						
=NOH oxime																			200			
HCHO (possibly hydrate)																						
—SH										0.05												
>PH										0.2												
>C=O																			3			
—C≡N										0.1												

Fig. 4.21. Spectra-structure correlations and average molar absorptivity data for the near-infrared region. [Reprinted from *Analytical Chemistry*, **32**, 140 (1960). Copyright 1959 by the American Chemical Society and reprinted by permission of the copyright owner.]

O—H, S—H, etc., and carbonyl overtones. (The carbonyl fundamental is extremely intense and thus gives rise to more intense overtones.)

The combination bands are often the most intense bands in the near-infrared region. The combination bands represent the sum or the difference of two fundamentals, or a fundamental and a first overtone, generally involving the vibrational and bending deformation modes of the same functional group.

The value of near-infrared spectroscopy lies in the greater resolution obtainable with quartz prism or grating near-infrared spectrophotometers than can be obtained with the salt optics or grating infrared instruments (except in the more refined and expensive instruments). Near-infrared spectrophotometers provide a resolution of approximately 5 Å, allowing one to determine band positions to ± 0.001 μ. This is in contrast to the resolution of infrared instruments which is about ± 0.02 μ for small laboratory instruments and slightly less than ± 0.01 μ for the better grating instruments. With reasonably complex molecules, the congestion in the 3 to 4 μ region limits the amount of useful information which can be obtained from this region, whereas the greater resolution provided in the near-infrared region permits greater distinction of the bands for diagnostic purposes.

Figure 4.21 presents a spectra-correlation chart, compiled by Goddu and Delker in 1960, of characteristic group frequencies in the near-infrared region. It appears that more comprehensive correlations have not been compiled. Near-infrared spectroscopy has been shown to be very useful for the detection of the presence of cyclopropanes.[3] A more complete discussion of individual absorbing chromophores will not be presented here owing to the limited current use of near-infrared spectroscopy. However, the potential utility of near-infrared spectroscopy should not be ignored, and the practicing chemist should be aware of its potential uses.

4.7 FAR-INFRARED SPECTROSCOPY

The far-infrared region contains absorption bands arising from heavy atom bond stretching and bending deformations, functional group skeletal deformations, ring torsional deformations, and lattice mode vibrations (in the solid state only). Considerable research effort is currently being devoted to the study of far-infrared spectra of molecules. Figure 4.22 presents a spectra-structure correlation for the far-infrared region. Reference 17, Sec. 4.11, may be referred to for a more detailed description of far infrared spectroscopy.

---

[3] P. G. Gassman and F. V. Zalar, *J. Org. Chem.*, **31**, 166 (1966).

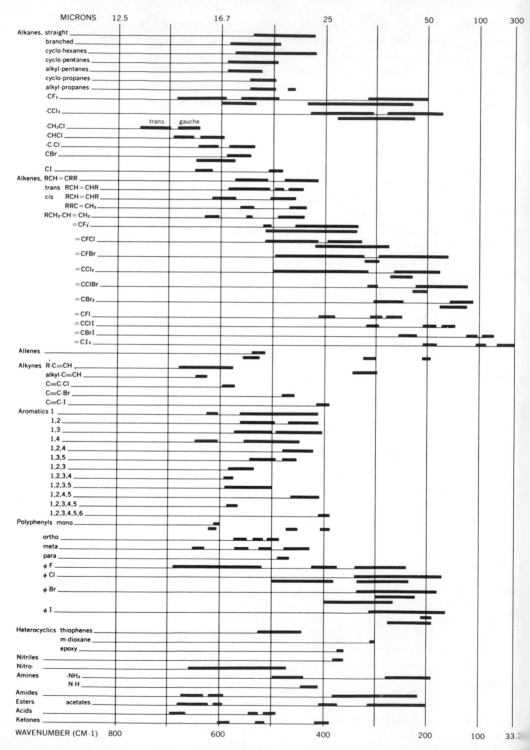

BECKMAN FAR-INFRARED VIBRATIONAL FREQUENCY CORRELATION CHART

Fig. 4.22. Spectra-structure correlation for the far-infrared region

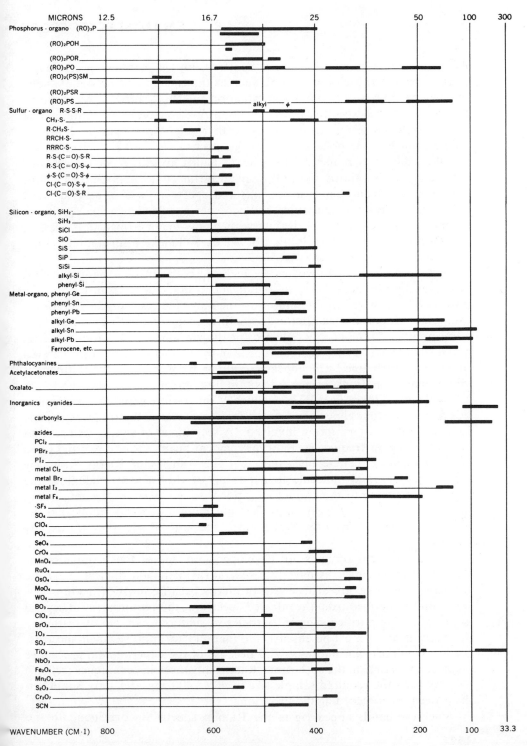

An important adjunct to infrared spectroscopy is Raman spectroscopy. Highly symmetrical chromophores, such as the carbon-carbon double bond of ethylene, do not absorb infrared radiation and thus cannot be detected by means of infrared spectroscopy. To be infrared active, the chromophore must have a dipole moment; whereas with Raman spectroscopy, however, only a change in the polarizability of the electrons of the system on vibrational excitation is required. In general, bands appearing in the infrared spectrum will also appear in the Raman spectrum, but the intensities will be reversed; i.e., strong bands in the infrared spectrum will appear as weak bands in the Raman spectrum, and vice versa. The principal utility of Raman spectroscopy lies in dealing with highly symmetrical molecules, or chromophores within a molecule, for which sufficient information cannot be gained from the infrared spectrum.

Raman spectroscopy involves the inelastic collision of a photon with a molecule, approximately 1% of the incident light, with transferral of a portion of the energy of the photon to the molecule resulting in vibrational excitation. The scattered photon has had its energy reduced by an amount corresponding to the energy required for the vibrational excitation process, and it will appear at a lower frequency than the incident beam. In general, a number of such lines will appear, resulting from a number of different vibrational excitations occurring within the molecule. These lines are referred to as *Stokes lines*. The frequency separation between the incident beam and the Stokes lines, the Raman frequency $\Delta \nu$, is the frequency of the fundamental of the vibration involved. A second set of much weaker lines appears at higher frequencies than the incident beam and are referred to as *anti-Stokes lines*. These lines represent the transfer of energy from the vibrationally excited molecule to a photon of the incident radiation, the molecule returning to the ground vibrational state. Because the anti-Stokes lines are usually very weak, and the information derived duplicates the information gained from the Stokes lines, the anti-Stokes lines are usually ignored.

Raman spectroscopy has not developed into one of the major tools of the structural chemist, as have infrared, ultraviolet, and nuclear magnetic resonance spectroscopy. This is true primarily because most of the information could be derived from the infrared spectrum, the requirement of several grams of highly purified sample (particles of material capable of scattering the incident beam must be absent), the inability to apply Raman spectroscopy to materials that fluoresce, and because of the experimental procedure involved in recording the spectrum. Recent advances in techniques (ref. 15) have reduced the required sample size to 40 to 400 mg and have simplified the procedure of recording the spectrum.

Since the bands appearing in the Raman spectrum correspond to the bands normally occurring in the infrared region, the spectra-structure correlations outlined in the previous sections on infrared spectroscopy also apply to the interpretation of Raman spectra.

4.9   INFRARED SPECTRAL PROBLEMS

Identify each of the following compounds from their infrared spectra.

1. $C_4H_6O_2$; infrared spectrum of unknown recorded as a neat liquid film.

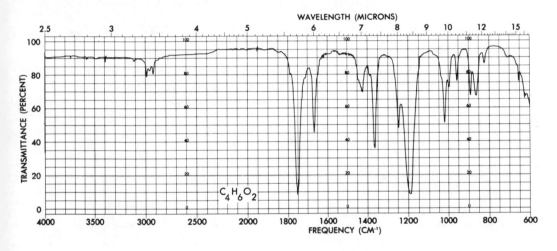

2. C_5H_8O; infrared spectrum of unknown recorded as a neat liquid film.

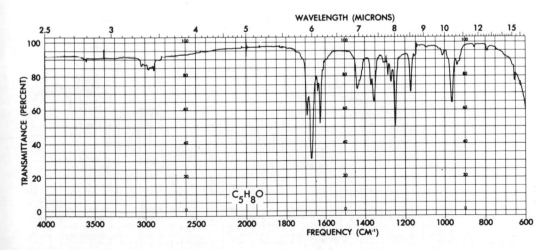

3. $C_8H_{10}O$; infrared spectrum of unknown recorded in carbon tetrachloride (4000 to 900 cm^{-1}) and carbon disulfide (1000 to 600 cm^{-1}) solution.

WAVELENGTH (MICRONS)

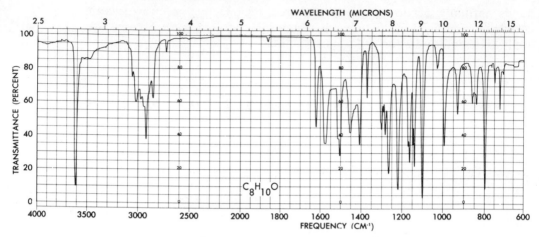

4. $C_9H_{10}O$; infrared spectrum of unknown recorded in carbon tetrachloride (4000 to 900 cm^{-1}) and carbon disulfide (1000 to 600 cm^{-1}) solution.

WAVELENGTH (MICRONS)

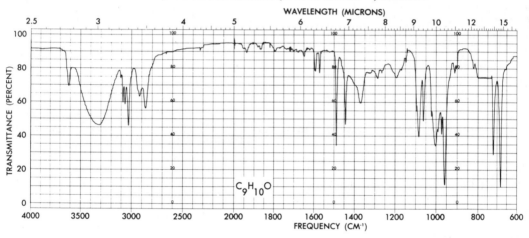

5. $C_4H_6O_2$; infrared spectrum of unknown recorded as a neat liquid film.

WAVELENGTH (MICRONS)

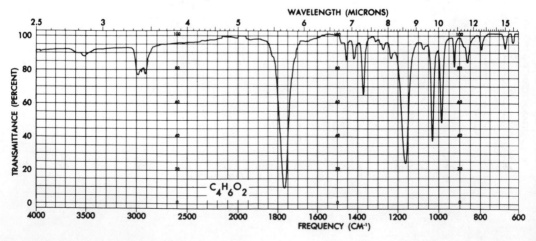

6. $C_7H_7NO_3$; infrared spectrum of unknown recorded as a neat liquid film.

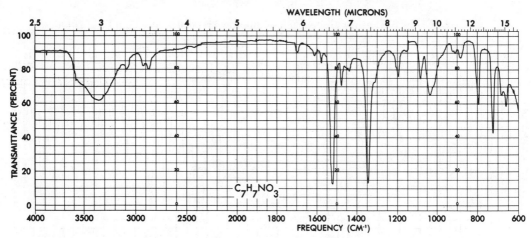

7. C_4H_7Br; infrared spectrum of unknown recorded as a neat liquid film.

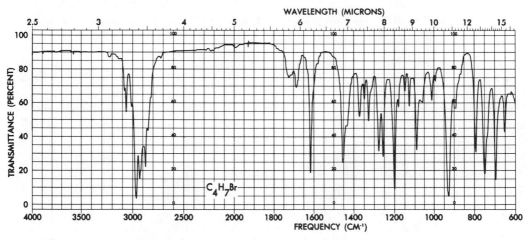

8. C_6H_8O; infrared spectrum of unknown recorded as a neat liquid film.

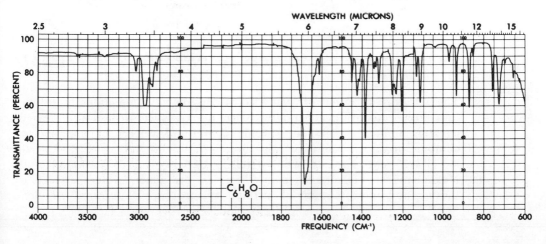

9. $C_7H_7NO_3$; infrared spectrum of unknown recorded as a neat liquid film.

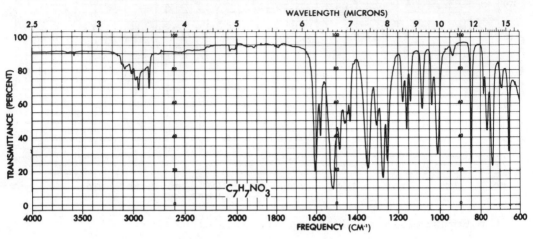

10. C_6H_5BrS; infrared spectrum of unknown recorded in carbon tetrachloride (4000 to 900 cm^{-1}) and carbon disulfide (1000 to 600 cm^{-1}) solution.

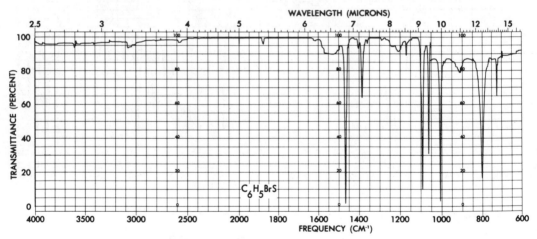

4.10 *LITERATURE SOURCES OF INFRARED SPECTRAL DATA*

The literature covering infrared spectroscopy can be conveniently grouped into three categories: general reference books, specific literature reviews covering limited areas, and specific compound references and spectral compilations.

The reference books by Conley (ref. 5), Bellamy (ref. 3), Lawson (ref. 9), Nakanishi (ref. 11), and Szymanski (ref. 15) are general reference texts on the use and interpretation of infrared spectroscopy. *Infrared—A Bibli-*

ography, by Brown et al. (ref. 4), is a compilation of original literature references on instrumentation, techniques, and general compound classifications covering the years from 1935 to 1951.   Specific literature reviews covering limited areas are too numerous to cite in this text.   The reader is referred to abstracting and current literature services to locate specific reviews, for example, ref. 8.

Literature references for specific compounds can be found in ref. 2, 7, and 8 in Sec. 4.11. Published spectra of individual compounds can be found in the Sadtler Standard Spectra files (ref. 13), the American Petroleum Institute infrared spectra files (ref. 1), the Manufacturing Chemists' Association Research Project infrared spectra files (ref. 10), and the atlases of infrared absorption spectra of steroids (ref. 6).

Instructions on the use and applications of the various compilations of literature references and spectra accompany each compilation.

4.11   REFERENCES

INFRARED SPECTROSCOPY

1. *American Petroleum Institute Research Project 44 Infrared Files.*   Petroleum Research Laboratory, Carnegie Institute of Technology, Pittsburgh, Pennsylvania.

2. Ministry of Aviation Technical Information and Library Services, ed., *An Index of Published Infra-Red Spectra*, Vol. 1 and 2.   London: Her Majesty's Stationery Office, 1960.

3. Bellamy, L. J., *The Infra-red Spectra of Complex Molecules.* New York: John Wiley & Sons, Inc., 1958.

4. Brown, C. R., Ayton, M. W., Goodwin, T. C., and Derby, T. J., *Infrared—A Bibliography.* Washington, D.C.: Library of Congress, Technical Information Division, 1954.

5. Conley, R. T., *Infrared Spectroscopy.* Boston: Allyn and Bacon, Inc., 1966.

6. Dobriner, K., Katzenellenbogen, E. R., and Jones, R. N., *Infrared Absorption Spectra of Steroids—An Atlas*, Vol. 1. New York: Interscience Publishers, Inc., 1953.

7. Hershenson, H. M., *Infrared Absorption Spectra Index.* New York: Academic Press, 1959.

8. *IR, Raman, Microwave Current Literature Service*, Butterworth and Co., Ltd., London, England; and Verlag Chemie, Gmbh., Weinheim, Germany.

9. Lawson, K. E., *Infrared Absorption of Inorganic Substances.* New York: Reinhold Publishing Corporation, 1961.

10. *Manufacturing Chemists' Association Research Project Infrared Files*, Chemical Thermodynamic Properties Center, Department of Chemistry, Texas A & M University. College Station, Texas.

11. Nakanishi, K., *Infrared Absorption Spectroscopy*. San Francisco: Holden-Day, Inc., 1962.

12. Roberts, G., Gallagher, B. S., and Jones, R. N., *Infrared Absorption Spectra of Steroids—An Atlas*, Vol. 2. New York: Interscience Publishers, Inc., 1958.

13. *Sadtler Standard Spectra*, Sadtler Research Laboratories, Philadelphia, Pennsylvania.

14. Szymanski, H. A., *Infrared Band Handbook*. New York: Plenum Publishing Company, 1962, and supplements 1 and 2, (1964).

15. Szymanski, H. A., *Interpreted Infrared Spectra*, Vol. 1. New York: Plenum Publishing Company, 1964.

RAMAN SPECTROSCOPY

16. Jones, R. N., DiGiorgio, J. B., Elliott, J. J., and Nonnenmacher, G. A. A., *J. Org. Chem.*, **30,** 1822 (1965).

FAR-INFRARED SPECTROSCOPY

17. Stewart, J. E., "Far Infrared Spectroscopy" in *Interpretive Spectroscopy*, Freeman, S. K., ed., New York: Reinhold Publishing Corporation, 1965, p 131.

5

Nuclear Magnetic Resonance

5.1 INTRODUCTION

Nuclear magnetic resonance spectroscopy involves the transition of a nucleus from one spin state to another spin state, with the resultant absorption of energy. The interaction of a nucleus with an applied magnetic field is illustrated in Fig. 5.1.

Nuclei possess a mechanical spin which, in conjunction with the charge of the nucleus, produces a magnetic field whose axis is directed along the spin axis of the nucleus. If this nucleus is placed in a magnetic field, H_0, the nucleus may assume $2I + 1$ spin orientations (spin states) where I is the nuclear spin quantum number, with respect to the applied field. These spin states are labeled $-I$, $-I + 1$, ..., $+I$, for example, with I of $\frac{1}{2}$, spin states of $-\frac{1}{2}$ and $+\frac{1}{2}$ are possible; with I of 1, spin states of -1, 0, and $+1$ are possible. Figure 5.1 illustrates the two spin-state orientations possible ($-\frac{1}{2}$ and $+\frac{1}{2}$) for the hydrogen nucleus.

The magnetic and spin axis of the nucleus in a given spin state rotates, precesses, about the direction of the applied field H_0. The precessional frequency (Larmor frequency) of the nucleus is exactly equal to the frequency of the electromagnetic radiation required to induce transitions from one spin state to another, for example, the $-\frac{1}{2} \longrightarrow +\frac{1}{2}$ spin transition for the hydrogen nucleus.

In the absence of a magnetic field at room temperature the energies of the spin states are degenerate. As a magnetic field is impressed on the nucleus, the spin states lose their degeneracy and become separated by an energy difference (ΔE) which is a function of the strength of the applied magnetic field (see Fig. 5.2).

The relationship between the ΔE and field strength (H_0) for a bare

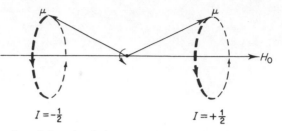

$I = -\frac{1}{2}$ $I = +\frac{1}{2}$

Fig. 5.1. Precession of the axis of the magnetic moment of a nucleus about the lines of an applied magnetic field.

nucleus is given by Eq. (5.1)

$$\Delta E = h\nu = \frac{\mu \beta_N H_0}{I} \tag{5.1}$$

where h is Planck's constant, ν is the frequency of the exciting radiation, μ is the magnetic moment of the nucleus expressed in multiples of the nuclear magneton[1] (see Table 5.1), and β_N is a constant called the nuclear magneton constant. For nuclei with more than two spin states, radiation-induced spin transitions are allowed only between adjacent spin states, the energies between any two spin states being equal (see later discussion).

Because various nuclei differ in their value of μ and I, they will undergo nuclear spin transitions at different frequencies in the same applied magnetic field. Table 5.1 lists several of the more common nuclei with their nuclear spin quantum number I, μ, resonance frequency in a 10,000 gauss field, and natural abundance. Since the frequency and field strength are directly proportional, the data given in Table 5.1 can be scaled up or down to correspond to other applied field strengths. It should also be pointed out at this time that the sensitivity, or response, of an equal number of different nuclei will vary (see Eq. (5.2)) as a function of the angular momentum m

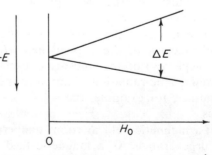

Fig. 5.2. Spin energy level separation for a nucleus with I of $\frac{1}{2}$ as a function of the applied field H_0.

---

[1] The magnetic moment μ may possess either a $+$ or $-$ sign and is due to the fact that the I and μ vectors may be parallel or antiparallel ($\mu = CI$).

Table 5.1. Resonance Frequencies at 10,000 Gauss, and Related Data for Selected Nuclei*

Nucleus	Spin quantum number, I	Magnetic moment, $\frac{eh}{4\pi}$	Resonance frequency MHz	Natural abundance, %
^{1}H	$\frac{1}{2}$	2.79268	42.5759	99.9844
^{2}H	1	0.857386	6.53566	0.0156
^{7}Li	$\frac{3}{2}$	3.2560	16.547	92.57
^{9}Be	$\frac{3}{2}$	−1.1773	5.983	100
^{10}B	3	1.8005	4.575	18.83
^{11}B	$\frac{3}{2}$	2.6880	13.660	81.17
^{13}C	$\frac{1}{2}$	0.70220	10.705	1.108
^{14}N	1	0.40358	3.076	99.635
^{15}N	$\frac{1}{2}$	−0.28304	4.315	0.365
^{17}O	$\frac{5}{2}$	−1.8930	5.772	0.037
^{19}F	$\frac{1}{2}$	2.6273	40.055	100
^{23}Na	$\frac{3}{2}$	2.2161	11.262	100
^{25}Mg	$\frac{5}{2}$	−0.85471	2.606	10.05
^{27}Al	$\frac{5}{2}$	3.6385	11.094	100
^{29}Si	$\frac{1}{2}$	−0.55477	8.458	4.70
^{31}P	$\frac{1}{2}$	1.1305	17.236	100
^{33}S	$\frac{3}{2}$	0.64274	3.266	0.74
^{35}Cl	$\frac{3}{2}$	0.82091	4.172	75.4
^{37}Cl	$\frac{3}{2}$	0.68330	3.472	24.6
^{39}K	$\frac{3}{2}$	0.39094	1.987	93.08
^{79}Br	$\frac{3}{2}$	2.0991	10.667	50.57
^{81}Br	$\frac{3}{2}$	2.2626	11.499	49.43
free electron	$\frac{1}{2}$	−1836	28024.6	—

*In multiples of the nuclear magneton.

and spin quantum number; the sensitivity decreases with decreasing m and increasing I. In addition, the natural abundances of many of the nuclei appearing in Table 5.1 are very low, increasing the difficulty of detecting the resonance signals. One may increase the atom concentration (usually a very expensive process) by synthesis, or he may employ computer averaging of transients techniques (CAT). CAT is a process in which the spectrometer scans the resonance region many times, summing the resonance signals to enhance signal-to-noise ratios in order to observe the resonance signals.

The population distribution between two spin states is given by Eq. (5.2), where m is the angular momentum.

$$\frac{State\ 1}{State\ 2} = \left(1 - \frac{m\mu H_0}{IkT}\right)\frac{1}{2I+1} \tag{5.2}$$

For hydrogen, the population distribution is 0.99999/1.00001 at 10,000 gauss and at room temperature. This ratio is rather small, demanding sensitive detection techniques.

Absorption of energy results in the excitation of the nucleus to a higher-energy spin state. The probability of spontaneous emission from the higher spin state is negligible, although induced emission is just as probable as radiation induced excitation. Thus a net initial absorption of energy is observed until equal populations are attained, whereupon further absorption will cease (*saturation*). The rate at which saturation is attained depends on the intensity (amplitude) of the exciting radiation and on the rate at which spin excited nuclei return to the spin ground state (*relaxation*). In the nuclear magnetic resonance experiment, it is most desirable to maintain sustained absorption of the radio-frequency energy and to avoid saturation. It is important, then, that we inquire as to how a nucleus in a higher spin state can return to a lower spin state (*relaxation*), enabling a sustained absorption of the radio-frequency energy.

Relaxation processes can be divided into two categories, spin-spin and spin-lattice relaxation. Spin-spin relaxation involves a transfer of spin energy of one nucleus to a neighboring nucleus, the rate of transfer being denoted as the relaxation time T_2 (the mean half-life required for a perturbed system to return to equilibrium). Spin-lattice relaxation (lattice being defined as the aggregate of atoms or molecules under study) involves the conversion of the spin energy into thermal energy. The mean half-life of spin-lattice relaxation (T_1) is affected by small fluctuations of the magnetic field throughout the sample. These fluctuations may arise from (1) individual aggregates of molecules, (2) by the presence of paramagnetic substances (oxygen and magnetic materials), or (3) by the presence of an atom possessing an electric quadrupole moment (atoms with $I > \frac{1}{2}$). (The reader is referred to more complete texts for more detailed descriptions of these processes.) The first effect is greatly influenced, in a complex manner, by the viscosity of the system. Generally, an increase in the viscosity decreases T_1. Similarly, the presence of paramagnetic materials or atoms possessing an electric quadrupole moment causes T_1 to decrease. Both T_1 and T_2 affect the width of the absorption line, the width increasing as T_1 and T_2 decreases.

From the foregoing discussion, certain precautions must be exercised in the preparation of the sample and in recording of the spectrum. The solvent must be free of magnetic inpurities, the sample must be free from paramagnetic species, and for very careful measurements the sample should be freeze-degassed to remove dissolved oxygen.

Adequate spin-lattice relaxation is required to maintain the unequal population of spin states for continuous radio-frequency absorption at a given excitation amplitude. Should this condition not exist, saturation will occur. Saturation interferes with peak shapes and, most importantly, with the integrated peak areas; it should be avoided if at all possible. Generally, reducing the radio-frequency power input until a constant absorption response is obtained will alleviate this problem.

5.2   ORIGIN OF CHEMICAL SHIFT

Let us now consider a nucleus contained in a molecule which is in the vicinity of various other atoms, or groups of atoms. The magnetic field "felt" at the nucleus of our atom will not be equal to the applied field H_0, but may be less than or greater than the applied field H_0, and will be represented by H_{net} (net magnetic field). H_{net} is a function of the applied field strength H_0 and may be represented by Eq. (5.3), in which σ_A is the shielding constant for nucleus A which we are considering.

$$H_{net} = H_0 (1 - \sigma_A) \qquad (5.3)$$

The shielding constant σ_A is composed of contributions as illustrated by Eq. (5.4).

$$\sigma_A = \sigma_{AA}^{dia} + \sigma_{AA}^{para} + \sum_{B \neq A} \sigma_{AB} + \sigma_A^{deloc} \qquad (5.4)$$

The term σ_{AA}^{dia} is the contribution to the shielding constant of atom A by induced diamagnetic currents on atom A under the influence of the applied field H_0. The magnitude of A is a function of the electron density about the nucleus of atom A; hence, it is an indirect function of the electronegativity of the atom or functional groups bonded to A. The term σ_{AA}^{para} is the contribution to σ_A by induced paramagnetic currents on atom A, and it arises as a consequence of the mixing of ground and excited electronic states (p or d-orbitals only; hence σ_{AA}^{para} is zero when the electrons of A are localized in pure s-orbitals, for example, the hydrogen atom) under the influence of the applied field.

The term σ_{AB} is the contribution to σ_A due to local induced currents, either diamagnetic (positive contribution) or paramagnetic (negative contribution), on the atom or functional group B, and it is sometimes referred to as the neighbor anisotropy effect. The magnitude of σ_{AB} is a function of the atom or functional group B, the distance between A and B, and the spatial orientation of B with respect to A. If the functional group B is axially symmetric, for example, a single atom or a methyl group, the magnitude of σ_{AB} may be calculated by use of Eq. (5.5), generally referred to as the McConnell equation:

$$\sigma_{AB} = \frac{1}{3N_0 R_{AB}^3} \Delta \chi_B (1 - 3 \cos^2 \theta) \qquad (5.5)$$

where N_0 is Avogadro's number, R_{AB} is the internuclear distance, $\Delta \chi_B = \chi^{\parallel}{}_B - \chi^{\perp}{}_B$ (difference in the longitudinal or parallel and transverse or perpendicular magnetic susceptibilities), and θ is the angle between the symmetry axis of B and the A—B internuclear vector. In certain instances Eq. (5.5) does not satisfactorily correlate with experimental data, and caution must be exercised in its use.

The final term of Eq. (5.4), σ_A^{deloc}, arises from induced currents involving delocalized electrons, for example, the π-electron system of aromatics. The effects of remote functional groups, giving rise to σ_{AB} and σ_A^{deloc}, will be discussed with respect to individual functional groups later in this chapter.

The shielding constant σ_A may result in an increase (*shielding*) or a decrease (*deshielding*) in the resultant field felt by the nucleus.

Substitution of the net field (H_{net}) felt at the nucleus of atom A in Eq. (5.1) gives Eq. (5.6).

$$\Delta E_A = h\nu = \frac{\mu \beta_N H_{net}}{I} \tag{5.6}$$

If one considers other nuclei in the same molecule, the remainder of the molecule will display different σ values with respect to the various nuclei unless the nuclei are identical with respect to the remainder of the molecule. The different shieldings of the other nuclei in the molecule result in different frequencies to induce spin transitions with these nuclei. The change in frequency of resonance with respect to an arbitrarily chosen position is termed *chemical shift*. With nuclear magnetic resonance, we have an extremely sensitive probe to investigate the environment of various nuclei in a single molecule.

5.3   DETERMINATION OF CHEMICAL SHIFT

Equation (5.1) contains two potential variables, the frequency ν and the magnetic field strength H_0. Experimentally, we might maintain one of these variables constant, while varying the other. The frequencies required for resonance in a magnetic field of 10,000 gauss are in the region of 10^6 Hz. With proton magnetic resonance, differences in chemical shift are of the order of 1 to several hundred Hz demanding an extremely fine control and measurement of frequency if the frequency is to be used as the measurable variable. Electronically, it is easier to maintain a constant radio frequency and vary the field strength. This is accomplished by summing a high-intensity, constant magnetic field with a weak, varying field controlled by a small flow of current through the sweep coils (see Fig. 5.3). The sample is placed in the magnetic field, and a constant-energy radio frequency is impressed on the sample. As the flow of current is changed in the sweep coils,

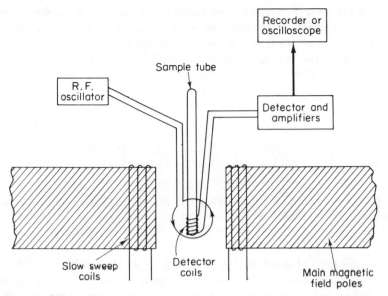

Fig. 5.3. Nuclear induction system of a nuclear magnetic resonance spectrometer.

the magnetic field changes until the correct conditions for resonance occur with the absorption of radio-frequency energy. The reorientation of the spin angular momenta of the nuclei undergoing spin excitation induces a signal in the receiver coils (the plane of the receiver coil is maintained at a right angle with respect to the plane of the radio-frequency coil), which is amplified and sent to a recorder, giving rise to an induction spectrum. Such a setup is referred to as a *double-coil* system and is used only for high-resolution nuclear magnetic resonance spectroscopy. Another experimental arrangement involves the use of only a single coil for the radio-frequency coil and receiver coil; this is referred to as a *single-coil* system. Single-coil systems give rise to true absorption spectra and are used for both low- and high-resolution nuclear magnetic resonance spectroscopy.

At present, there are two basically different instrument designs for recording nuclear magnetic resonance spectra. One involves recording the absorption or induced signal vs. a variable—field strength, time, or chart speed; the two quantities are essentially independent of one another. Spectra recorded on such instruments do not provide direct indications of the resonance positions. Examples of such instruments include the Varian Associates HR-40, HR-60, and HR-100, in which the number following the HR indicates the basic spectrometer frequency. To accurately pinpoint the peak positions, an internal standard is added to the sample; the position of absorption of the reference standard is used as a reference field strength. The peak positions are then determined by extrapolation or interpolation techniques, using a side band from the standard impressed at a known number of Hz from the standard by means of an audio-frequency oscillator. The side bands used to determine peak positions should be positioned close to the absorption peaks to reduce errors introduced by drift in the rate of sweeping the magnetic field. Figure 5.4 illustrates the use of the side-band technique to calculate peak positions.

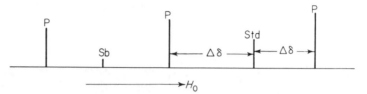

Fig. 5.4. Use of the side-banding technique to determine chemical shifts. The lines labeled P represent peaks of the spectrum, Std the internal standard, and Sb the sideband peak displaced at a known number of Hz from the reference peak.

A second type of instrumentation involves an interlocking of the resonance signal output with the rate of sweeping of the magnetic field; the Varian Associates A-series instruments (the A representing proton stabilization control) are examples. The spectra thus obtained have the signal intensity recorded directly vs. the field strength. An internal standard is still required to position the peaks within the magnetic field, but side-banding techniques are not required to determine the peak positions. The field-strength scale may be expanded with both types of instruments; however, the A-type instruments are limited in the extent of expansion of the field-strength scale, indicated as the full-chart sweep width in Hz. To keep the resonance peaks on the chart during scale expansion recordings, the region of the magnetic field being swept may be adjusted; this is indicated as the sweep offset in Hz. Specific illustrations of such spectra, with interpretations, will be shown later in this chapter.

Several nomenclature conventions have been employed to indicate the position of resonance of a nucleus. Although all these conventions will be described, the usage of a single convention is being recommended.[2] Individual peak positions, reserved for use in cases in which a complete analysis of the spectrum is not possible (see Secs. 5.12 and 5.13), are reported in Hertz (Hz) relative to the standard used. Occasionally cycles per second (abbreviated as cps or c/s; note that 1 Hertz is equivalent to 1 cycle per second) is used, although the use of these dimensions is not recom-

---

[2]The nomenclature conventions recommended in this text are consistent with the preliminary recommendations of the *ad hoc* Committee on Nuclear Magnetic Resonance (Dr. B. L. Shapiro, Chairman), advisory to the National Bureau of Standards Office of Standard Reference Data (through the Office of Critical Tables of the National Research Council), as well as the recommendations of the A.S.T.M. Committee E-13.7 (Dr. E. G. Brame, Jr., Chairman).

mended. *In hydrogen magnetic resonance, peaks appearing at a field lower than the standard are assigned a positive sign, and peaks appearing at a higher field are assigned a negative sign* (the standard being assigned a chemical shift of 0 Hz). It is not uncommon to encounter a sign assignment just the opposite when dealing with hydrogen magnetic resonance, and one must be wary. *In all other forms of magnetic resonance, for example,* ^{11}B, ^{13}C, ^{19}F, *etc., peaks appearing at a field lower than the standard are assigned a negative sign and peaks appearing at higher field are assigned a positive sign.* When recording peak positions in Hz the spectrometer frequency and the standard used must be specified. Tetramethylsilane (TMS) is the recommended standard for hydrogen magnetic resonance. Other standards may be used and corrections applied to change the chemical shifts relative to TMS.

The *delta* (δ) system is recommended when accurate resonance positions are known. δ values are calculated using Eq. (5.7).

$$\delta = \frac{(\text{chemical shift in Hz}) \times 10^6}{\text{spectrometer frequency}} \tag{5.7}$$

δ is a dimensionless number as calculated by Eq. (5.7). The δ system is used as follows for a hydrogen appearing at 72 Hz to lower field than TMS: $\delta = 1.2$. The sign of δ is the same as that of the chemical shift in Hz. Differences in chemical shifts ($\Delta\delta$) are assigned a dimension of parts per million (ppm). When the δ system is used, TMS is the standard and the spectrometer frequency need not be specified.

The *tau* (τ) scale[3] of chemical shifts (Eq. 5.8) has found considerable use. The use of τ values implies the use of TMS as a standard, and hence is applicable only to hydrogen magnetic resonance ($\tau = 10 - \delta$).

$$\tau = 10 - \frac{(\text{chemical shift relative to TMS}) \times 10^6}{\text{spectrometer frequency}} \tag{5.8}$$

5.4 HYDROGEN MAGNETIC RESONANCE

We shall now discuss the trends in the chemical shift of hydrogens as a function of their environment. Chemical shift correlations for other nuclei will not be included, owing to a limitation of space and a less general use by the organic chemist. References to collections of chemical shift data for other nuclei are given at the end of this chapter.

The several factors contributing to the chemical shift of a nucleus in a molecule were discussed in Sec. 5.2.

The chemical shift is highly dependent on the electron density about the

[3]G. V. D. Tiers, *J. Phys. Chem.*, **62**, 1151 (1958).

nucleus being observed, or associated with the atom to which it is bonded. As the electron density decreases, the nucleus experiences a greater deshielding and absorbs at lower field positions. The position of resonance shifts to lower field as one proceeds from left to right across the periodic table (due to the space-inductive effect) and also down the periodic table (due to increased polarizability) except for the hydrogen halides. In fact, the chemical shifts of the methyl hydrogens of CH_3X-type compounds show a nearly

Table 5.2. Average Chemical Shifts (δ) of α-Hydrogens in Substituted Alkanes*

Functional group X	CH_3X	$-CH_2X$	$-CH-X$
H	0.233	0.9	1.25
CH_3 or CH_2	0.9	1.25	1.5
F	4.26	4.4	—
Cl	3.05	3.4	4.0
Br	2.68	3.3	4.1
I	2.16	3.2	4.2
OH	3.47	3.6	3.6
O—Alkyl	3.3	3.4	—
O—Aryl	3.7	3.9	—
OCO—Alkyl	3.6	4.1	5.0
OCO—Aryl	3.8	4.2	5.1
SH	2.44	2.7	—
SR†	2.1	2.5	—
SOR	2.5	—	2.8
SO_2R	2.8	2.9	3.1
NR_2	2.2	2.6	2.9
NR—Aryl	2.9	—	—
NCOR	2.8	—	3.2
NO_2	4.28	4.4	4.7
CHO	2.20	2.3	2.4
CO—Alkyl	2.1	2.4	2.5
CO—Aryl	2.6	3.0	3.4
COOH	2.07	2.3	2.6
CO_2R	2.1	2.3	2.6
$CONH_2$‡	2.02	2.2	—
$CR=CR^1R^2$	2.0–1.6	2.3	2.6
Phenyl	2.3	2.7	2.9
Aryl§	3.0–2.5	—	—
$C\equiv CR$	2.0	—	—
$C\equiv N$	2.0	2.3	2.7

*The data appearing in this table were taken from various published compilations and original literature references. The tabulated values are average values for compounds which do not contain another functional group within two carbon atoms from the indicated hydrogens. For methylene and methine hydrogens most values fall within ±0.15 ppm of the tabulated values.
†S-aryl derivatives absorb at somewhat lower fields.
‡Replacement of NH_2 by N-alkyl$_2$ results in a slight shift upfield.
§Includes polycyclic and many heterocyclic aromatics. Values for $-CH_2-X$ and $-CH-X$ probably appear at lower fields also.

linear correspondence with the electronegativity of the X atom from lithium to oxygen. The positions of absorption for the α-hydrogens in a variety of acyclic compounds are tabulated in Table 5.2. The shielding and deshielding effects by X of hydrogens decreases as one goes from α- to the β- to the γ-position, the γ-position being relatively unaffected.

Hydrogens bonded to carbon in cyclic structures absorb at only slightly different positions, except for cyclopropane which absorbs at very high field ($\delta = 0.222$). Substituted cyclopropanes absorb in the $\delta = 0.5$ to 2.0 region. Cyclopentane absorbs at $\delta = 1.51$, cyclohexane at $\delta = 1.44$, while cycloheptane absorbs at $\delta = 1.53$.

The presence of an additional substituent Y on the carbon, to give X—CH$_2$—Y type compounds, leads to further deshielding of the hydrogens. The effect of the second substituent is not always additive since electron withdrawal becomes more difficult as the number of electron withdrawing substituents increases. However, Shoolery has developed an empirical method of calculating the chemical shifts of the hydrogens in CH$_3$X and X—CH$_2$—Y by averaging the shifts caused by successive substitutions of hydrogen in methane, and then designating these shifts as effective shielding constants, $\sigma_{i_{\text{eff}}}$. Table 5.3 lists the $\sigma_{i_{\text{eff}}}$'s for a number of functional groups. These constants are used with Eq. (5.9) to calculate the expected

Table 5.3. Shoolery's Effective Shielding Constants

Functional group	$\sigma_{i_{\text{eff}}}$
—Cl	2.53
—Br	2.33
—I	1.82
—OH	2.56
—O—Alkyl	2.36
—O—Aryl	3.23
$\overset{\text{O}}{\overset{\|}{-\text{OCR}}}$	3.13
—SR	1.64
—N(Alkyl)$_2$	1.57
—CH$_3$	0.47
—CR^1=CR^{2}R^3	1.32
—C$_6$H$_5$	1.85
—C≡C—R	1.44
—RC=O	1.70
$\overset{\text{O}}{\overset{\|}{-\text{C}-\text{OR}}}$	1.55
$\overset{\text{O}}{\overset{\|}{-\text{C}-\text{NR}_2}}$	1.59
—CF$_3$	1.14
—C≡N	1.70

chemical shifts of various hydrogens. [Schoolery's calculations were based on the τ scale; Eq. (5.9) has been converted to the δ scale.]

$$\delta = 0.233 + \Sigma \, \sigma_{i_{\text{eff}}} \tag{5.9}$$

This approach is quite successful in predicting the chemical shifts of hydrogens in CH_3X and CH_2XY type compounds, but it tends to be quite poor in predicting the chemical shifts of methine (CHXYZ) hydrogens. Several examples of the use of Schoolery's rules, including good and poor correspondence with experimentally observed values, are tabulated in Table 5.4.

**Table 5.4. Comparison of Observed and Predicted Chemical Shifts
Using Shoolery's Rules**

Compound	Calculated δ	Observed δ
$BrC\underline{H}_2Cl$	5.09	5.16
$IC\underline{H}_2I$	3.87	4.09
$C_6H_5C\underline{H}_2OR$	4.44	4.41
$C_6H_5C\underline{H}_2CH_3$	2.52	2.55
$C_6H_5C\underline{H}_2C_6H_5$	3.91	3.92
$C{=}C{-}C\underline{H}_2OH$	3.92	3.91
$C{=}C{-}C\underline{H}_2{-}C{=}C$	2.87	2.91
$CH_3C\underline{H}_2\overset{\displaystyle O}{\overset{\|}{C}}{-}R$	2.40	2.47
$(C_2H_5O)_3C\underline{H}$	7.31	4.96
$(CH_3)_2C\underline{H}I$	2.99	4.24

Extension of the inductive effect explanation from aliphatic to aromatic, olefinic, and acetylenic compounds leads to a prediction of chemical shift trends in opposition to the experimentally observed trend. Before discussing the chemical shifts of protons attached to unsaturated centers, it is appropriate to discuss the shielding properties of various systems and the importance of these effects on nuclei bonded to these systems and on nuclei in other portions of the molecule.

Doubly and triply bonded systems possess a more polarizable and delocalized electronic structure than singly bonded systems, and thus they produce significant induced fields when placed in a magnetic field. The magnitude and sign of the induced fields are highly directional with respect to the axes of the unsaturated center.

Hydrogens bonded to C=C and C=O absorb at much lower field positions than one might expect if the deshielding were due to local inductive effects of the doubly bonded system. The additional deshielding, approximately 4 to 5 ppm, of these hydrogens must be due to a long-range deshielding by the C=C and C=O. Theoretical and experimental evidence has indicated that the space about the C=O can be divided into two regions, one producing paramagnetic shifts (indicated by a minus sign) and the other region producing diamagnetic shifts (indicated by a positive sign). The division of the space about the carbonyl group into the paramagnetic and diamagnetic shielding regions is not clear. Earlier investigators proposed that the space about the carbonyl group and the carbon-carbon double bond should be divided as illustrated in Fig. 5.5a. Recently, Karabotsos and co-workers[4] have proposed that division of the space about the carbonyl group as illustrated in Fig. 5.5b better explains most of the long range shielding effects of the carbonyl group. The two models predict the same general trends, the differences appearing only in very specific compounds.

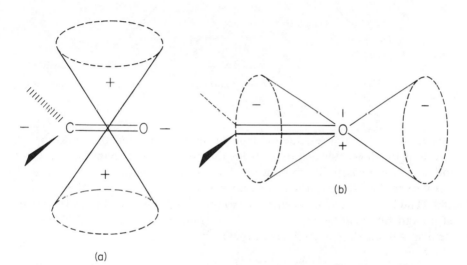

Fig. 5.5. Regions of diamagnetic (+) and paramagnetic (−) shielding by the carbonyl group.

Nuclei in other parts of the molecule may experience long range shielding and deshielding if the geometry of the molecule is such that a nucleus resides in the shielding or deshielding regions of the carbonyl group. In non-rigid molecules, one or more molecular conformations may place the absorbing nucleus in these regions, resulting in diamagnetic or paramagnetic shifts. The magnitude of these shifts increases as the distance between the carbonyl group and the absorbing nucleus decreases, and as the angle the nucleus makes with respect to the central axis of the $C\!=\!O$ approaches 0 or 90°. The long range shielding associated with the $C\!=\!C$ is similar to that of the carbonyl group.

The regions of diamagnetic and paramagnetic shielding produced by an aromatic ring are illustrated in Fig. 5.6. The effects are generally much

[4]G. J. Karabatsos, G. C. Sonnichsen, N. Hsi, and D. J. Fenoglio, *J. Amer. Chem. Soc.*, **89**, 5067 (1967).

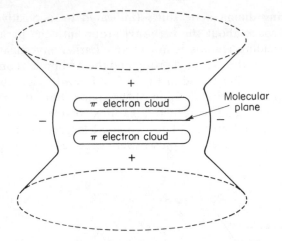

Fig. 5.6. Regions of diamagnetic (+) and paramagnetic (−) shielding of the benzene ring.

greater owing to the presence of strong ring currents, and they may be observed at much greater distances from the benzene ring than for the C=C and C=O systems. The maximum diamagnetic shielding is observed directly over the π-electron system and not over the center of the aromatic ring. The low-field chemical shift of aromatic hydrogens is due to the strong paramagnetic shielding in the plane of the ring. Structures **1** and **2** illustrate the diamagnetic shielding effects of the aromatic ring.

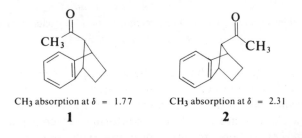

CH$_3$ absorption at δ = 1.77   CH$_3$ absorption at δ = 2.31

1   **2**

Carbon-carbon triple bonds present a situation quite different from the C=C and C=O systems. The regions of diamagnetic and paramagnetic shielding are illustrated in Fig. 5.7. The regions of diamagnetic and paramagnetic shielding of the C≡C are the opposite of the C=C. This results in the apparently anomalous chemical shift, if based on local shielding when compared with saturated C—H and vinyl C—H, of acetylenic hydrogens in the δ = 2.35 region.

Long range shielding effects may also be caused by electron currents in saturated bonds. In cyclic compounds axial and pseudoaxial hydrogens appear at higher field than their equatorial and pseudoequatorial counterparts

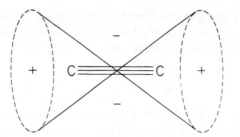

Fig. 5.7. Regions of diamagnetic (+) and paramagnetic (−) shielding of the carbon-carbon triple bond.

do. This has been attributed to an anisotropic shielding of the axial hydrogen by the electrons of the β,γ bonds.

From the preceding paragraphs it is obvious that a certain degree of caution must be exercised in assigning hydrogen resonances based on the general regions of absorption. One must keep in mind that the geometry of the molecule, contributions by various molecular conformations, and the presence of functional groups may have a profound effect on the chemical shift of a given hydrogen. In many cases chemical evidence must be integrated with the spectral evidence in order to derive the correct structure.

Table 5.5 lists typical chemical shifts of hydrogens bonded to various unsaturated systems.

Pascual, Meier, and Simon[5] have developed an empirical correlation of

Table 5.5. Chemical Shifts of Hydrogens Bonded to Unsaturated Centers

Type	Unconjugated, δ	Conjugated, δ
$C{=}CH_2$	4.6–5.0	5.4–7.0*
C=C with H	5.0–5.7	5.7–7.3*
Aromatic	6.5–8.3†	——
Nonbenzenoid aromatic	6.2–9.0	——
Acetylenic	2.3–2.7	2.7–3.2
Aldehydic	9.5–9.8	9.5–10.1*
$H{-}\overset{O}{\underset{\|}{C}}{-}N{\big\langle}$	7.9–8.1	——
$H{-}\overset{O}{\underset{\|}{C}}{-}O{-}$	8.0–8.2	——

*The position depends on the type of functional group in conjugation with the unsaturated group.
†The position of aromatic hydrogen absorption depends on the type of substituent attached to the aromatic ring (see Table 5.6).

[5]C. Pascual, J. Meier, and W. Simon, *Helv. chim. acta.*, **49**, 164 (1966).

chemical shifts of hydrogens bonded to olefinic systems similar to Shoolery's rules. The chemical shift of ethylenic hydrogens in carbon tetrachloride solution is calculated by use of Eq. (5.10)

$$\delta = \delta_{ethylene} + \sum_i Z_i \tag{5.10}$$

where $\delta_{ethylene}$ is the chemical shift of the hydrogens of ethylene and is taken as $\delta = 5.28$ and Z_i is the $\Delta\delta$ that results on substitution of an ith functional group for hydrogen in ethylene. The values of Z_i depend on the stereochemical relationship of the substituent with respect to the hydrogen in question in the substituted ethylene.

Table 5.6. Substituent Constants for Substituted Ethylenes*

Substituent	gem	cis	trans	Substituent	gem	cis	trans
—H	0	0	0	—CHO	1.03	0.97	1.21
—Alkyl	0.44	−0.26	−0.29	—C(=O)—N<	1.37	0.93	0.35
—Cycloalkyl	0.71	−0.33	−0.30				
—CH₂O, —CH₂I	0.67	−0.02	−0.07	—C(=O)—Cl	1.10	1.41	0.99
—CH₂S	0.53	−0.15	−0.15	—OR§	1.18	−1.06	−1.28
—CH₂Cl, —CH₂Br	0.72	0.12	0.07	—OR‖	1.14	−0.65	−1.05
—CH₂N	0.66	−0.05	−0.23	—OCOR	2.09	−0.04	−0.67
—C≡C	0.50	0.35	0.10	—Aryl	1.35	0.37	−0.10
—C≡N	0.23	0.78	0.58	—Cl	1.00	0.19	0.03
—C=C†	0.98	−0.04	−0.21	—Br	1.04	0.40	0.55
—C=C‡	1.26	0.08	−0.01				
—C=O†	1.10	1.13	0.81	—N(R§)(R)	0.69	−1.19	−1.31
—C=O‡	1.06	1.01	0.95				
—COOH†	1.00	1.35	0.74	—N(R‖)(R)	2.30	−0.73	−0.81
—COOH‡	0.69	0.97	0.39				
—COOR†	0.84	1.15	0.56	—SR	1.00	−0.24	−0.04
—COOR‡	0.68	1.02	0.33	—SO₂R	1.58	1.15	0.95

*For carbon tetrachloride solutions.
†For a single functional group in conjugation with the first C=C.
‡For a functional group which is also conjugated to a further substituent (e.g., 1,3,5-hexatriene).
§ R = aliphatic.
‖ R = unsaturated group.

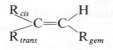

The values for Z_i are given in Table 5.6; they give values of δ within 0.15 of the experimental values in 73.3% of the 1070 compounds correlated. Serious deviations generally occur in the more highly substituted systems. An example of the use of Eq. 5.10 is illustrated with *cis*- and *trans*-stilbene, using the Z_i values in Table 5.6.

$$\delta_{cis} = 5.28 + Z_{i_{gem}} + Z_{i_{trans}} = 5.28 + 1.35 + (-0.10) = 6.53$$

$$\text{Observed } \delta = 6.55$$

$$\delta_{trans} = 5.28 + Z_{i_{gem}} + Z_{i_{cis}} = 5.28 + 1.35 + 0.37 = 7.00$$

$$\text{Observed } \delta = 6.99$$

π-Electron systems are capable of transmitting inductive and resonance effects owing to substituents bonded to the systems, and may result in diamagnetic or paramagnetic shifts. For example, with compounds **3** and **4**, the terminal vinyl hydrogens of **3** appear at lower field ($\delta = 6.20$ and 6.38), owing

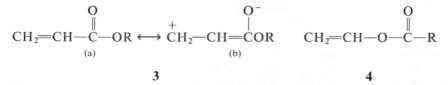

to deshielding caused by contribution of structure **3**(b), than do the similarly situated hydrogens in **4** ($\delta = 4.55$ and 4.85).

The chemical shift of aromatic hydrogens is quite sensitive to the type of substituent attached to the ring. Electron withdrawing groups result in a

Table 5.7. Chemical Shifts of Hydrogens in Monosubstituted Benzenes

Functional group	Chemical shifts (δ) relative to benzene (at $\delta = 7.27$)		
	ortho	meta	para
NO_2	−0.97	−0.30	−0.42
CO_2R	−0.93	−0.20	−0.27
$COCH_3$	−0.63	−0.27	−0.27
$COOH$	−0.63	−0.10	−0.17
CCl_3	−0.80	−0.17	−0.23
CH_3	0.10	0.10	0.10
Cl	0.00	0.00	0.00
OCH_3	0.23	0.23	0.23
OH	0.37	0.37	0.37
NH_2	0.77	0.13	0.40
$N(CH_3)_2$	0.50	0.20	0.50

deshielding of the hydrogens, particularly in the *ortho* position, and electron-donating groups result in shielding of the hydrogens. The chemical shift of aromatic hydrogens can be used to determine the type of functional group bonded to the aromatic ring. In many cases, the substitution pattern can be determined from the ratios of the various hydrogens observed at different field positions. Table 5.7 lists the chemical shifts for ortho-, meta- and para-hydrogens produced by a few substituents in monosubstituted benzene derivatives.

5.4.1 Chemical Shifts of Hydrogen Attached to Atoms Other than Carbon

5.4.1a Hydrogen bonded to oxygen

The chemical shifts of hydrogens bonded to oxygen are extremely sensitive to structural and environmental changes. The hydroxyl hydrogens of alcohols generally absorb in the $\delta = 0.5$ to 5.0 region (see Table 5.8), the position being dependent on the concentration. This is due to facile intermolecular hydrogen bonding and exchange, and results in a deshielding of the hydroxyl hydrogen. Distinction between intermolecular and intramolecular hydrogen bonding can be made by use of dilution studies. On dilution, a plot of the chemical shift vs. concentration results in a line with a significant slope for intermolecularly bonded hydrogen and very small slope for intramolecularly bonded hydrogen (theoretically, one would predict a zero slope). The resonances of monomeric hydroxyl hydrogens are near $\delta = 0.5$ in carbon tetrachloride. Phenolic hydroxyl hydrogens absorb at lower fields, approximately $\delta = 4.5$ for monomeric species. Their chemical shift is also a function of the concentration.

The acidic hydrogens of enols and carboxylic acids appear at very low fields. Enolic hydroxyl hydrogens absorb near $\delta = 15.5$, the enhanced deshielding being due to the positive nature of the hydrogen and strong intramolecular hydrogen bonding. Carboxylic acid hydrogens absorb in the $\delta = 9$ to 12 region and are not affected by concentration changes. This is due to the fact that carboxylic acids normally exist in dimeric form in nonpolar solvents.

The resonance signal of hydroxyl hydrogens is further complicated by the possibility of chemical exchange of hydrogens between different molecules. The peak shape, both in broadness and splitting by spin-spin interaction with the carbinol proton (see following discussion on spin-spin coupling, Sec. 5.7 and 5.11) of the hydrogen resonance peaks depends on the extent of chemical exchange and will be discussed in greater detail under time-averaging of chemical shifts (Sec. 5.6). Since the hydrogen exchange is acid catalyzed, it can be substantially reduced by careful purification of the compound and solvent, or by recording the spectrum in dimethyl sulfoxide solution.[6]

An alternative method of characterizing the hydroxyl hydrogen resonance is to remove the hydrogen and replace it with deuterium. This is accomplished by dissolving a small portion of the sample in ether or chloroform and shaking the solution several times with deuterium oxide. The organic phase is dried over a drying agent, and the organic compound is recovered by evaporation.

5.4.1b Hydrogen bonded to nitrogen

The chemical shifts of hydrogens bonded to nitrogen are also quite sensitive to structural and environmental changes. The resonance signals are generally very broad, owing to an interaction with the electric quadrupole moment of the nitrogen, and they may not be readily apparent. Hydrogen exchange

Table 5.8. **Chemical Shifts of Hydrogen Bonded to Oxygen, Nitrogen, and Sulfur**

Function group		Chemical shift, δ	
OH	aliphatic alcohols	0.5	(monomeric)
		0.5–5	(associated)
	phenols	4.5	(monomeric)
		4.5–8	(associated)
	enols	15.5	
	carboxylic acids	9–12	(dimeric)
	hydrogen bonded to carbonyl systems	13–16	
NH_2	alkylamine	0.6–1.6	
	arylamine	2.7–4.0	
	amide	7.8	
NH	alkylamine	0.3–0.5	
	arylamine	2.7–2.8	
$R_3\overset{+}{N}H$	ammonium salts	7.1–7.7	(in trifluoroacetic acid)
SH	aliphatic	1.3–1.7	
	aromatic	2.5–4	

[6]The exchange process is slowed considerably owing to hydrogen bonding between the hydroxyl hydrogen and the sulfoxide group. The hydroxyl hydrogen of a primary alcohol appears as a triplet (see Sec. 5.9 on spin-spin coupling), of a secondary alcohol as a doublet, and of a tertiary alcohol as a singlet [O. L. Chapman and R. W. King, *J. Amer. Chem. Soc.*, **86**, 1257 (1964)].

also occurs. Table 5.8 lists the resonance ranges for various types of N—H containing compounds.

5.4.1c Hydrogen bonded to sulfur

The chemical shifts of hydrogens bonded to sulfur in aliphatic thiols appear in the $\delta = 1.2$ to 1.6 region. Aromatic thiols absorb at a somewhat lower field near $\delta = 3.0$ to 3.5 (see Table 5.8.)

5.5   CHEMICAL SHIFTS OF OTHER NUCLEI

Many other nuclei have been studied by nuclear magnetic resonance techniques, including ^{11}B, ^{13}C, ^{14}N, ^{19}F, ^{31}P, etc. A discussion of the dependence of structure on the chemical shift of these nuclei is beyond the scope of this text, and the reader is referred to other more comprehensive texts listed under Sec. 5.18.

5.6   NUCLEAR SPIN-SPIN INTERACTIONS

The correlation of nuclear resonance frequencies with structure provides very useful information for determining the structures of molecules. However, the value of nuclear magnetic resonance is greatly increased by the occurrence of an additional phenomenon, that of nuclear spin-spin interaction leading to a splitting of the resonance lines. A nucleus with spin quantum number I greater than zero possesses $(2I + 1)$ equally separated spin states when placed in a magnetic field. The effects of these individual spin states is transferred, generally through the bonding electrons by spin angular momentum polarization of the bonding electrons, to adjacent nuclei. As a result, these nuclei experience different effective magnetic environments, depending on the orientation of the spin state of the coupling nucleus relative to the applied magnetic field. This interaction results in a number of resonance lines for a given absorbing nucleus instead of a single resonance line. This phenomenon is referred to as *spin-spin coupling* and is illustrated in the spin-energy diagram of Fig. 5.8 for a two-spin system.

Since only transitions in which the total change in the spin angular momentum equals 1 are allowed, only transitions $1 \longrightarrow 2$, and $3 \longrightarrow 4$ involving spin inversion of the X nucleus only, and $1 \longrightarrow 3$, and $2 \longrightarrow 4$ involving spin inversion of the A nucleus only, are allowed. If there is no spin-spin interaction, the energies corresponding to the $1 \longrightarrow 2$ and $3 \longrightarrow 4$

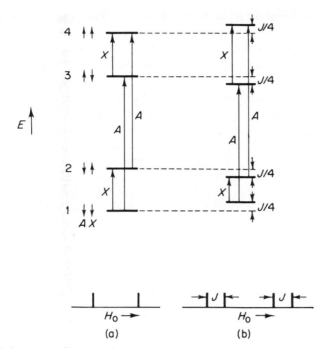

Fig. 5.8. Spin energy diagrams for a non-spin-spin coupled system (a) and spin-spin coupled system (b) with the resulting spectra.

transitions are equal, as are the $1 \longrightarrow 3$ and $2 \longrightarrow 4$ transitions, giving only a 2-line spectrum. If we now allow spin-spin interaction of $+J$ to occur, levels 1 and 4 are destabilized by $J/4$, and levels 2 and 3 are stabilized by $J/4$ relative to the uncoupled states (see Fig. 5.8b). The only allowed transitions are still the $1 \longrightarrow 2$, $3 \longrightarrow 4$, $1 \longrightarrow 3$, and $2 \longrightarrow 4$ transitions. The energy of the $1 \longrightarrow 2$ and $1 \longrightarrow 3$ transitions have been decreased by $J/2$ relative to the non-spin-spin interaction diagram, and the transitions $2 \longrightarrow 4$ and $3 \longrightarrow 4$ have increased by $J/2$. The resulting spectrum now displays four lines, the lines for the individual nuclei being separated by J. Since the magnetic moments of nuclei are independent of the applied field strength, the size of this coupling interaction, referred to as the *coupling constant*, J, is field strength independent. This is in distinct contrast to the chemical shift which is field dependent.

An alternative situation is possible in which a spin-spin interaction of $-J$ occurs. Levels 1 and 4 are stabilized by $J/4$, and levels 2 and 3 are destabilized by $J/4$. When this occurs it is apparent that an inversion in the energies of the $1 \longrightarrow 2$ and $3 \longrightarrow 4$, and the $1 \longrightarrow 3$ and $2 \longrightarrow 4$ transitions will occur relative to Fig. 5.8b; however the appearance of the nuclear magnetic resonance spectra of the two systems will be identical. The determination of the relative sign of the coupling constant is quite difficult and requires the application of special techniques. The reader is referred to the standard texts on nuclear magnetic resonance listed in Sec. 5.18.

The construction of spin energy diagrams may be used for more complicated spin systems involving a greater number of interacting nuclei. However, in the more complex spin systems, additional transitions become allowed, leading to very complex spectra. Instead of continuing this approach in the more complex systems, we shall develop a very simplified, *first-order* approach which is much easier to apply in simple systems.

Spin-spin interaction in a simple system can be illustrated by means of simple molecule HD, in which I for hydrogen is $\frac{1}{2}$ and for deuterium is 1. The hydrogen possesses two spin states, $-\frac{1}{2}$ and $+\frac{1}{2}$, and the deuterium three spin states, -1, 0, and $+1$ (see Fig. 5.9). The spin states of hydrogen have equal probability and thus will be equally populated, as will the three spin states of deuterium.

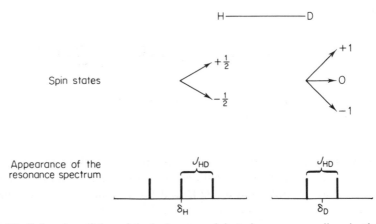

Fig. 5.9. Spin-spin splitting of the hydrogen and deuterium resonance lines in the HD molecule.

Interaction of the spin states of deuterium with the hydrogen nucleus will result in a splitting of the hydrogen resonance into three distinct lines, one line being effectively deshielded, one remaining unchanged (interaction with the zero spin state of deuterium), and one line being effectively shielded. The intensities of the three lines will be equal, owing to the equal population of the three spin states of the deuterium (line intensity will be discussed later). The separation between the individual lines is the same and is the coupling constant J_{HD} given in Hz. (See footnote, p. 166.) The deuterium resonance is split into two lines, one deshielded and one shielded relative to a noncoupled deuterium resonance. The resonance lines of the deuterium are again of equal intensity. The separation of the deuterium resonance lines is the same as the separation of the hydrogen resonance lines (J_{HD}). When spin-spin coupling is present, the chemical shift of the absorbing nucleus is taken as the center of gravity of the resonance lines resulting from that nucleus (see Fig. 5.10).

Let us now consider the effect of spin-spin coupling between hydrogens of differing chemical shift in a more complex organic molecule. If we consider the appearance of the H_A resonance in the fragment **5** in Fig. 5.10, the hydrogen H_A will interact with the two spin states of the adjacent hydrogen, giving a two-line spectrum similar to the deuterium resonance lines in HD. Hydrogen H_A in fragment **6** in Fig. 5.10 is coupled to the two adjacent methylene hydrogens (assumed to have identical chemical shifts, and widely separated from H_A in chemical shift). The net spin interaction with the two

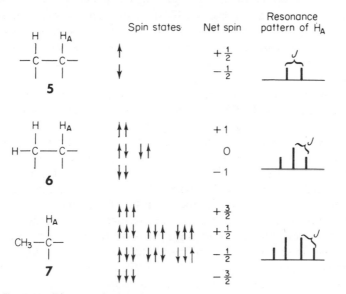

Fig. 5.10. Diagram of spin-spin interactions between adjacent hydrogens.

methylene hydrogens may be such that the methylene hydrogens give one spin state in which both hydrogens have $+\frac{1}{2}$ spins, two spin states of one hydrogen $+\frac{1}{2}$ with the other $-\frac{1}{2}$, and one spin state with both spins $-\frac{1}{2}$ (illustrated by means of the arrows in Fig. 5.10). All four net spin states are of equal probability and will thus produce a three-line pattern for H_A in which the intensities of the lines will be 1:2:1, the distance between two adjacent lines being the coupling constant J. Finally, if we consider the fragment **7** in Fig. 5.10, the three chemically equivalent methyl hydrogens produce the net spin states indicated to the right of **7**, in which one net spin state is $+1\frac{1}{2}$, three states with net spin of $+\frac{1}{2}$ and $-\frac{1}{2}$, and one net spin state of $-1\frac{1}{2}$. Again, since all net spin states are of equal probability, the absorption pattern for H_A will appear as four lines of relative intensity 1:3:3:1 with a spacing of J.

In general, the multiplicity of an absorption pattern can be represented by Eq. (5.11) in which I is the spin quantum number of the coupled nucleus, and n is the number of chemically equivalent coupled nuclei.

$$\text{no. of peaks} = 2In + 1 \qquad (5.11)$$

The relative intensity of the peaks, produced by spin-spin coupling between hydrogens which have widely differing chemical shifts, approaches the co-efficients of r in the expanded form of Eq. (5.12), in which n is the number of adjacent, chemically equivalent hydrogens.

$$(r + 1)^n \qquad (5.12)$$

For example, the use of Eqs. (5.11) and (5.12) predicts a four-line pattern with relative intensities 1:3:3:1 (coefficients of r in the expanded form of $(r + 1)^3 = r^3 + 3r^2 + 3r + r^0$) for an interaction with a methyl group. The multiplicity of hydrogen absorption patterns can be referred to as doublets, triplets, quartets, etc., if the relative intensities of the peaks follow the coefficients of the expanded form of Eq. (5.12).

In a molecule of any complexity, we generally have fragments present of the type $>CH-CH_A-CH<$, in which the H_A hydrogen may be coupled to one or more hydrogens of differing chemical shift. The coupling constants involved usually have different values, and complex patterns may result. Furthermore, the assumption that the hydrogens have greatly different chemical shifts (implying $\Delta\delta \gg J$), as employed in the previous paragraphs, is hardly ever applicable when dealing with typical organic molecules. As the difference in chemical shifts of hydrogens decreases with respect to their coupling constants, the absorption patterns described in the previous paragraph become highly distorted, and additional absorption lines may arise owing to spin level interactions which allow forbidden spin-spin transitions to become "less" forbidden. The number of theoretical transitions in a three-spin system is 15; in a four-spin system it is 56, and in a five-spin system, it is 210. The handling of such systems will be discussed later.

The value of the coupling constant is a function of the number of intervening bonds, the type of intervening bonds, the orientation of one nucleus with respect to the other nucleus, and the type of functional group(s) present in the system. Table 5.9 lists typical ranges for a variety of hydrogen-hydrogen spin-coupling constants. Several of these systems will be discussed in greater detail in the subsequent paragraphs.

The geminal coupling constant, $^2J_{HH}$, (the superscript designates the number of bonds between the coupled nuclei) varies greatly, from approximately -17 to $+40$ Hz. Independent correlations of J as a function of the H—C—H angle and the electronegativity of substituents attached to the C atom have been published. However, Pople and Bothner-By[7] recently

---

[7] J. A. Pople and A. Bothner-By, *J. Chem. Phys.*, **42**, 1339 (1965).

Table 5.9. Hydrogen-Hydrogen Coupling Constants

Structure	J_{HH}, range in Hz
H—H	280
$\mathrm{C}\!\!<^H_H$ (geminal hydrogens of a saturated methylene)	$\mp 12\text{–}15^{*}$
6-membered ring	11–13
3-m. ring	4–8
epoxides	3.5–5.5
$-\!\overset{\mathrm{H}}{\underset{\mathrm{I}}{\mathrm{C}}}\!-\!\overset{\mathrm{H}}{\underset{\mathrm{I}}{\mathrm{C}}}\!-$ (vicinal)	$\mp 0\text{–}9\dagger$
6-m. ring (diaxial)	10–13
6-m. ring (axial-equatorial)	4–7
3-m. ring (*cis*)	7–11
3-m. ring (*trans*)	4–8
epoxides (*cis*)	2–3.5
epoxides (*trans*)	3–5
$-\mathrm{C}-\mathrm{C}-\mathrm{C}-$ (1,3)	$\mp 0\text{–}7\ddagger$
$-\mathrm{C}-(\mathrm{C})_n\mathrm{C}-$ ($n > 1$)	$\longrightarrow 0$
$=\mathrm{C}\!\!<^H_H$ (*geminal*)	1–3.5§
$\mathrm{C}\!\!=\!\!\mathrm{C}$ (*cis*)	6–14§
$\mathrm{C}\!\!=\!\!\mathrm{C}$ (*trans*)	11–18§
$\mathrm{C}\!\!=\!\!\mathrm{C}\!\!<^{C-H}_{H}$	4–10
$\mathrm{C}\!\!=\!\!\mathrm{C}$ (1,3)	0.5–3.0
$\mathrm{C}\!-\!\mathrm{C}\!\!=\!\!\mathrm{C}\!-\!\mathrm{C}$ (1,4)	0–1.6
$\mathrm{C}\!\!=\!\!\mathrm{C}\!\!=\!\!\mathrm{C}\!-\!\mathrm{H}$ (1,3)	5–6
$\mathrm{C}\!-\!\mathrm{C}\!\!=\!\!\mathrm{C}\!\!=\!\!\mathrm{C}\!-\!\mathrm{H}$ (1,4)	2–3
benzene ring (ortho)	7–10
(meta)	2–3
(para)	~1

Table 5.9. Hydrogen-Hydrogen Coupling Constants (Continued)

Structure		J_{HH}, *range in Hz*
	X=O (1,2)	1–2
	X=N (1,2)	2–3
	X=S (1,2)	5.5
	X=O, N, S (2,3)	3.5
	(1,3)	1–2
	(5-m. ring)	~3.5
	(7-m. ring)	10–13
		1–3
C—C≡C—H	(1,3)	2–4
C—C≡C—C	(1,4)	2–3
C—C≡C—C≡C—H	(1,5)	1–2
C—(C≡C)₂—C	(1,6)	~1
C—(C≡C)₃—C	(1,8)	~0.4

*This coupling constant is a function of the H—C—H angle and the electronegativity of attached substituents.

†*J* is a function of the dihedral angle and subsubstituents.

‡*J* is highly dependent on the spatial orientation of the two coupled nuclei.

§*J* is a function of bond angle and substituent electronegativities.

suggested a more unified correlation based on molecular orbital calculations and the consideration of inductive and hyperconjugative effects, and on the contribution of the factors as a function of the symmetry of the remainder of the molecule. These rules can be summarized as follows:

1. The geminal coupling constant becomes more positive as the hybridization on the C atom becomes more *s*-like.

2. Withdrawal of electron density from orbitals symmetric between hydrogen atoms (*sigma* orbital inductive effect), by groups directly bonded to the C atom, leads to a positive change in the coupling constant, and similar groups on the *beta* C atom lead to a negative change in the coupling constant.

3. Withdrawal of electron density from orbitals antisymmetric between hydrogen atoms (hyperconjugative effects) should lead to a negative change in the coupling constant.

4. Opposite changes occur in rules 2 and 3 when electron donation occurs.

The following examples will be used to illustrate the application of these rules. The values for the geminal coupling constants for the series of compounds CH_4, CH_3OH, CH_3Cl, and CH_3F are -12.4, -10.8, -10.8, and -9.6 Hz (the determination of the coupling constant between chemically identical nuclei will be considered in Sec. 5.10); the coupling constant becomes more positive as the substituent becomes more electronegative and capable of inductive electron withdrawal. In contrast, the geminal coupling constants for acetone and acetonitrile are -14.9 and -16.9 Hz and are indicative of extensive hyperconjugative electron withdrawal. The geminal coupling constant in ethylene (sp^2 carbon) is in the region of $+2.0$ to $+2.5$ Hz, whereas in propene, chloroethylene, and fluoroethylene, the coupling constants are $+2.08$, -1.4, and -3.2 Hz, respectively. The variations displayed in these constants are the results of inductive electron withdrawal. The geminal coupling constant in formaldehyde is $+40$ to $+42.2$ Hz. The magnitude of this coupling constant is attributed to electron withdrawal by induction by the oxygen, and to hyperconjugative donation by the non-bonded electrons on the oxygen.

Several theoretical relationships have been developed for calculating the vicinal coupling constant ($^3J_{HH}$) as a function of the dihedral angle ϕ between the two C—H bonds (see **8**).

8

The original relationships were derived by Karplus and are given by Eqs. (5.13) and (5.14).

$$J_{vicinal} = \begin{cases} 8.5 \cos^2 \phi - 0.28 & (0° < \phi < 90°) \quad (5.13) \\ 9.5 \cos^2 \phi - 0.28 & (90° < \phi < 180°) \quad (5.14) \end{cases}$$

Karplus later refined the relationship of J with ϕ as given in Eq. (5.15).

$$J_{vicinal} = 4.22 - 0.5 \cos \phi + 4.2 \cos^2 \phi \qquad (5.15)$$

Williamson and Johnson have proposed Eq. (5.16) and (5.17) for the relationship between J and ϕ.

$$J_{vicinal} = \begin{cases} 10 \cos^2 \phi & (0 < \phi < 90°) \quad (5.16) \\ 16 \cos^2 \phi & (90 < \phi < 180°) \quad (5.17) \end{cases}$$

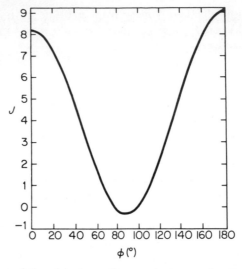

Fig. 5.11. Variation of the vicinal coupling constant as a function of the dihedral angle ϕ.

The angular dependence as calculated from Eq. (5.15) is presented in Fig. 5.11. Experimental values, derived from conformationally rigid systems in which the bond angles are reasonably well known, are in good agreement with the calculated values. However, the effect of electronegative substituents has not been completely evaluated. In systems in which free rotation exists between the central carbon atoms, the observed J values will be the time-averaged values of the various coupling constants from all possible conformations.

Long-range coupling, through more than three saturated bonds, is observed when the bonding system exists in the M, or W, conformation $\left(\begin{smallmatrix} & C & & C & \\ H & & C & & H \end{smallmatrix}\right)$. Long-range coupling has been observed between the indicated hydrogens in compounds **9** and **10**.

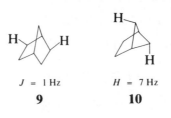

$J = 1\,\text{Hz}$ $H = 7\,\text{Hz}$

9   **10**

The coupling constant is also a function of the electronegativity of any functional groups contained in the system. Table 5.10 lists data tabulated by Bhacca and Williams[8] for the system **11**, in which J_{AX} and J_{BX} decrease as the electronegativity of X increases.

Table 5.10. Hydrogen Coupling Constants as a Function of the Substituent in 11 (Hz)

X	J_{AX}	J_{BX}	J_{AB}
CN	4.6	9.3	-12.6
COOH	4.4	8.5	-12.6
C_6H_5	4.2	8.9	-12.7
Cl	3.2	8.0	-13.2
OH	2.4	7.4	-12.6
OCOCH$_3$	2.5	7.6	-13.3

$$\begin{array}{ccc} H_A & & H_X \\ | & & | \\ -C & \!\!-\!\!-\!\! & C- \\ | & & | \\ H_B & & X \end{array}$$

11

The size of the coupling constant between olefinic hydrogens also depends on many factors. As indicated in Eqs. 5.13 through 5.17, the size of the coupling constant should be a function of the stereochemistry about the double bond; thus it plays an important role in allowing the chemist to distinguish between *cis*- and *trans*-disubstituted olefins (see entries in Table 5.9). The coupling constant between the olefinic hydrogens of cyclic olefins is quite sensitive to the ring size, the coupling constant decreasing markedly as the size of the ring decreases (see Table 5.11).

Table 5.11. Coupling Constants of Olefinic Protons in Cyclic Olefins

Ring size (no. of carbon atoms)	Coupling constant, Hz
7	10–13
6	8.8–10.5
5	5.1–7.0
4	2.5–4.0
3	0.5–2.0

The coupling constant is also dependent on the electronegativity of functional groups attached to the carbon-carbon double bond. For the system **12**, the relationships given in Eqs. 5.18 through 5.23 have been proposed in which $\Delta\chi$ is the difference in electronegativities between H and X on the Pauling scale, and ΔE is the difference in electronegativities on the Huggins scale.

12

[8] See ref. 1, Sec. 5.18.

Although these equations are for monosubstituted ethylenes, these correlations can be used to calculate the coupling constants in *cis-* and *trans-*disubstituted olefins by incorporating an additional electronegativity factor for the second group.

$$J_{gem} = 2.5(1 - 0.056(\Delta\chi)) \tag{5.18}$$

$$J_{trans} = 19.1(1 - 0.18(\Delta\chi)) \tag{5.19}$$

$$J_{cis} = 11.6(1 - 0.34(\Delta\chi)) \tag{5.20}$$

$$J_{gem} = 2.5 - 3.2(\Delta E) \tag{5.21}$$

$$J_{trans} = 19.1 - 3.3(\Delta E) \tag{5.22}$$

$$J_{cis} = 11.7 - 4.0(\Delta E) \tag{5.23}$$

5.7   VIRTUAL COUPLING

Molecules containing —$CHCH_3$ groups generally give rise to AB_3 patterns. However, occasionally, the methyl signal for such a system will not appear as a simple doublet, but will appear as a much more complex pattern, barely recognizable as being due to a methyl group. This may happen when there is strong coupling between the methine hydrogen and another chemically and magnetically similar hydrogen, or group of hydrogens, in the molecule. Even though this third hydrogen, or group of hydrogens, does not couple directly with the methyl group, $J = 0$, the spectrum appears as if strong long-range coupling does exist. This interaction is termed *virtual coupling*. In general, virtual coupling may occur in any system **14** between H_1 and H_3, when H_1 and H_2, and H_2 and H_3 are strongly coupled and all are of similar chemical shift.

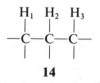

14

5.8   SPIN-SPIN COUPLING
BETWEEN OTHER
NUCLEAR COMBINATIONS

As indicated in the opening paragraphs of this chapter, spin-spin coupling may exist between nuclei having I not equal to zero. It should now be obvious that the organic chemist is extremely fortunate that the predominant isotopes of carbon and oxygen, ^{12}C and ^{16}O, possess I values of zero and do not spin-spin couple with other nuclei. If this were not true, the complexity of hydrogen magnetic resonance spectra would be formidable. However, there are many other nuclei encountered in organic molecules which possess I values not equal to zero.

Table 5.12 lists typical ranges of coupling constants for a variety of different nuclear combinations. These coupling constants also vary with bond angle, the number of intervening bonds, and the electronegativity of attached atoms or functional groups in a manner described for hydrogen-hydrogen coupling.

The natural occurrence of more than one isotope of a particular atom, of which one, or both, may spin couple with other nuclei, leads to the superpositioning of two coupling patterns in the sample spectrum. The spectrum of chloroform, Fig. 5.12, in which ^{12}C is accompanied by a small natural abundance of ^{13}C (see Table 5.1) of spin $\frac{1}{2}$, displays an intense singlet for the uncoupled absorption of the hydrogen in $^{12}CHCl_3$, and two weak outer bands (*satellites*) for the coupled absorption of the hydrogen in $^{13}CHCl_3$. The relative intensity of these ^{13}CH coupling peaks is directly equal to the $^{13}C/^{12}C$ ratio. (Substitution of one isotope for another in a molecule

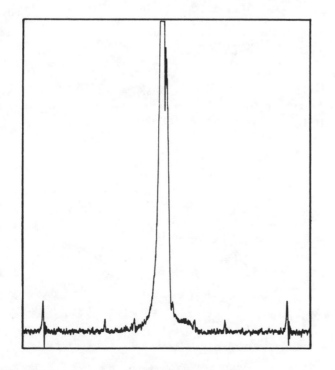

Fig. 5.12. Hydrogen resonance spectrum of chloroform displaying an intense, uncoupled central peak corresponding to $^{12}CHCl_3$ with the coupled peaks of $^{13}CHCl_3$ appearing as the weak outer peaks. (Recorded on a Varian Associates HR-60 Spectrometer.)

Table 5.12. Coupling Constants for Various Nuclei Combinations

Coupling constant	System	Coupling constant range (Hz)
J_{HD}	General	$0.154 J_{HH}$
$J_{^{10}BH}$	$^{10}B-H$	$\sim 30-50$
$J_{^{11}BH}$	$^{11}B-H$	75–182
	$^{11}B-\!\!-\!\!-H-\!\!-\!\!-\,^{11}B$ (bridged)	30–50
	$^{11}B-C-H$	~ 0
$J_{^{13}CH}$	$^{13}CH_3X$	110–155
	$^{13}C-H$	160–235
	$^{13}C-C-H$	4–6
	$^{13}C-C-C-H$	3–4
$J_{^{14}NH}$	NH_4^+	52.6
$J_{^{15}NH}$	NH_4^+	73.7
$J_{^{19}FH}$	$\diagdown C(H)(F)\diagup$	44–81
	$\diagup CF-CH\diagdown$	0–30
	$=C(H)(F)$	70–80
	$F\,C=C\,H$ (trans)	30–50
	$F\,C=C\,H$ (cis)	2–20

	(ortho)	8–10
	(meta)	5–8
	(para)	~ 2

Coupling constant	System	Coupling constant range (Hz)
$J_{^{13}C^{19}F}$	$^{13}C-{}^{19}F$	250–300
	$^{13}C-C-{}^{19}F$	30–40
$J_{^{19}F^{19}F}$	$\diagdown C(F)(F)\diagup$	155–225
	$\diagup CF-CF\diagdown$ (trans)	≈ 16
	(gauche)	≈ 18
	$=C(F)(F)$	28–87
	$F\,C=C\,F$ (cis)	20–58
	$F\,C=C\,F$ (trans)	95–120

	(ortho)	~ 20
	(meta)	2–4
	(para)	11–15

Table 5.12. Coupling Constants for Various Nuclei Combinations (Continued)

Coupling constant	System	Coupling constant range (Hz)
$J_{^{29}SiH}$	^{29}Si—H	190–380
$J_{^{31}PH}$	^{31}P—H	179–700
	^{31}P—C—H	~4
	^{31}P—C—C—H	~12–14
$J_{^{199}HgH}$	^{199}Hg—C—H	80–235
	^{199}Hg—C—C—H	115–200

produces only very slight changes in the chemical shifts of attached nuclei, deuterium producing the greatest change of about 1 to 3 Hz at 60 MHz.)

The ^{13}CH coupling constant has been used as a probe to determine the percent s character of a carbon-hydrogen bond, particularly in strained, small ring compounds. The quantitative aspects of this method has been questioned, owing to the occurrence of other phenomena which also affect the size of the coupling constant.

5.9   SPIN-SPIN COUPLING BETWEEN CHEMICALLY IDENTICAL NUCLEI

Spin-spin coupling occurs between nuclei having the same chemical shift, for example, the hydrogens of methyl chloride and the 1- and 2-hydrogens of 1,1,2,2-tetrachloroethane, although the effects of this coupling are not observable in the hydrogen resonance spectra of these compounds. Indirect methods must be used to determine the coupling constants between identical nuclei. Isotopic substitution, followed by determination of the coupling constant between the isotopic atoms, allows one to calculate the coupling constant between nuclei of identical chemical shift by multiplying the coupling constant observed between the isotopic atoms by the ratio of their gyromagnetic ratios. For example, in the compounds just cited, the J_{HD}'s are determined directly from the resonance spectra of the monodeuteriomethyl chloride and monodeuterio-1,1,2,2-tetrachloroethane, and are multiplied by 1/0.154 or 6.5 to derive the J_{HH}'s.

Another technique for determining the coupling constant between hydrogens of identical chemical shift on different carbon atoms involves analysis of the ^{13}C satellite patterns. The hydrogen on ^{13}C in the fragment H—^{12}C—^{13}C—H is only very slightly different in chemical shift from the hydrogen on ^{12}C, but the peak(s) due to the hydrogen on ^{13}C are well displaced from the peak(s) arising from the hydrogen on ^{12}C due to the large ^{13}CH coupling constant. The ^{13}C satellite pattern will thus resemble an

AMX pattern (see Sec. 5.13) in which the hydrogen resonance is split by coupling with the ^{13}C and the hydrogen on the adjacent ^{12}C. This H—H coupling constant is thus taken as the coupling constant between the hydrogens of identical chemical shift in the system H—^{12}C—^{12}C—H.

5.10   SPIN DECOUPLING

In spectra of relatively complex molecules, it may be quite difficult to determine which nuclei are spin-coupled leading to the observed patterns. This is particularly true if several of the coupling constants are of similar size, if one or more nuclei are buried in a complex absorption multiplet, or if long-range coupling is present.   In such cases, it would be desirable to destroy the spin-spin coupling interaction between various nuclei in the system.   Spin-spin splitting is observed when the nucleus which is responsible for the splitting remains in a given spin state for a period of time which is long in comparison with the reciprocal of the difference in chemical shifts, in Hz, of the coupled nuclei. If we can cause the lifetime of a spin state to decrease sufficiently so that the absorbing nucleus "sees" only a time average of the various spin states possible, a single resonance line will result. Experimentally, it has been found possible to do this by irradiating the nucleus responsible for the spin-spin interaction at its resonance frequency. This is referred to as *double resonance*.

It is difficult to obtain experimentally the precise resonance frequency in the MHz region to effect double resonance. A technique has been de-

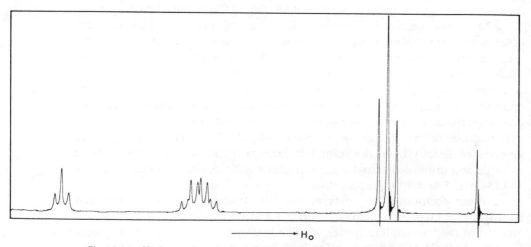

Fig. 5.13. Hydrogen resonance spectrum of purified ethanol showing the spin-spin coupling between the hydroxyl and methylene hydrogens.   The high field resonance peak is TMS.

veloped which involves recording the resonance spectrum in the first side-band region separated from the center-band region by a given number of Hz. A side-banding technique is then utilized, similar to that discussed for the determination of chemical shifts, which impresses a side band from the center band, on the resonance line of the nucleus one wishes to decouple. The frequency used to irradiate the coupled nucleus is the number of Hz from the center band to the first side band where the spectrum is being recorded, plus or minus the difference between the chemical shifts of the coupled nuclei in Hz in the center band.

Examples of double resonance are illustrated in Figs. 5.14 and 5.15 for purified ethanol (Fig. 5.13). In Fig. 5.14, the coupling of the methyl group

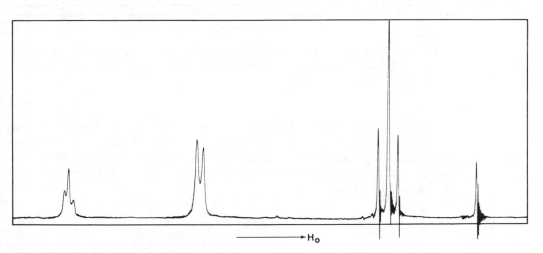

Fig. 5.14. Hydrogen double resonance spectrum of ethanol in which the spin-spin coupling interaction between the methyl and methylene hydrogens has been destroyed by saturation of the methyl hydrogens.

with the methylene group has been removed so that the methylene group now appears as a doublet due to spin-spin splitting only by the hydroxyl hydrogen. Note that the effect of the spin-spin interaction of the methylene group on the methyl group has not been destroyed. In Fig. 5.15, the methylene group has been irradiated at a frequency which destroys the spin-spin coupling to the methyl group only. The coupling between the methylene group and the hydroxyl hydrogen has not been destroyed.

The technique of double resonance can be extended by the use of two audio oscillators to accomplish the saturation of two different sets of nuclei (*triple resonance*). Figure 5.16 displays the resonance spectrum of ethanol in which both the methyl and hydroxyl hydrogens were irradiated, destroying nearly all of the coupling interactions to the methylene group.

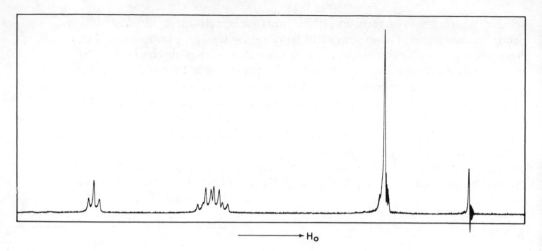

Fig. 5.15. Hydrogen double resonance spectrum of ethanol in which the spin-spin coupling between the methylene and methyl hydrogens has been destroyed by saturation of the methylene hydrogens.

Heteronuclear double resonance requires a different technique because the side-band techniques do not provide sufficient energy to irradiate the coupled nucleus ($\Delta \sigma \approx 10^6$ Hz). It is then necessary to irradiate the coupled nucleus independently at its own resonance frequency. Systems have been designed to accomplish this.

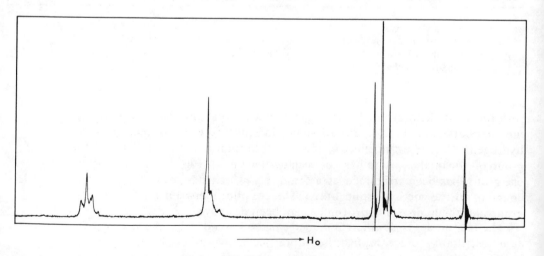

Fig. 5.16. Hydrogen triple resonance spectrum of ethanol in which the spin-spin coupling between the methyl and hydroxyl hydrogens with the methylene hydrogens has been destroyed by irradiating simultaneously the methyl and hydroxyl hydrogens.

Our discussions of chemical shift and spin-spin coupling phenomena have thus far been quite general. We have not stopped to consider in detail both the dynamic processes that may be occurring within or between molecules and the effect of these dynamic processes on the appearance of the nuclear magnetic resonance spectrum. Two dynamic processes are of great interest to the spectroscopist. These are (1) rotations about bond axes and (2) intramolecular and intermolecular exchange of nuclei between functional groups (for example, hydroxyl hydrogen exchange). Both of these dynamic processes result in changes of the chemical environment of a given nucleus. Exchange of chemical environments involving bond rotations are illustrated with structures **13** and **14**.

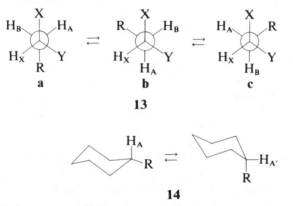

13

14

In **13**, rotation about the central carbon-carbon bond produces three rotational conformations in which the chemical environment of H_A, and also of H_B, is different in each rotational conformation. Because the dihedral angle H_X—C—C—H_A, and H_X—C—C—H_B, also changes in the three conformations, the coupling constant between H_A and H_X, and H_B and H_X, is also different in **13a, b,** and **c.**

The appearance of the nuclear magnetic resonance spectrum of a system such as **13** is a function of the rate of interconversion of the various conformations of the molecule, the difference in chemical shifts of a given nucleus in the available conformations (for example, H_A in **13a, b,** and **c**), and the population of the various conformations. If the rate, or frequency, of interconversion of conformations is much greater than the difference in chemical shifts in Hz of H_A in conformers **13a, b,** or **c**, a sharp, time-averaged resonance will result. If the frequency of interconversion approximates the difference in chemical shifts, a broad resonance line will result; and if the frequency of interconversion is much slower than the difference in chemical shifts, sharp, individual resonances will be obtained for the nuclei in each of the available conformations.[9] This phenomenon is illustrated in Fig. 5.17 for a nucleus equilibrating between two different sites in a molecule. P is the exchange frequency (given in exchanges per second when $(\delta_A - \delta_{A'})$ is expressed in cps).

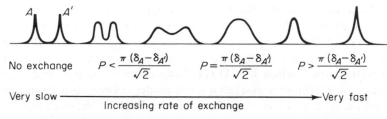

No exchange $P < \dfrac{\pi\,(\delta_A - \delta_{A'})}{\sqrt{2}}$ $P = \dfrac{\pi\,(\delta_A - \delta_{A'})}{\sqrt{2}}$ $P > \dfrac{\pi\,(\delta_A - \delta_{A'})}{\sqrt{2}}$

Very slow ──────────────────────────────────→ Very fast
Increasing rate of exchange

Fig. 5.17. Line shapes of resonance peaks of an equilibrating system.

The chemical shift of a nucleus in a rapidly interconverting system is given by Eq. (5.24)

$$\delta_{A_{av}} = \sum_i N_i \delta_{A_i} \tag{5.24}$$

in which N_i is the mole fraction of the ith conformer and δ_{i_A} is the chemical shift of H_A in the ith conformation. Similarly, the observed coupling constant is given by Eq. (5.25)

$$J_{AX_{av}} = \sum_i N_i J_{AX_i} \tag{5.25}$$

in which J_{AX_i} is the coupling constant between H_A and H_X in the ith conformation.

The nuclear magnetic resonance spectra of equilibrating systems may provide a great deal of kinetic and thermodynamic information about the system. The equilibrium constant for a given system may be calculated by using Eq. (5.26), providing the chemical shifts of nucleus A in the two equilibrating species (δ_A and $\delta_{A'}$) can be independently determined.

$$K_{eq} = \frac{\delta_A - \delta_{A_{av}}}{\delta_{A_{av}} - \delta_A} \tag{5.26}$$

The equilibrium constant is readily converted in ΔG for the reaction, which in turn when determined as a function of temperature allows calculation of ΔH and ΔS.

Determination of the values of δ_A and $\delta_{A'}$ may be carried out in two ways: (1) the temperature of the sample may be lowered until the frequency of the interconversion of A to A' is less than $\Delta\delta_{AA'}$ (see the following para-

---

[9]The nmr spectrometer may be compared to a camera equipped with a relatively slow shutter. A clear picture will be obtained of a nearly stationary object, whereas a blurred picture will be obtained of a rapidly moving object.

graph); or an additional functional group may be incorporated in the molecule which restricts the nucleus as A or A'. The latter method demands that the new functional group does not contribute to the shielding or deshielding of the nucleus of interest. It has recently been demonstrated that the latter condition is very difficult to achieve and that δ_A and $\delta_{A'}$ should be determined by low-temperature nuclear magnetic resonance measurements.[10]

Kinetic information is obtained by analysis of the line shapes of the spectrum when $P \approx \dfrac{\pi(\delta_A - \delta_{A'})}{\sqrt{2}}$; however, this requires being able to record the resonance spectrum under these conditions. Associated with each transformation within a molecule is an activation energy required to affect the transformation. For most transformations, the thermal energy at room temperature is sufficient to overcome the activation energy, and one generally observes only time-averaged spectra. Lowering the temperature of the sample will reduce the available thermal energy, and, if the activation energy lies in the accessible thermal region, the rate of the exchange may be reduced until the peaks broaden and eventually separate into distinct resonance lines (transversing right to left in Fig. 5.16 as a result of lowering the temperature). The temperature at which the individual resonance lines merge into a broad resonance line is referred to as the *coalescence temperature*.

The temperature of the sample is controlled by placing the sample in a small Dewar through which a stream of heated or cooled nitrogen is passed. The entire Dewar fits into a probe assembly which is held in the magnetic field. The accessible temperature range is approximately 200 to $-100°C$, and several reports give as low as $-160°C$. Care must be exercised in using high temperatures in that the sample tube should be carefully sealed and annealed, and placed in an oven or sand bath and heated for a short period of time to at least $15°$ above the maximum temperature at which the spectrum is to be recorded. This helps to avoid sample tube failure in the probe.

An excellent example of the effect of lowering the temperature of a sample is illustrated with compound **15**, which may exist in either conformation **15a** or **15b**. The resonance spectra of **15** recorded at various temperatures are shown in Fig. 5.18. At room temperature only broad, time-averaged peaks are observed, whereas at approximately $-16°$ individual peaks for both **15a** and **15b** are discernable.

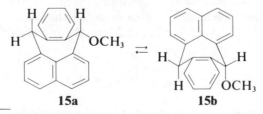

15a   **15b**

[10]S. Wolfe and J. R. Campbell, *Chem. Comm.*, **872** (1967); E. L. Eliel and R. J. L. Martin, *J. Amer. Chem. Soc.*, **90**, 682 (1968); F. R. Jensen and B. H. Beck, *Ibid.*, **90**, 3251 (1968).

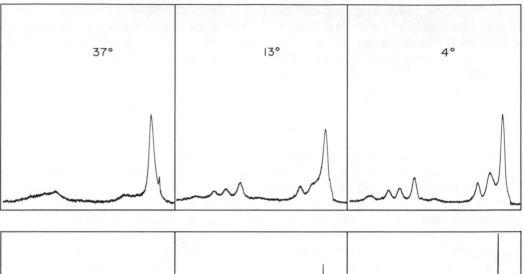

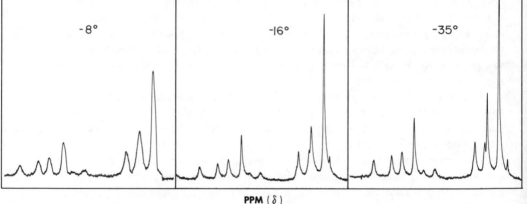

PPM (δ)

Fig. 5.18. Hydrogen resonance spectrum of compound **15** showing only the resonance peaks of the indicated aliphatic hydrogens in the δ = 3.0 to 6.5 region as a function of the temperature of the sample. At $-35°$ the predominant isomer present is **15b.**

    The complete analysis of the line shapes of a nuclear magnetic resonance spectrum requires computer techniques[11] and provides data on the rate constants for the reaction and the activation enthalpy and entropy.

    It is instructive to consider in greater detail the chemical shifts of H_A and H_B in structure **13.** Applying Eq. (5.24) to H_A and H_B gives

$$\delta_A = N_a\delta_{A_a} + N_b\delta_{A_b} + N_c\delta_{A_c}$$

and

$$\delta_B = N_a\delta_{B_a} + N_b\delta_{B_b} + N_c\delta_{B_c} \; .$$

[11]A computer program for the complete analysis of the line shapes has been developed by Prof. Gerhard Binsch, Department of Chemistry, University of Notre Dame.

H_A and H_B are diastereotopic in all three of the rotamers **13a, b**, and **c**. (A *diastereotopic* relationship exists between two atoms or groups of atoms when the two are in diastereomeric environments; in example **13** the diastereomeric environments are provided by the asymmetric carbon $-CXYH_X$.) H_A and H_B are *not* chemically equivalent despite the fact that both hydrogens are bonded to the same carbon! The chemical shifts of H_A and H_B are consequently different (although the difference may be quite small) in each of the three rotamers; $\delta_A \neq \delta_B$ under any conditions of rapid rotation and rotamer populations. Similarly $J_{AX} \neq J_{BX}$. Occasionally it is useful to modify a functional group to increase the difference between the chemical shifts of diastereotopic hydrogens. For example the difference in the chemical shifts of the diastereotopic hydrogens of the 1,2-diphenylethyl system is greatly enhanced on converting the alcohol to the benzoate (see Fig. 5.19) in which the diastereotopic hydrogens appear at $\delta = 3.26$ as an *AB* portion of an *ABX* system (see pp. 205–206).

Just as a pair of diastereotopic hydrogens or groups in the same molecule may have different chemical shifts, corresponding hydrogens or groups in two diastereomers, in principle, will have different chemical shifts. When the chemical shifts of hydrogens in diastereomers are of sufficient difference, nuclear magnetic resonance provides a very accurate method for the analysis of mixtures of diastereomers.

When R = H in **13,** the back carbon becomes a methyl carbon. The mole fractions of the three possible conformations will be equal, and all three methyl hydrogens will have exactly the same chemical shift and the same coupling constant with H_X.

The exchange of nuclei between functional groups is best illustrated by the hydroxyl hydrogen exchange encountered with hydroxyl containing

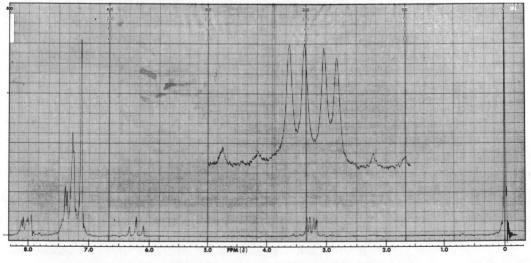

Fig. 5.19. Hydrogen resonance spectrum of 1,2-diphenylethyl benzoate.

compounds. The rate of this exchange is dependent on the concentration of the sample, the type of solvent used, the temperature of the sample, and the possible presence of acid or base catalysts. As the concentration of the sample decreases, the rate of intermolecular exchange decreases, as well as a change in the chemical shift. The use of solvents which are excellent hydrogen bond acceptors, for example, dimethyl sulfoxide, drastically reduces hydrogen exchange. Reduction of the temperature of the sample decreases the rate of exchange. The presence of acid or base catalysts facilitates the exchange reaction.

Not only does the rate of chemical exchange affect the chemical shift, but rapid exchange also destroys the spin-spin coupling of the hydroxyl hydrogen with the carbinol hydrogen(s). This is illustrated in the resonance spectrum of an acidified sample of ethanol shown in Fig. 5.20 in which the hydroxyl hydrogen coupling is not apparent (compare with Fig. 5.13).

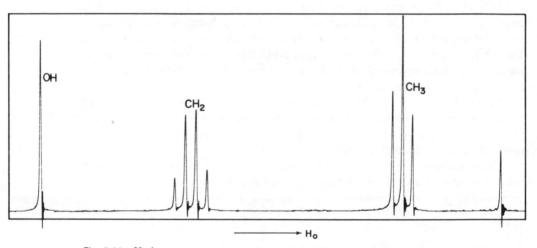

Fig. 5.20. Hydrogen resonance spectrum of acidified ethanol displaying only a singlet for the low field hydroxyl signal.

5.12 DESIGNATION OF SPIN SYSTEMS

In a spectrum containing the absorption peaks of a variety of differing nuclei, a system of designating the types of spin systems present is desirable. By a spin system, we mean sets of nuclei which are spin-spin coupled giving rise to recognizable absorption patterns. A single molecule may contain several such distinguishable spin systems.

The designation of the various nuclei in a spin system is based on the relative chemical shifts and the size of the chemical shift difference with respect to the coupling constant J. The symbols used to designate individual

Table 5.13. **Designation of Individual Nuclei and Spin Systems**

Symbol	Description*
A	H_A with δ_A.
A_n	$n\,H_A$ of same δ_A and magnetically identical.
AA'	Hydrogens H_A and $H_{A'}$ with same chemical shift, but magnetically different with respect to other nuclei.
AB	Hydrogens H_A and H_B with δ_A and δ_B, where $J_{AB} \geq (\delta_A - \delta_B)$.
AX	Hydrogens H_A and H_X with δ_A and δ_X, where $J_{AX} < (\delta_A - \delta_B)$.
AMX	Hydrogens H_A, H_M, and H_X with δ_A, δ_M, and δ_X, where $J_{AM} < (\delta_A - \delta_M)$ and $J_{XM} < (\delta_M - \delta_X)$
ABX	Hydrogens H_A, H_B, and H_X with δ_A, δ_B, and δ_X, where $J_{AB} \geq (\delta_A - \delta_B)$, $J_{AX} < (\delta_A - \delta_X)$ and $J_{BX} < (\delta_B - \delta_X)$.
ABC	Hydrogens H_A, H_B, and H_C with δ_A, δ_B, and δ_C, where all $\Delta\delta$'s $< J$'s

*Although all entries in this table utilize the hydrogen atom, these systems can be used with any other nuclear system or mixed nuclear system.

nuclei and spin systems are tabulated in Table 5.13. Examples of the use of these symbols appear in the following paragraphs.

Most of the entries in Table 5.13 are self-explanatory and will be illustrated by the use of actual resonance spectra. In general, spectra are presented in the literature with increasing field strength from left to right. The symbols designating the spin systems are employed such that the first letters of the alphabet (A, B, C, etc.) are used to represent nuclei appearing at lowest fields, while the latter letters (M, X, Y, etc.) represent nuclei appearing at progressively higher fields. For example, in an ABX system, the A nucleus would appear at the lowest field strength, the X nucleus appearing at the highest field strength.

The use of primed symbols, for example, in AA', also requires further explanation. Up to now we have considered only the chemical identity, or nonidentity, of nuclei in the resonance region. The term *magnetic identity* implies an equal spin-spin interaction, that is, an equal coupling constant between each nucleus of a given set of coupled nuclei and the absorbing nucleus. This condition is not always met. In many instances, two chemically identical nuclei interact with different coupling constants with another nucleus and, thus, are not magnetically identical with respect to that nucleus. The prime is used to indicate this magnetic nonidentity. The following examples illustrate this phenomenon.

In the cyclopropane derivative **16**, we have three chemically different types of hydrogens, H_A, H_B, and H_X. Depending on the substituent X, the methylene hydrogens may be designated as above, or they may be reversed. If we further differentiate the two H_A's and H_B's as H_A and H_B on one atom, and $H_{A'}$ and $H_{B'}$ on the other carbon atom, we can readily see that the two H_A's are not magnetically identical with the individual B hydrogens, or vice versa; yet they are chemically identical. Hydrogen H_A is spin-spin coupled with H_B with a geminal coupling constant and is spin-spin coupled with hydrogen $H_{B'}$ with a vicinal coupling constant. Since the geminal and vicinal coupling constants would be expected to have different values, we have two sets of nuclei which are chemically identical, AA' and BB', but magnetically dissimilar with respect to a third nucleus.

A similar situation exists in 1, 3-butadiene (**17**). It is obvious that hydrogens H_A and $H_{A'}$ are chemically identical, as are also H_B and $H_{B'}$, and H_C and $H_{C'}$. However, the spin-spin interactions between the C' and A', and C' and B' hydrogens are different from those between the C' and A, and C' and B hydrogens. Again magnetic nonidentity is indicated. Magnetic nonidentity may be obvious in some spectra, whereas in others, the complexity of the absorption patterns may not allow a straightforward analysis.

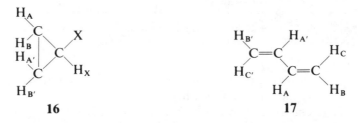

         16   **17**

Certain ambiguities in the designation of spin systems may still arise when applying the nomenclature rules outlined in the foregoing paragraphs. For example, if we were to designate a system as $AA'BB'CC'$ (butadiene), we are not clearly indicating the sequence of nuclei with respect to the carbon-atom framework of the molecule. To avoid such confusion, the symbols designating the types of nuclei bonded to one carbon atom may be separated by a hyphen from the symbols designating the types of nuclei on the adjacent carbon atom. Thus the preferred designation for butadiene would be $BC—A—A'—B'C'$.

5.13   INTERPRETATION OF SIMPLE SPIN SYSTEMS

5.13.1 The Two-spin Systems

We shall begin our discussion of simple spin systems by considering the two-spin system, the one-spin system being trivial and giving rise to a single absorption line. The simplest two-spin system is the A_2 system, which gives rise to a single absorption line. If we allow the chemical shifts of the two nuclei to become slightly different, and then introduce coupling between the nuclei such that $J_{AB} > \Delta\delta_{AB}$, a number of lines appear in the resonance spectrum. If we were to split the resonance lines of nuclei A and B in a symmetrical fashion about their respective chemical shifts, the lower field line of the B nucleus doublet would appear at lower field than the higher field line of the A nucleus (crossing of energy states). However, the energy states of one nucleus, produced by spin-spin interaction with a neighboring nucleus, may not overlap (cross) the energy states of that nucleus. To avoid the crossing of states in the AB system, the inner lines of the A and B doublets are more intense than the outer lines so that the centers of gravity of the A and B doublets are at their respective chemical shifts. The spin-spin coupling interaction is diagrammed in Fig. 5.21. As the chemical shift difference ($\Delta\delta$) between the nuclei A and B increases relative to J_{AB}, the outer lines increase in intensity at the expense of the inner lines. Finally, as the system graduates to the AX system, the lines become more equal in intensity.

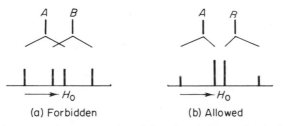

Fig. 5.21. Diagrammatic representation of the spin-spin interaction in the AB system.

The relative intensities of the outer-to-inner lines in an AB or AX system, and the difference in chemical shifts ($\Delta\delta$) are calculated using Eq. (5.27) and (5.28), respectively.

$$\text{Relative intensity}\left(\frac{\text{line 1}}{\text{line 2}}\right) = \frac{Q - J}{Q + J} \tag{5.27}$$

$$\Delta\delta = (Q^2 - J^2)^{1/2} \tag{5.28}$$

The coupling constant J is the distance between lines 1 and 2, or 3 and 4, and the quantity Q is the distance between lines 1 and 3, or lines 2 and 4.

Examples of spectra containing the AB and AX systems are illustrated in Figs. 5.22 and 5.23. The hydrogen resonance spectrum of cis-β-phenylmercaptostyrene, Fig. 5.22, shows an AB pattern centered at $\delta = 6.48$ for the two ethylenic hydrogens. The observed coupling constant is 10.7 Hz, and $\Delta\delta_{AB}$ is 0.095 ppm. The observed intensity ratio is 0.056, compared to 0.061 as calculated from Eq. 5.27. The complex absorption pattern at lower field is due to the aromatic hydrogens.

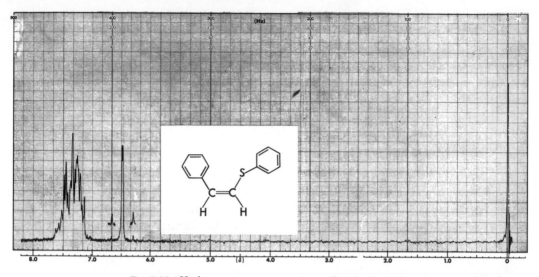

Fig. 5.22. Hydrogen resonance spectrum of *cis*-β-phenylmercaptostyrene.

Figure 5.23 shows the hydrogen resonance spectrum of *cis*-β-ethoxy-styrene in which the ethylenic hydrogens appear as an AX system, with $\delta_A = 5.18$ and $\delta_X = 6.06$ and a coupling constant of 7.2 Hz. The observed intensity ratio is 0.77, with a calculated value of 0.77. The triplet at $\delta = 1.21$ and the quartet at $\delta = 3.77$ represent the A_2X_3 pattern of the CH_3CH_2 group, while the aromatic hydrogens give rise to the complex multiplet centered at $\delta = 7.3$.

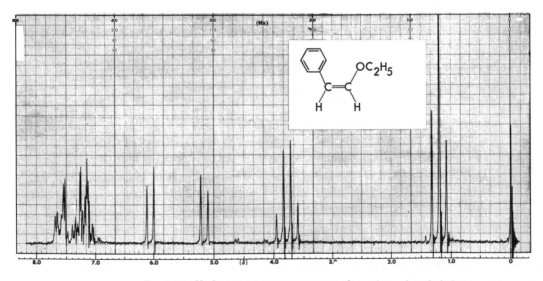

Fig. 5.23. Hydrogen resonance spectrum of *cis*-ethoxyphenylethylene.

5.13.2 Three-spin Systems

The simplest three-spin system is the A_3 system, for example, an isolated methyl group, which gives rise to a single absorption line. If one of the nuclei becomes slightly different in chemical shift relative to the other two nuclei, such that $\Delta\delta$ is less than the coupling constant, we have an A_2B or AB_2 system. The spectra of such systems do not possess any symmetry and cannot be readily analyzed as described for the AB and AX systems. A similar situation is encountered with the ABC system in which all three nuclei are slightly different in chemical shift. In spectra where there appears to be a three-spin system of the A_2B or ABC type, the student is referred to Wiberg and Nist's *The Interpretation of NMR Spectra* in order to determine the chemical shifts and coupling constants. Wiberg and Nist have calculated the absorption patterns for various spin systems, maintaining a constant chemical shift difference (6 Hz) while varying the coupling constant. The student selects the calculated spectrum which most closely resembles the portion of the spectrum of his sample in peak shape, relative positions, and intensities. The difference in chemical shifts and coupling constants may then be scaled up to correspond to the observed pattern. Numerous computer programs are available for the calculation of spectra and for comparison with observed spectra.

The A_2X or AX_2 system, in which one nucleus is of a greatly different chemical shift, may be represented in a straightforward fashion as indicated in Fig. 5.24. The theoretical intensities of the lines in the A and X portions

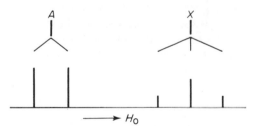

Fig. 5.24. Representation of spin-spin splitting in the A_2X system.

of the spectrum, as calculated from Eq. (5.12), apply only when $\Delta\delta_{AX}/J_{AX} \longrightarrow \infty$. As the ratio of $\Delta\delta_{AX}$ to J_{AX} decreases, the absorption patterns become distorted, with the inner lines increasing in intensity at the expense of the outer lines, as in the $AX \longrightarrow AB$ transformation.

Another three-spin system is the ABX system where $\Delta\delta_{AB} < J_{AB}$, and $\Delta\delta_{AX}$ and $\Delta\delta_{BX} > J_{AX}$ and J_{BX}. The spin-spin coupling interaction is diagrammed in Fig. 5.25. If we first couple the A and B nuclei independently of the X nucleus, we derive the typical AB pattern. Spin-spin interaction between the A and X nuclei doubles the A and X resonance lines, and inter-

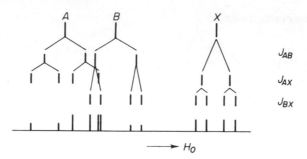

Fig. 5.25. Representation of the spin-spin splitting in the ABX system.

action between the B and X nuclei doubles the B and again the X lines. It should be pointed out that coupling of the A and B nuclei with the X nucleus leads to an overlapping of the A and B resonance lines. This is permissible except in cases when direct coupling between two nuclei might lead to crossing of states.

Occasionally, the X portion of the ABX spectrum may appear as an apparent triplet if $J_{AX} \approx J_{BX}$ which results in an overlap of the two central lines of the X portion. In such cases, the term triplet should not be used, but the pattern should be described as "the X portion of an ABX system." A typical example of a spectrum containing an ABX system is shown in Fig. 5.19 for 1, 2-diphenylethyl benzoate. The AB portion arises from the dissimilarity of the diastereotopic methylene hydrogens, as described in Sec. 5.6.

The final three-spin system is the AMX system. The absorption pattern is predicted in the straightforward manner as illustrated in Fig. 5.26. The relative sizes of the coupling constants will vary, depending on whether we are dealing with an AM—X or an A—M—X system. A typical example

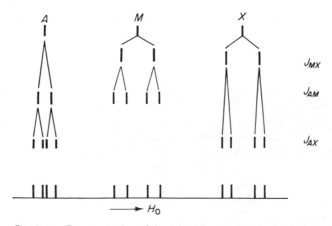

Fig. 5.26. Representation of the spin-spin coupling in the AMX system.

of an *AMX* pattern is illustrated in the hydrogen resonance spectrum of styrene (Fig. 5.27).

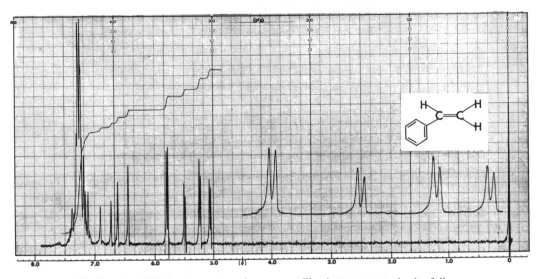

Fig. 5.27. Hydrogen resonance spectrum of styrene. The bottom trace is the full spectrum, total sweep width (total chart scan) of 500 Hz, which displays the aromatic hydrogen and the *AMX* vinyl regions with TMS appearing at $\delta = 0$. The *M* and *X* hydrogen patterns are displayed in the upper right-hand scan with a total sweep width of 100 Hz and sweep offset (displacement of the field at the O Hz chart position) of 296 Hz. The upper left-hand trace is the integral line indicating the relative intensity of the various peaks.

5.13.3 Higher-spin Systems

Higher-spin systems of the general type $A_n X_m$ give rise to patterns as described in Sec. 5.7. For spectra containing AB_3, A_2B_2, AB_4, and A_2B_3, the reader is referred to Wiberg and Nist's *The Interpretation of NMR Spectra.*

5.14   QUANTITATIVE APPLICATIONS OF NMR

The total integrated absorption of the same number of nuclei (*sensitivity*) of different chemical shift, but of the same isotopic species, is the same. Different nuclear species, however, possess different sensitivities and will not lead to equal integrated absorptions for an equal number of nuclei. This is of little consequence because the absorptions of two different nuclei do not occur within the same spectral region and, thus, never appear in the same spectrum.

The equal sensitivity, regardless of the chemical environment, provides the chemist with a very powerful quantitative tool. Measurement of the areas of individual absorption patterns allows one to derive the relative ratio of nuclei of different chemical environments within the molecule. If a definite structural assignment can be made for any one absorption pattern, and if the number of such functional groups within the molecule is known, the relative ratios of hydrogens can be transformed into an absolute ratio.

The analysis of mixtures may also be accomplished, provided that the absorption of one nucleus, or set of nuclei, appearing in each molecule present in the mixture can be distinguished and resolved from the absorption patterns of other nuclei in the other molecules. This technique provides one of the easiest methods for the analysis of diastereomeric mixtures.

Quantitative methods have been devised for the determination of the percent of hydrogen contained in organic molecules by adding a weighed portion of a standard, of known percent hydrogen, to a weighed portion of the unknown, and then integrating the resonance spectrum. The percent hydrogen in the unknown is then calculated in a straightforward manner.

The integration of resonance spectra can be accomplished by manual or electronic methods. Manual methods involve the measurement of areas by means of a planimeter or by cutting out the peak areas (usually a very difficult process) and weighing the cutouts. Electronic integration is generally achieved by accumulating the voltage input to the recorder and plotting this accumulated voltage as a function of time or chart distance. Figure 5.27 illustrates the use of this method. The line appearing above the resonance peaks is the integration line, and the height of the vertical steps is proportional to the number of nuclei appearing under that resonance peak. Digital readout devices are also available to print out the accumulated voltages at desired intervals.

5.15   SAMPLE PREPARATION

One of the most important factors in the production of good spectra is care in the preparation of the sample. The sample should be of high purity and free from traces of lint, dust, or other foreign material, particularly impurities having magnetic properties. The presence of such materials causes extensive line broadening.

Nonviscous liquids may be run neat (without solvent) with approximately 3% by weight tetramethylsilane added as an internal standard for hydrogen magnetic resonance. Other suitable standards are available for other nuclear species. Viscous liquids and solids must be dissolved in a suitable solvent, generally 10 to 25% by weight, to give a solution of low viscosity. The choice of the solvent will depend on the solubility characteristics of the sample. The solvent should not absorb in regions which will obscure the absorption patterns arising from the sample. Solvent absorption interferences may be avoided by the use of fully deuterated or halogenated molecules, for example, $CDCl_3$ and CCl_4. A great many deuterated solvents suitable for nuclear magnetic resonance studies are commercially available. The solvent should be of high purity and protected from contamination.

The neat liquid sample, or solution, is placed in a carefully cleaned sample tube. Special tubing is available specifically for use in nuclear magnetic resonance spectrometers. The wall diameter should be relatively thin and kept within close thickness tolerances. This provides for as great a volume of sample within the magnetic field as is possible, and a minimum of field distortion due to variances in wall thickness. The sample tube should be sealed. For very careful work the tube should be sealed under reduced pressure, using degassing techniques to remove dissolved oxygen which may lead to line broadening. The sample is then ready to have its nuclear magnetic resonance spectrum recorded.

5.16 NUCLEAR MAGNETIC RESONANCE SPECTRAL PROBLEMS

Identify each of the following unknowns on the basis of their nuclear magnetic resonance spectra and other information provided.

1. $C_5H_{12}O_2$; nmr spectrum, Fig. 5.28; displays broad absorption in the 3300 cm^{-1} (3.0μ) region of the infrared region.

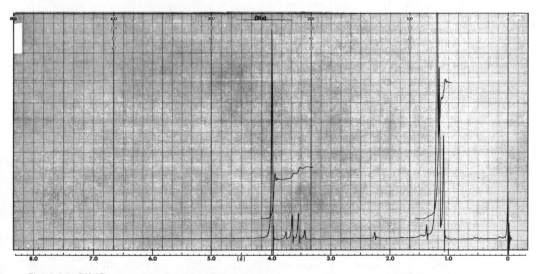

Fig. 5.28. NMR spectrum of unknown for Problem 1 recorded as a 30% solution in deuteriochloroform with sweep width 500 Hz.

2. $C_8H_8O_2$; nmr spectrum, Fig. 5.29; absorbs at 1692 cm^{-1} (5.91μ) transparent in the 3300 cm^{-1} (3.0 μ) region.

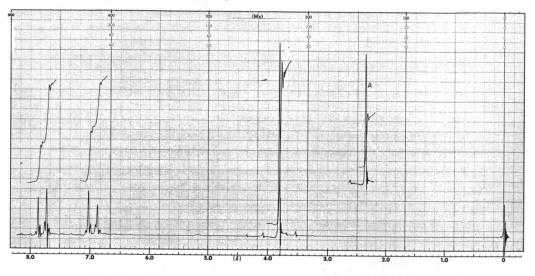

Fig. 5.29. NMR spectrum of unknown for Problem 2 recorded as a 30% solution in deuteriochloroform with sweep width 500 Hz; trace A sweep width 500 Hz, sweep offset 450 Hz.

3. C_8H_8O; nmr spectrum, Fig. 5.30; transparent in the 3300 cm^{-1} (3.0 μ) and 2000 to 1650 cm^{-1} (5 to 6 μ) regions.

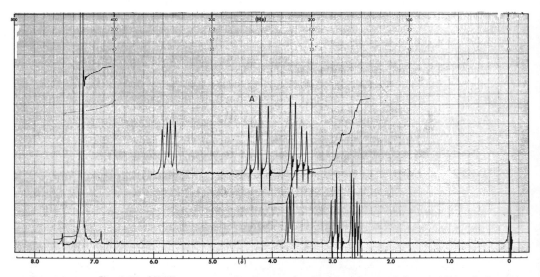

Fig. 5.30. NMR spectrum of unknown for Problem 3 recorded as a 30% solution in deuteriochloroform with sweep width 500 Hz; trace A sweep width 250 Hz, sweep offset 50 Hz.

4. C_5H_8O; nmr spectrum, Fig. 5.31; absorbs at 1684 cm^{-1} (5.93 μ).

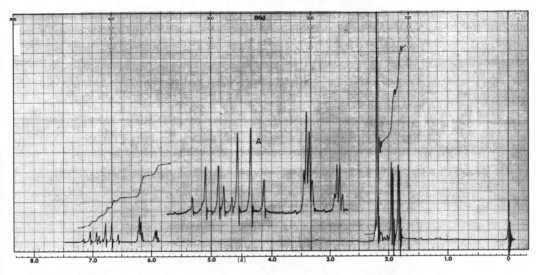

Fig. 5.31. NMR spectrum of unknown for Problem 4 recorded as a 30% solution in deuteriochloroform with sweep width 500 Hz; trace A sweep width 250 Hz, sweep off-set 270 Hz.

5. $C_6H_{12}O$: nmr spectrum, Fig. 5.32; transparent in the 3300 cm^{-1} (3.0 μ) and 2000 to 1650 cm^{-1} (5 to 6 μ) regions.

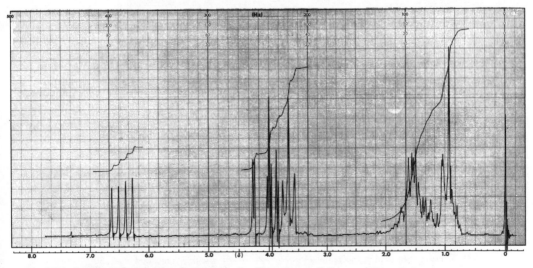

Fig. 5.32. NMR spectrum of unknown for Problem 5 recorded as a 30% solution in deuteriochloroform with sweep width 500 Hz.

6. $C_6H_{14}O_2$; nmr spectrum, Fig. 5.33; displays broad absorption in the 3300 cm^{-1} (3.0 μ) region.

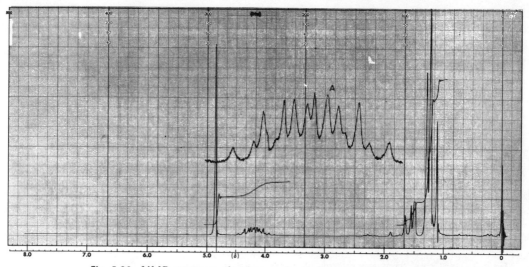

Fig. 5.33. NMR spectrum of unknown for Problem 6 recorded as a 30% solution in deuteriochloroform with sweep width 500 Hz; trace A sweep width 100 Hz, sweep offset 210 Hz.

7. $C_{11}H_{16}O$; nmr spectrum, Fig. 5.34; displays broad absorption in the 3300 cm^{-1} (3.0 μ) region.

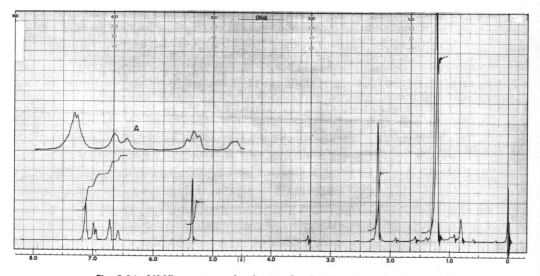

Fig. 5.34. NMR spectrum of unknown for Problem 7 recorded as a 30% solution in deuteriochloroform with sweep width 500 Hz; trace A sweep width 100 Hz, sweep offset 340 Hz.

8. $C_5H_{12}O_2$; nmr spectrum, Fig. 5.35; displays broad absorption in the 3300 cm^{-1} (3.0 μ) region.

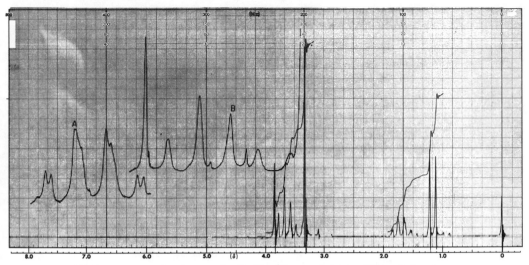

Fig. 5.35. NMR spectrum of unknown for Problem 8 recorded as a 30% solution in deuteriochloroform with sweep width 500 Hz; trace A sweep width 100 Hz, sweep offset 20 Hz; trace B sweep width 100 Hz, sweep offset 160 Hz.

9. $C_9H_{14}O$; nmr spectrum, Fig. 5.36; absorbs at 1682 cm^{-1} (5.94 μ) and 1626 cm^{-1} (6.15 μ).

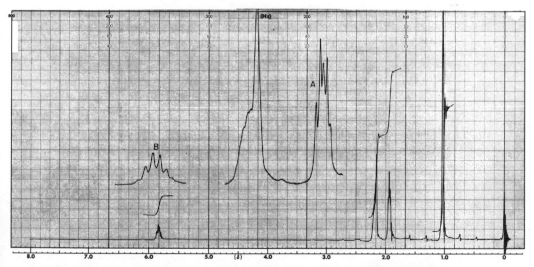

Fig. 5.36. NMR spectrum of unknown for Problem 9 recorded as a 30% solution in deuteriochloroform with sweep width 500 Hz; trace A sweep width 100 Hz, sweep offset 80 Hz; trace B sweep width 100 Hz, sweep offset 280 Hz.

10. $C_6H_{11}ClO$; nmr spectrum, Fig. 5.37; transparent in the 3300 cm^{-1} (3.0 μ) and 2000 to 1650 cm^{-1} (5 to 6 μ) regions.

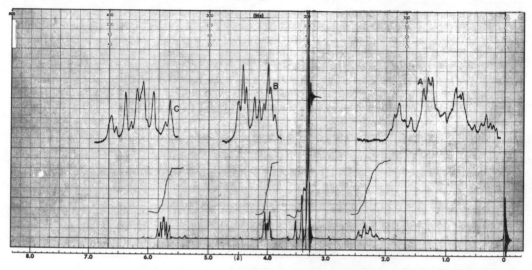

Fig. 5.37. NMR spectrum of unknown for Problem 10 recorded as a 30% solution in deuteriochloroform with sweep width 500 Hz; trace A sweep width 100 Hz, sweep offset 190 Hz; trace B sweep width 100 Hz, sweep offset 190 Hz; trace C sweep width 100 Hz, sweep offset 270 Hz.

5.17 NMR LITERATURE

The most recent comprehensive text on nuclear magnetic resonance is that by Emsley, Feeney, and Sutcliffe (ref. 2). Volume 1 discusses the theoretical aspects of nuclear magnetic resonance and the analysis of high-resolution spectra. Volume 2 contains discussions of correlations of resonance spectral parameters with molecular structure for various nuclear species. Both volumes contain a great many references to the original literature. Another comprehensive text is that by Pople, Schneider, and Bernstein (ref. 4), although this text is outdated by the recent advances in nmr spectroscopy with respect to structure-spectra correlations. The smaller texts by Bhacca and Williams (ref. 1), Jackman (ref. 3), and Roberts (ref. 5 and 6) are recommended for the occasional user of nmr.

The analysis of resonance spectra (the determination of chemical shifts and coupling constants) in complex spin systems is greatly facilitated by Wiberg and Nist's *The Interpretation of NMR Spectra* (ref. 7), in which line positions and intensities have been calculated as functions of the coupling constant and the difference in chemical shifts.

The spectra of individual compounds have been compiled in references 11, 12, 14, and 15. *NMR, EPR, and NQR Current Literature Service,* published periodically by Butterworth (ref. 13), provides a continuing survey and compilation of literature citations in various areas of nuclear magnetic resonance.

5.18   REFERENCES

GENERAL

1. Bhacca, N. S., and Williams, D. H., *Applications of NMR Spectroscopy in Organic Chemistry.* San Francisco: Holden-Day, Inc., 1964.

2. Emsley, J. W., Feeney, J., and Sutcliffe, L. H., *High Resolution Nuclear Magnetic Resonance Spectroscopy, Vol. 1 and 2.* New York: Pergamon Press, 1965.

3. Jackman, L. M., *Applications of Nuclear Magnetic Resonance Spectroscopy in Organic Chemistry.* New York: Pergamon Press, 1959.

4. Pople, J. A., Schneider, W. G., and Bernstein, H. J., *High Resolution Nuclear Magnetic Resonance.* New York: McGraw-Hill Book Company, 1959.

5. Roberts, J. D., *Nuclear Magnetic Resonance. Application to Organic Chemistry.* New York: McGraw-Hill Book Company, 1959.

6. Roberts, J. D., *An Introduction to the Analysis of Spin-Spin Splitting in High-Resolution Nuclear Magnetic Resonance Spectra.* New York: W. A. Benjamin, Inc., 1962.

7. Wiberg, K. B., and Nist, B. J., *The Interpretation of NMR Spectra.* New York: W. A. Benjamin, Inc., 1962.

8. Mathieson, D. W., ed., *Nuclear Magnetic Resonance for Organic Chemists.* New York: Academic Press, Inc., 1967.

9. Mislow, K., and Raban, M., "Stereochemical Relationships of Groups in Molecules," in *Topics in Stereochemistry,* E. L. Eliel and N. L. Allinger, eds., Vol. I, Chap. 1. New York: Interscience Publishers, Inc.

10. Emsley, J. W., Feeney, J., and Sutcliffe, L. H., *Progress in Nuclear Magnetic Resonance Spectroscopy,* Vol. 1–3. New York: Pergamon Press, 1966–1967.

COMPILATIONS OF NMR SPECTRA AND LITERATURE REFERENCES

11. *A Catalogue of the Nuclear Magnetic Resonance Spectra of Hydrogen in Hydrocarbons and Their Derivatives,* Humble Oil and Refining Co., Baytown, Tex.

12. *High Resolution NMR Spectra Catalogue,* Vol. 1 (1962) and Vol. 2 (1963). Varian Associates, Palo Alto, Calif.

13. *NMR, EPR and NQR Current Literature Service,* Butterworth Inc., Washington, D.C.

14. *Sadtler NMR Spectra,* Sadtler Research Laboratories, Philadelphia, Pa.

15. *Nuclear Magnetic Resonance Spectral Data.* American Petroleum Institute, Project No. 44, Chemical Thermodynamic Properties Center, Agricultural and Mechanical College of Texas, College Station, Tex.

16. Bovey, F. A., *NMR Data Tables for Organic Compounds, Vol. 1.* New York: Interscience Publishers, Inc., 1967.

17. Brügel, W., *Nuclear Magnetic Resonance Spectra and Chemical Structures.* New York: Academic Press, 1967.

6

Electron Paramagnetic Resonance

6.1   INTRODUCTION

Electron paramagnetic resonance (epr)[1] is similar to nuclear magnetic resonance in many respects. A free electron possesses a spin quantum number of $\frac{1}{2}$ and, hence, may exist in one of two states, $+\frac{1}{2}$ or $-\frac{1}{2}$, of equal energy (degenerate). As in nuclear magnetic resonance, this degeneracy of the energy states is removed upon application of an external magnetic field, and transitions from one state to the other can be induced by irradiation of the electron with the appropriate amount of energy. The energy required is given by Eq. (6.1)

$$E = h\nu = g\mu_0 H_0 \tag{6.1}$$

where g, the Landé g factor, is the proportionality constant [(magnetic moment)/(angular momentum)] with a value of 2.002319 for an unbound electron, H_0 is the applied field, and μ_0 is defined by Eq. (6.2).

$$\mu_0 = \frac{eh}{4\pi m_e c} \tag{6.2}$$

In Eq. (6.2), e is the electronic charge, m_e is the mass of the electron, and c is the speed of light. Typical energies involved in electron spin transitions are in the region of 1 calorie (cal). As in the case of nuclear magnetic resonance, the energy required to induce a spin transition is proportional to the field strength H_0. The field strength normally used is 3400 gauss, thus requiring an exciting frequency of 9.5 GHz (gigaHertz or 10^9 Hz).

The intensity of electron spin resonance lines depends on several factors, some of which are indicated in Eq. (6.3) which gives the ratio of the populations of the $-\frac{1}{2}/+\frac{1}{2}$ states.

$$\frac{n(-\frac{1}{2})}{n(+\frac{1}{2})} = \exp\left(\frac{g\mu_0 H_0}{kT}\right) \tag{6.3}$$

---

[1] Also referred to as electron spin resonance (esr).

The intensity of the resonance line increases with H_0 (or frequency) and the g factor (to be discussed further later), and is inversely proportional to the absolute temperature. In addition to the factors indicated in Eq. (6.3), the intensity of the resonance line is a function of the concentration of the paramagnetic species. The steady-state concentration of paramagnetic species generally required is in the 10^{-10} to 10^{-12} mole/liter range, although very sensitive spectrometers can detect paramagnetic species at as low a concentration as 10^{-14} mole/liter. The various experimental factors involved in determining the intensity of the resonance line places an experimental uncertainty of generally $\pm 50\%$ on the determination of the absolute concentration of the paramagnetic species.

The electron paramagnetic resonance lines are almost always recorded as a derivative spectrum (the first derivative of the equation describing the trace of the absorption line) for experimental reasons. The total absorption of energy compared with the total energy input is exceedingly small and, hence, difficult to determine accurately. To amplify the absorption signal, the magnetic field is modulated by adding a small oscillating magnetic field to the main applied field H_0 which alternately adds to and subtracts from the main applied field H_0. Figure 6.1a illustrates a typical absorption curve, and Fig. 6.1b illustrates the derivative curve of the absorption curve.

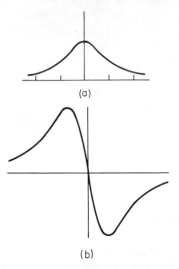

(a)

(b)

Fig. 6.1. Typical absorption (a) and derivative (b) curves.

6.2   g VALUES

As in nuclear magnetic resonance, electron paramagnetic resonance provides two basic kinds of data: the position of absorption, and the coupling constants of the unpaired, or unbound, electron with other nuclei possessing $I \neq 0$. The position of resonance is indicated as a g value which is a function of the ratio of the frequency to the H_0 at resonance conditions. For the free electron, this g value is the Landé g factor, or 2.002319. Variations of the g value from 2.002319 are due to magnetic interactions involving the orbital angular momentum of the unpaired electron; hence, it is a function of the chemical environment. For liquids, in which molecular collisions lead to randomization of the interactions with other nuclei, a specific paramagnetic species will possess a single g value (*isotropic*). However, in solids, a paramagnetic species may exhibit different g values along the x, y, and z axes depending on the symmetry of the molecule (*anisotropic*). (It should be noted that epr spectra of solids can be obtained; such is not true in high-resolution nmr). Organic radicals in solution all possess g values very close to the g value for the unbound electron. In general, hydrocarbon free radicals possess lower g values than hydrocarbon free radicals containing oxygen or nitrogen, which are, in turn, lower than radicals containing halogen or the peroxy ($-00\cdot$) group. Table 6.1 contains a few representative g values.

Table 6.1. g **Values for Organic Radicals**

Radical	g Value
Vinyl	2.00220
Anthracene cation	2.00249
Allyl	2.00254
Methyl	2.00255
Ethyl	2.00260
Naphthalene anion	2.00263
Anthracene anion	2.00266
Benzene anion	2.00276
trans-Stilbene anion	2.00285
Benzophenone anion	2.00359
1,4-Benzosemiquinone	2.00468
2-Chloro-benzosemiquinone	2.00486
2,3-Dichlorobenzosemiquinone	2.00512
2-Bromobenzosemiquinone	2.00507
Triphenylphosphine cation	2.00554
Cumylperoxy radical	2.0155

6.3 HYPERFINE SPLITTING

Although the unpaired electron may exist in only two spin states, $-\frac{1}{2}$ and $+\frac{1}{2}$, and is thus expected to give rise to only a single resonance line, these two spin states may interact with the magnetic moments of nuclei with which the unpaired electron may be wholly or partially associated. This interaction causes a further splitting of the resonance line into several lines (almost identical with spin-spin coupling in nuclear magnetic resonance). The transitions between these sublevels of the $-\frac{1}{2}$ and $+\frac{1}{2}$ spin states of the unpaired electron results in a number of resonance lines. This interaction is termed *hyperfine splitting* and is the distance between associated peaks of a subset, as measured in gauss, and is indicated as a_x, where x is the atom leading to the hyperfine splitting. The number of lines and the relative intensities of the lines are given by Eq. (6.4) and (6.5), in which n is the number of chemically identical nuclei interacting with the unpaired electron, and I is the spin quantum number of the nth atom (these equations are identical with the equations used to predict the number of lines and their relative intensities in nuclear magnetic resonance).

$$\text{No. of peaks} = (2In + 1) \qquad (6.4)$$

$$\text{Relative intensities} = (r + 1)^n \quad \text{(coefficients of the expanded form)} \qquad (6.5)$$

Two kinds of interactions lead to hyperfine splitting interactions; the first is referred to as a dipole-dipole interaction, and the second is referred to as Fermi contact interaction. Dipole-dipole hyperfine interactions occur only in solids and are caused by the presence of a nucleus ($I \neq 0$) near the unpaired electron, but not right at the electron (intermolecular interaction), and is highly anisotropic. In liquids and gases, the dipole-dipole hyperfine splitting is zero because of the time averaging of the fields due to other nuclei in the sample.

The Fermi contact interaction arises from a close, direct interaction (intramolecular interaction) of the unpaired electron with the nucleus and is orientation independent (*isotropic*). The value of the hyperfine splitting constant arising from the Fermi contact interaction is proportional to the free electron spin density associated with the interacting nucleus. In some cases, a further contribution arises owing to the electron spin density on the neighboring atoms.

The unpaired electron in a π-electron system will have finite spin density associated with each atom in the conjugated π-system; thus it will show hyperfine splitting with these atoms and substituents bonded to these atoms. Values of a_H for such π systems may be approximated by Eq. (6.6)

$$a_{\mathrm{H}} = Q\rho \qquad (6.6)$$

where Q is a constant with a value between 22.5 and 30 gauss, usually taken as 24 gauss, and ρ is the unpaired electron spin density on the carbon attached to the coupled hydrogen. Values of ρ may be calculated theoretically (see ref. 3) and compared with the experimentally observed values to gain insight as to the orbital occupied by the unpaired electron and as an aid in assignment of the structure of the paramagnetic species. The *alpha*-hydrogens on alkyl groups bonded to π systems give rise to hyperfine splitting even though they are not bonded directly to the π system. This interaction is attributed to a hyperconjugative interaction of the alkyl group with the

π system, and the values of a_H are also approximated by Eq. (6.6) using the same Q value as used for the π-electron system.

The *alpha* hydrogen ($-HC\cdot$) hyperfine splitting constants in alkyl radicals average 21–23 gauss (see ref. 4). The *beta-a_H* values are generally larger (see Table 6.2) and decrease with increasing substitution on the β-carbon. The values of the *beta-a_H* are also found to be rotationally dependent (refs. 4 and 7) and thus resemble the trends observed with the vicinal spin-spin coupling constants in nuclear magnetic resonance. When

Table 6.2. Selected Hyperfine Splitting Constants*

Radical	Hyperfine splitting constant (gauss)			
$CH_3 \cdot$	23.04			
$CH_3CH_2 \cdot$	22.38(α)	26.87(β)		
$CH_3CH_2CH_2 \cdot$	22.08(α)	33.2(β)	0.38(γ)	
$(CH_3)_2CH \cdot$	22.11(α)	24.68(β)		
$(CH_3)_3C \cdot$	—	22.72(β)		
Benzene $^+$	2.89			
Benzene $^-$	3.75			
Pyrazine $^+$	7.6(a_N)	3.26(a_H)		
Pyrazine $^-$	7.1(a_N)	2.6(a_H)		
	5.12(2)	4.45(3)	0.59(4)	0.79(CH_3)
	9.70(a_N)	3.36(2)	1.07(3)	4.03(4)
	14.35(a_N)	3.40(2)	1.05(3)	
	3.78(a_F)	1.62(2)	0.54(3)	3.43(CH_3)
	4.90(1)	1.83(2)		
	4.95(1)	1.87(2)		

*Values are selected from refs. 2 and 4. The $\cdot$ indicates a neutral free radical, $+$ represents a radical cation formed by removal of an electron, and $-$ represents a radical anion formed by the addition of an electron. The values in this table have been measured in a variety of different solvents. Some paramagnetic species show slightly different hyperfine splitting constants in different solvents.

progressing from the *beta* to the *gamma* carbon, a_H diminishes very rapidly, typical values of *gamma*-a_H being in the 0.4 to 1.1 gauss range.

Isotopic substitution of deuterium for hydrogen leads to a reduction of the hyperfine splitting constant according to Eq. (6.7) in which g_I are the g factors for hydrogen (5.5855) and deuterium (0.8574).

$$a_D = \frac{g_{I_D} \times a_H}{g_{I_H}} \tag{6.7}$$

As in nuclear magnetic resonance, deuterium substitution is often employed to reduce the complexity of a spectrum to facilitate interpretation.

The values of $a_{^{13}C}$ do not follow a simple $Q\rho$ relationship. Contributions to $a_{^{13}C}$ arise from unpaired electron spin density on the adjacent carbon atoms. Resonance lines arising from ^{13}C hyperfine splitting appear as very weak lines in the epr spectrum, similar to the ^{13}C satellites in nmr spectra. The $a_{^{14}N}$, however, appears to follow the $Q\rho$ relationship, the contributions from the spin density on adjacent atoms being negligible.

For more detailed discussions of the various factors involved in electron paramagnetic resonance, such as relaxation phenomena, see the references listed in Sec. 6.6.

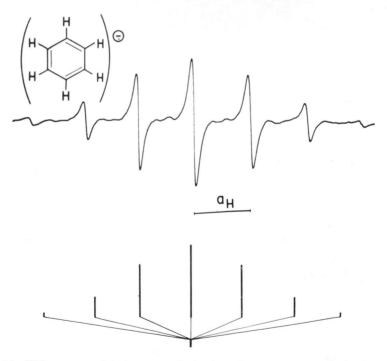

Fig. 6.2. EPR spectrum of the benzene radical anion. (From F. Gerson, *Hockauflösende ESR-Spektroskopie*, Verlag-Chemie, Weinheim, 1967.)

Figure 6.2 shows the epr spectrum of the benzene radical anion. Equations (6.4) and (6.5) predict a seven-line spectrum with relative intensities of 1:6:15:20:15:6:1, as illustrated below the spectrum. The hyperfine splitting constant is 3.75 gauss.

Figure 6.3 shows the epr spectrum of the naphthalene radical anion in which hyperfine splitting occurs due to four equivalent *alpha* hydrogens

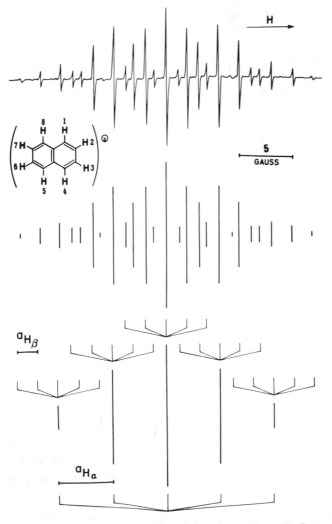

Fig. 6.3. EPR spectrum of the naphthalene radical anion. (From F. Gerson, *Hock-auflösende ESR-Spektroskopie*, Verlag-Chemie, Weinheim, 1967.)

(a_H = 4.95 gauss) and four equivalent *beta* hydrogens (a_H = 1.87 gauss). The derivation of the splitting pattern is illustrated below the spectrum.

Figure 6.4 shows the epr spectrum of the 1,4,5,8-tetraazanaphthalene in which hyperfine splitting occurs due to four equivalent nitrogen (^{14}N) atoms (I = 1) and four equivalent hydrogens ($I = \frac{1}{2}$).   The derivation of the splitting pattern is illustrated below the spectrum.

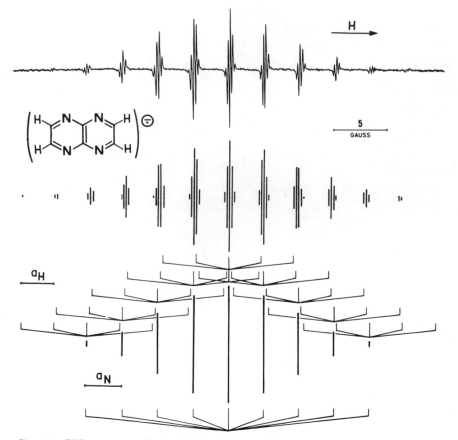

Fig. 6.4. EPR spectrum of the 1,4,5,8-tetraazanaphthalene radical anion.   (From F. Gerson, *Hochauflösende ESR-Spektroskopie*, Verlag-Chemie, Weinheim, 1967.)

6.5   APPLICATIONS
OF EPR SPECTROSCOPY
IN ORGANIC CHEMISTRY

Applications of epr spectroscopy in organic chemistry are somewhat limited by several factors.   The foremost limitation is the necessity of having a paramagnetic species (having an unpaired electron).   Such species are relatively rare in organic chemistry. A second problem is encountered in the generation of the radicals and in maintaining a steady-state concentration high enough to be detected. Radicals are generally prepared in one of four ways: (1) homolytic fission of a weak bond (for example, the thermally induced fission of the central carbon-carbon bond in hexaphenylethane), (2) addition of an electron to a substrate molecule (reduction) by reaction with a powerful electron donor (e.g., sodium or potassium) or by electrolytic means, (3) abstraction of an electron (oxidation) by electrolytic or chemical means, and (4) by disproportionation between a dianion and a neutral molecule (for example, olefin plus an olefin dianion to give an olefin mono-anion radical). Finally, the use of epr spectroscopy in structure determination is limited because structural information can be derived concerning only the portion of the molecule in which the spin density of the unpaired electron is delocalized.

As indicated earlier, the occurrence of paramagnetic species in organic chemistry is quite rare, and in those cases where such species are generated, the half-lives of the species are so short that samples cannot be prepared and handled as are nmr samples. Organic radicals are often formed as reactive intermediates in chemical reactions. To generate a sufficient steady-state concentration of radicals in the probe of the epr spectrometer, flow systems have been developed in which the reaction producing the radicals occurs in the probe. Any signal thus produced can be interpreted in terms of the structure of the paramagnetic species, employing the approach to hyperfine splitting outlined in the foregoing paragraphs. The lack of an epr signal, however, does not necessarily mean that the reaction does not proceed via radical intermediates, since the steady-state concentration of the radicals may have been below the detection limit of the spectrometer. The detection of radical intermediates in reactions is one of the most important applications of epr spectroscopy in organic chemistry; it has led to extensive revision of our concepts of certain reaction mechanisms.

Russell and co-workers (see ref. 8) have recently applied epr spectroscopy to problems of structure and conformation in organic chemistry. Treatment of compounds containing the —COCH$_2$— and —COCHOH— functional groups with potassium t-butoxide in dimethylsulfoxide in the presence of oxygen, of —COCO— with propiophenone enolate anion, or of —COCHBr— with potassium butoxide in dimethylsulfoxide results in the formation of *semidiones* (structure **1**). Hydrogens on the two carbon atoms attached to the semidione chromophore lead to hyperfine splitting. Analysis of the hyperfine splittings allows determination of the number of hydrogens

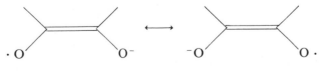

1

present, and the orientation of the hydrogens with respect to the nodal plane of the semidione chromophore (the reader is referred to the original article, ref. 9, for further details). In conformationally mobile systems, the epr spectra will be composites of the spectra of the two conformations; hence, both partial structures and conformations can be determined. (The epr time scale is of the order of 10^{-8} to 10^{-9} sec, $1/(\nu) = 2.8\ a_{\mathrm{H}}$ in MHz, compared with 1 to 10^{-2} sec for the nmr region, which is generally much greater than the frequency of interconversion of conformers.)

Further utilization of epr spectroscopy in organic structure determination will undoubtedly develop in the near future.

6.6   REFERENCES

1. Bersohn, M., and Baird, J. C., *An Introduction to Electron Paramagnetic Resonance.* New York: W. A. Benjamin, Inc., 1966.

2. Bowers, K. W., "Electron Spin Resonance of Radical Ions" in *Advances in Magnetic Resonance*, **1,** 317 (1965).

3. Carrington, A., "Electron-Spin Resonance Spectra of Aromatic Radicals and Radical Ions," *Quarterly Reviews*, **17,** 67 (1963).

4. Fessenden, R. W., and Shuler, R. H., *J. Chem. Phys.*, **39,** 2147 (1963).

5. Gerson, F., "Hochauflösende ESR Spektroskopie," Weinheim, Germany: Verlag-Chemie, 1967. (An English edition is in preparation: Wiley-Interscience, Inc., New York.)

6. Ingram, D. J. E., *Free Radicals as Studied by Electron Spin Resonance.* New York: Academic Press, Inc., 1958.

7. Pake, G. E., *Paramagnetic Resonance, An Introductory Monograph.* New York: W. A. Benjamin, Inc., 1962.

8. Russell, G. A., Review in *Science*, **161,** 423 (1968).

9. Talaty, E. R., and Russell, G. A., *J. Am. Chem. Soc.*, **87,** 4867 (1965).

7

Determination of Absolute

Stereochemistry

The previous chapters have outlined methods that provide useful information for the determination of the gross structure of a molecule. None of these techniques provides any information concerning the absolute stereochemistry of asymmetric centers in a molecule. The classical procedure for the determination of the absolute stereochemistry of a molecule involved the conversion to another molecule whose absolute stereochemistry was known. This procedure employs a sequence of chemical steps in which the asymmetric center was left undisturbed, or, in which the stereochemistry of each step was unambiguously known. Such correlations often involve a number of chemical steps and are quite tedious.

The use of only the sign of the rotation of plane polarized light at the sodium *D* line as an indication of the absolute stereochemistry is not valid, as we shall see in subsequent paragraphs. The comparison of changes in optical rotation when two similarly constituted chromophores are subjected to identical chemical transformations (known as the "rule of shift") does allow designation of the absolute stereochemistry. However, this procedure requires a series of model compounds of known absolute stereochemistry for the comparison.

Unambiguous assignment of absolute stereochemistry is possible employing X-ray or electron diffraction techniques. However, the equipment required for the measurements is not readily available to organic chemists, and the analysis of the diffraction patterns is extremely tedious, although the recent application of high-speed computers greatly facilitates such procedures. For these reasons, these techniques will not be discussed here.

Methods have been devised to calculate the sign and/or the magnitude of the rotation of plane polarized light; however, many of them do not appear to be generally applicable. Two recent approaches have made it possible to predict, with a high degree of accuracy, the absolute stereo-chemistry of an asymmetric center or molecule. Both of these approaches are limited in their scope, but they complement each other to a fair extent.

7.2 OPTICAL ROTATORY DISPERSION

A beam of plane polarized light may be considered to be composed of a left- and a right-handed circularly polarized component. The vectors representing these two components are illustrated in Fig. 7.1a (inner circle). The corresponding vector sum is also illustrated in Fig. 7.1a (outer circle). As

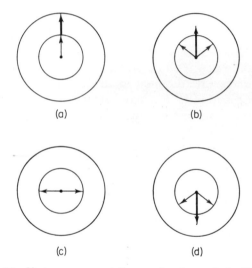

(a)   (b)

(c)   (d)

Fig. 7.1. Vector component diagram for plane polarized light.

the individual component vectors rotate, the vector sum traces a straight line, Fig. 7.1b, c, and d, characteristic of the plane polarized light beam. If the plane polarized light beam impinges on an optically active center, such as **1** in which the atom or group polarizabilities decrease in the order $A > B > C > D$, the velocity at which the two component vectors pass

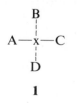

1

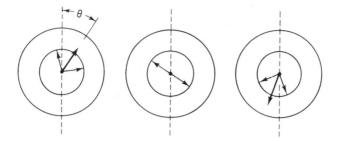

Fig. 7.2. Vector component diagram of the plane polarized light after passing through an optically active medium (the horizontal dotted line represents the original plane of polarization).

through the medium will be different. The property of such a medium is termed *circular birefringence.* (A physical analogy for this phenomenon is the attempted passage of a left- and a right-hand threaded bolt into a left-hand threaded nut; the former proceeds with little difficulty whereas the latter occurs with great difficulty.) The result will be a phase displacement by an angle θ from the original plane of polarization (see Fig. 7.2). The angle of rotation is given by Eq. (7.1) in which n_r and n_l are the refractive indices of the medium with respect to the right- and left-handed vector components, respectively, and λ_{vac} is the vacuum wavelength of the light employed.

$$\alpha = \frac{\pi}{\lambda_{vac}}(n_r - n_l) \tag{7.1}$$

If the optically active medium is also capable of absorbing light, an unequal absorption of the two component vectors will occur (*circular dichroism*). The resultant vector sum will trace an elongated ellipse (see Fig. 7.3), and the emergent light is said to be elliptically polarized. (The difference in the extinction coefficients k_r and k_l is of the order of 10^{-2} to 10^{-3}.)

The measurement of the optical rotation as a function of the wavelength is termed *optical rotatory dispersion* (ORD) and the measurement of the

Fig. 7.3. Elliptical polarization of light due to unequal refraction and absorption of the two component vectors.

unequal absorption of right and left circularly polarized light is termed
circular dichroism (CD). The two phenomena are intimately related; however, ORD and CD measurements are now commonplace. The same qualitative information is available from either type of measurement. Although
only ORD curves are discussed in this text, the application of CD data
parallels the application of ORD data.

Figure 7.4 illustrates two of many possible types of optical rotatory dis-
person curves. Optical rotatory dispersion curves may exhibit maxima,

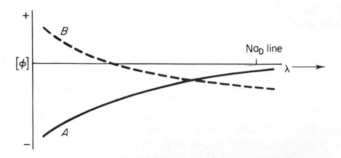

Fig. 7.4. Examples of typical optical rotatory dispersion curves: Curve A, normal
negative curve; and Curve B, plain positive curve.

minima (extrema), or inflections. Curve *A* (—) of Fig. 7.4 is referred to as
a normal negative dispersion curve. The curve is assigned a negative sign
because the molecular rotation increases in the negative direction in going
to shorter wavelengths. Curve *B* (---) in Fig. 7.4 is referred to as a plain
positive dispersion curve, even though it starts out with negative rotations
at longer wavelengths. (Actually, both curves are plain dispersion curves;
however, the term *normal* implies no extrema, inflections, or crossings of
the zero rotation line.)

Figure 7.5 illustrates two complex anomalous dispersion curves. Curve *A*
(—) is a positive, single, Cotton effect curve. The various points on the
curve are designated as peak (*P*) (instead of maximum, to avoid confusion
with the ultraviolet maximum which occurs at a different wavelength) and
trough (*T*). The vertical distance (*a*) between the peak and trough is the
amplitude, and the horizontal distance (*b*) is the breadth. A positive Cotton
effect curve is one in which the peak occurs at longer wavelength than does
a trough. If an optical rotatory dispersion curve displays several peaks,
troughs, or inflections, it is referred to as a multiple Cotton effect curve (see
curve *B* (---) in Fig. 7.5).

The optical rotatory dispersion curves can be described in terms of a
mathematical expression [Eq. (7.2)] known as the Drude equation:

$$[\phi] = \frac{K_1}{\lambda^2 - \lambda_0^2} + \frac{K_2}{\lambda^2 - \lambda_1^2} + \cdots + \frac{K_n}{\lambda^2 - \lambda_{n-1}^2} \qquad (7.2)$$

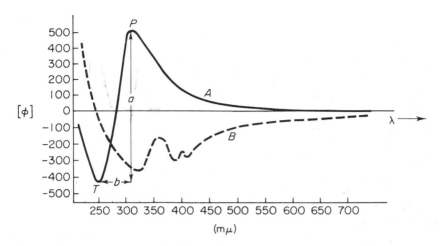

Fig. 7.5. Examples of typical anomalous dispersion curves.

where K_1, K_2, ..., K_n are constants for the absorbing chromophore, λ is the wavelength of the measurement, and λ_0, λ_1, ..., λ_{n-1} are the positions of absorption maxima of the compound in the ultraviolet region. The plain dispersion curves are represented by a single-term Drude equation in which λ_0 lies in the low end of the ultraviolet region. Multiple Cotton effect curves are represented by multiple-term Drude equations. In curve A of Fig. 7.5, the wavelength at the zero rotation point corresponds to the λ_{max} of the chromophore in the ultraviolet.

The description of an optical rotatory dispersion curve in the literature requires the reporting of sufficient information so that a curve can be reconstructed which accurately resembles the original curve. Also included are the solvent, concentration, and the temperature at which the measurements were taken. The specific or molar rotations at the longest wavelength recorded (usually ~ 700 mμ), at 589 mμ (the Na$_D$ line), and at peaks, troughs, and inflections should be recorded. The following is a typical example:

Compound A (of Fig. 7.5); ORD in methanol (C, 0.10),
$25°$, $[\alpha]_{700} + 30°$, $[\alpha]_{589} + 40°$, $[\alpha]_{320} + 450°$,
$[\alpha]_{300} 0°$, $[\alpha]_{275} - 390°$, $[\alpha]_{250} - 60°$.

The specific rotation is given by Eq. (7.3), where α is the rotation in degrees per centimeter path length, and C is the concentration in grams per cubic centimeter.

$$[\alpha] = \frac{\alpha \times 1800}{C \times \pi} \tag{7.3}$$

The molar rotation, $[\phi]$, is given by Eq. (7.4), in which M is the molecular weight of the compound.

$$[\phi] = [\alpha] \frac{M}{100} \tag{7.4}$$

In order to gain the maximum information in the accessible region of 700 to 215 mμ, the absorbing chromophore should have its maximum in this region; however, the extinction coefficient of the absorbing chromophore should be quite low, preferably below $\epsilon = 100$. The chromophore which has proven to be of greatest utility is the carbonyl group of ketones and alde-hydes giving rise to the $n \longrightarrow \pi^*$ band. Many other functional groups have also been used. These include the thione (C=S), disulfide, diselenide, nitro, nitrite, α-chloro-, α-bromo-, and α-azidoacids, polypeptides, and proteins which give rise to anomalous dispersion curves, and esters, lactones, alco-hols, amines, simple olefins, and hydrocarbons which give rise to plain dis-persion curves.

Optical rotatory dispersion data may be used in either qualitative or quantitative manners. Qualitative correlations relate the *sign* of the disper-sion curve with the absolute stereochemistry of molecules with similar gross structure. Again it is noted that model compounds are required, but Djerassi and co-workers have measured the optical rotatory dispersion curves for a great number of cyclic ketones which may be used as model systems. Quantitatively, the positions of the carbonyl groups, and a few other func-tional groups, in a complex molecule, such as the steroid or polycyclic ter-penoid series, have been correlated as a function of the sign, the amplitude and breadth, or the absolute $[\phi]$ values at various characteristic points on the dispersion curve. An empirical set of rules, the *octant rule*, has been developed to predict the absolute stereochemistry of a molecule based on the sign of its dispersion curve.

7.3   THE OCTANT RULE

The *octant rule* is a set of empirical rules which allows the assignment of the absolute stereochemistry about a five-, six-, or seven-membered ring ketone from the sign of the Cotton effect curve. The space about the carbonyl group is divided into octants about the *X*, *Y*, and *Z* axes as illustrated in Fig. 7.6. Plane *A* intersects the carbonyl oxygen atom and carbons 1 and 4. Plane *B* is perpendicular to plane *A* and intersects the oxygen and carbons 1, 2, and 6. Plane *C* is perpendicular to both planes *A* and *B*, and it intersects the carbon-oxygen double bond at the midpoint. The three planes thus divide the space about the carbonyl group into octants.

The four octants lying to the right of plane *C* in Fig. 7.6 are considered the rear octants; those to the left of plane *C* are the forward octants. *The*

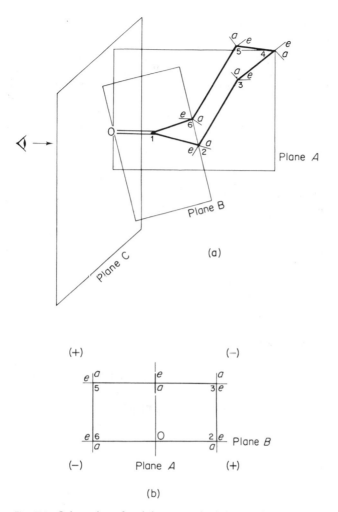

(a)

(b)

Fig. 7.6. Orientation of cyclohexanone in the octant framework.

octant rule states that the contribution by a substituent to the sign of the optical rotatory dispersion curve will be the sign product of its coordinate position.

Figure 7.6b illustrates the view, as one looks down the Y axis, of the carbonyl group. The bonds from each carbon atom are designated a for axial and e for equatorial. *The octant rule also states that any group lying in the A and B planes (this includes the 2e, 4a, 4e, and 6e bonds) will make no significant contribution to the sign of the dispersion curve.*

For the majority of the substituted cycloalkanones, the substituents will appear only in the four rear octants. Groups in these four octants will contribute to the sign of the dispersion curve, as indicated by the plus and minus

signs in Fig. 7.6b. The signs of the four forward octants will be just the
opposite.

   To illustrate the use of the octant rule, we shall apply it to (+)-*trans*-10-
methyl-2-decalone (**2**) which is known to have the absolute stereochemistry
as shown, and which exhibits a positive Cotton effect curve. The principal
challenge to the student is to correctly orient the three-dimensional representation of any molecule for incorporation into the octant framework. The
use of molecular models (Framework or Dreiding models) is highly recommended. The correct spatial orientation of **2** is shown in Fig. 7.7a, and the
placement in the octant framework is shown in Fig. 7.7b. Groups which are
similar in structure and equidistant from the carbonyl contribute equally to
the sign of the dispersion curve; the magnitude of the contribution decreases
as the distance between the group and the carbonyl group increases.[1] There-

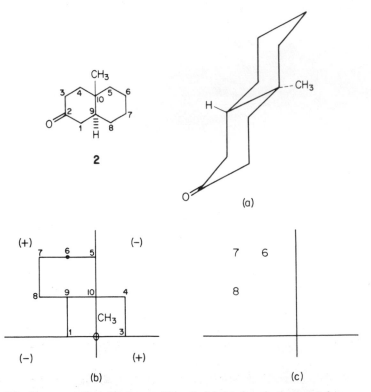

Fig. 7.7. Representation of (+)-*trans*-10-methyl-2-decalone in the octant framework.

---

[1]The α-axial-haloketone rule is a special application of the octant rule. α-Halocyclo-
hexanones exist primarily in an axial conformation in contrast to other α-substituted cyclohexanones due to a lesser unfavorable dipole-dipole interaction; hence they give rise to relatively large rotations, whereas other α-substituted cyclohexanones give rise to no perceptible
rotation since the substituent resides in the *B* plane.

fore, in Fig. 7.7b, the 4 and 9 carbons effectively cancel one another, leaving only carbon atoms 6, 7, and 8 to contribute to the sign of the dispersion curve, all of which give positive contributions in agreement with the observed sign. The representation of Fig. 7.7b can be abbreviated as shown in Fig. 7.7c, in which only the numbers of the contributing groups are shown. Groups appearing in the rear octants are represented in boldface numerals, while groups appearing in the forward octants are represented in italicized numerals.

An example of a molecule which possesses groups extending into the forward octants is illustrated with **3** in Fig. 7.8. The three-dimensional drawing, Fig. 7.8a, produces an octant representation as shown in Fig. 7.8b and c.

Conformationally mobile systems (e.g., a monosubstituted cyclohexanone in which the substituent is relatively small) provide added complications

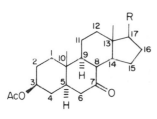

3

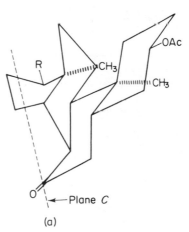

(a)

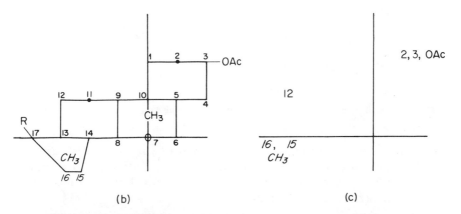

(b)   (c)

Fig. 7.8. Example of a molecule with substituent groups in the forward octants.

in that the substituent may be drawn as axial or equatorial. Normally, the group will prefer to be in the equatorial position. Such a situation can be illustrated with 3-methylcyclohexanone which can be represented by the two conformers, **4a** and **4b** in Fig. 7.9. The conformer with the equatorial

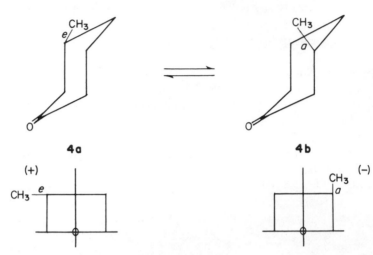

Fig. 7.9. Illustration of the application of the octant rule to a conformationally mobile system.

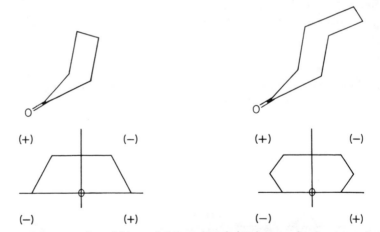

Fig. 7.10. Representation of five- and six-membered ring ketones in the octant framework.

methyl group will contribute more to the sign of the dispersion curve than will the axial conformer. Before the successful application of the octant rule can be made to polysubstituted systems, a decision must be made about which is the preferred conformation.[2]

---

[2] See: Eliel, E. L., Allinger, N. L., Angyal, S. J., and Morrison, G. A., *Conformational Analysis.* New York: Interscience Publishers, 1966.

The octant rule may also be applied to five- and seven-membered ring ketones, although the five-membered systems present conformational problems. These two ring systems are spatially oriented as shown in Fig. 7.10.

7.4 CORRELATIONS OF ORD DATA IN OTHER SYSTEMS

The application of the octant rule to heterocyclic ketones, in which the heteroatom is in the 3- or 5-position of the six-membered ring, has not been accomplished. Difficulties arise in assessing the contribution produced by the heteroatom vs. the opposite ring carbon atom and any other substituents which may be present. Additional correlations are required before applications to such systems can be safely made.

A preliminary report by Jennings, Klyne, and Scopes[3] on the correlation of ORD data of six-membered lactones (the heteroatom now appearing in the 2-position) with their absolute stereochemistry has appeared. The correlation is termed the *lactone sector rule* and is somewhat more complex to apply than the octant rule.

A modified octant rule has been proposed for cyclic α,β-unsaturated ketones and their derivatives.[4] The absolute stereochemistry of α-amino-acids has been correlated with the sign of the dispersion curve,[5] with **L**-α-aminoacids producing a steeply descending negative curve, and with **D**-α-aminoacids producing a steeply ascending positive curve. The absolute stereochemistry of sulfoxides and sulfinate esters has also been correlated with the observed Cotton effects.[6]

Optical rotatory dispersion data may also be used to predict the absolute skew, or chirality, of a cyclic cisoid diene.[7,8] If the skewed diene describes a right-hand helix, a positive dispersion curve will be obtained; if a left-hand helix is described, a negative dispersion curve will be obtained. Right- and left-handed helices are illustrated in Figs. 7.11a and 7.11b, respectively; the right-hand helix progresses "downhill" around the diene in a clockwise fashion and counterclockwise in the left-hand helix. Since the isopropyl prefers the pseudoequatorial conformation, Fig. 7.11a, a positive dispersion curve is obtained.

---

[3] Jennings, J. P., Klyne, W., and Scopes, P. M., *Proc. Chem. Soc.* (London), **412** (1964).

[4] Snatzke, G., *Tetrahedron*, **21**, 413, 421, 439 (1965).

[5] Craig, J. C., and Roy, S. K., *Tetrahedron*, **21**, 391 (1965).

[6] Mislow, K., Green, M. M., Lour, P., Melillo, J. T., Simmons, T., and Ternay, A. L., Jr., *J. Am. Chem. Soc.*, **87**, 1958 (1965).

[7] Burgstahler, A. W., Ziffer, H., and Weiss, U., *Ibid.*, **83**, 4660 (1961).

[8] Moscowitz, A., Charney, E., Weiss, U., and Ziffer, H., *Ibid.*, **83**, 4661 (1961).

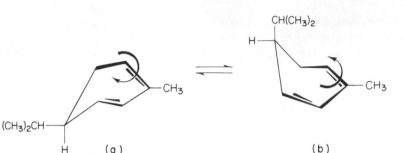

Right – hand helix
+ dispersion curve

Left – hand helix
– dispersion curve

Fig. 7.11. Illustration of the application of optical rotatory dispersion data with skewed dienes.

7.5   BREWSTER'S RULES

The octant rule cannot be applied to noncyclic compounds owing to the flexibility of the acyclic chain producing a great many molecular conformations. Brewster[9] has devised an empirical correlation which, not only predicts the sign of rotation at the sodium D line, but also predicts the magnitude of the rotation fairly closely. An asymmetric atom in the absolute configuration **5** is dextrorotatory when the polarizabilities of the substituent attachment atoms decrease in the order $A > B > C > D$. Table 7.1 lists the refractivities (R_D) of a number of attachment atoms.

Table 7.1. Rotational Rank of Common Substituents*

Substituent (A)	R_D of attachment atom (A)	Conformational rotatory powers† $k(C—H)(A—H)$
I	13.954	268
Br	8.741	192
SH	7.729	174
Cl	5.844	139
CN	3.580 (5.459)‡	87 (131)‡
C_6H_5	3.379 (6.757)‡	82 (158)‡
CO_2H	3.379 (4.680)‡	82 (114)‡
CH_3	2.591	60
NH_2	2.382	53
OH	1.518	23
H	1.028	0
D	1.004	—
F	0.81	−10

*Taken from the reference in footnote 9.
†Brewster's calculated values.
‡Value for the entire two-atom unsaturated unit ($C{\equiv}N$, $C{=}C$, $C{=}O$).

---

[9] Brewster, J. H., *J. Am. Chem. Soc.*, **81**, 5475, 5483, 5493 (1959).

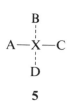

5

If we are dealing with a simple substituted methane, the prediction of the sign of rotation is relatively simple. However, if the substituents *A*, *B*, *C*, or *D* are complex side chains, the approach becomes more complicated.

In a complex molecule, many molecular conformations which contribute to the rotatory power of the molecule become possible. If we take a four-atom segment, for example, a disubstituted ethane, three conformations are possible, of which only one is illustrated below in **6**. The contribution to the sign of rotation between the substituents is indicated by the + and − signs in **6**.

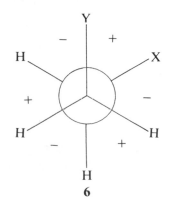

6

The rotatory power of each conformer is the sum of the individual inter-actions, for example, in **6**.

$$[M]_6 = k(XY - XH + HH - HH + HH - HY)$$
$$= k(XY - XH - HY + HH)$$
$$[M]_6 = k(X - H)(Y - H)$$

The reduced expression $k(X - H)(Y - H)$ is a constant which is taken from Table 7.1. The total rotatory power of a molecule is the summed average of all allowed molecular conformations.

In molecules containing more than a four-atom segment as illustrated above, we must consider all possible molecular conformations formed by combination of the various bond conformations. There are certain rules which must be followed in considering the various molecular conformations.

1. Only conformations of energy minima will be considered (staggered conformations).

2. The five-atom sequence 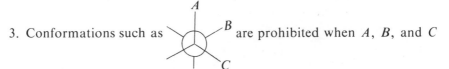 is prohibited when both A and A'
are larger than hydrogen.

3. Conformations such as ![conformation diagram with A, B, C substituents] are prohibited when A, B, and C
are all larger than hydrogen.

4. All allowed molecular conformations are considered equally probable.

We shall illustrate the application of Brewster's rules by calculating the
sign and magnitude of rotation of 2-pentanol of absolute stereochemistry **7**.
We have two four-atom sequences for which the bond conformations must
be considered. The bond conformations will be arbitrarily designated by the
central bond atom numbers and a letter as illustrated in Fig. 7.12. Con-

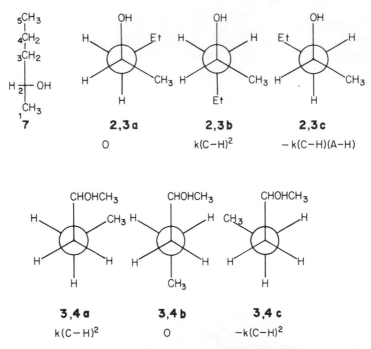

Fig. 7.12. Bond conformations for 2-pentanol 7.

formation **2,3a** violates rule 3 and is not allowed. Thus we have only five
allowed bond conformation combinations to consider. If we choose the
molecular conformation formed by combining bond conformations **2,3b**
and **3,4a**, and then sum their rotatory contributions (given below each

conformation number), we get

$$M_{2,3-b} + M_{3,4-a} = k(C—H)^2 + k(C—H)^2 = 2k(C—H)^2$$

Table 7.2 lists the rotatory contributions for each molecular conformation obtained in this manner [(A-H) indicates oxygen-hydrogen interaction]. The reader is advised to use molecular models to check for violations of rule 2.

Table 7.2. Contributions of Molecular Conformations to the Rotatory Power of 2-Pentanol

Bond conformations	Contribution
2,3-b + 3,4-a	$+2k(C—H)^2$
2,3-b + 3,4-b	$+k(C—H)^2$
2,3-b + 3,4-c	not allowed (rule 2)
2,3-c + 3,4-a	not allowed (rule 2)
2,3-c + 3,4-b	$-k(C—H)(A—H)$
2,3-c + 3,4-c	$-k(C—H)(A—H) - k(C—H)^2$

Sum $+2k(C—H)^2 - 2k(C—H)(A—H)$

$$[M] = \frac{+2k(C—H)^2 - 2k(C—H)(A—H)}{4} = \frac{2 \times 60° - 2 \times 23°}{4} = +18.5°$$

It must be kept in mind that assuming rule 4 to hold may not always be valid. Certainly, conformation populations vary greatly with the size of the attached groups (thus rules 1, 2, and 3), and serious deviations, even a change in sign of rotation, may occur in certain cases.

Brewster has also extended this approach to substituted five- and six-membered rings and to cyclic olefins. Available space does not allow a complete discussion of these approaches, and the reader is referred to the various literature references in Sec. 7.6.

Brewster's rules cannot be applied in cases in which functional groups enter into intramolecular complexing, for example, intramolecular hydrogen bonding.

7.6   REFERENCES

GENERAL

1. Crabbé, P., *Optical Rotatory Dispersion and Circular Dichroism in Organic Chemistry*. San Francisco: Holden-Day, Inc., 1965.

2. Djerassi, C., *Optical Rotatory Dispersion*. New York: McGraw-Hill Book Company, 1960.

3. Eliel, E. L., *Stereochemistry of Carbon Compounds*. New York: McGraw-Hill Book Company, 1962.

4. Velluz, L., Legrand, M., Grosjean, M., *Optical Circular Dichroism*. New York: Academic Press Inc., 1965.

8

Mass Spectrometry

8.1   INTRODUCTION

Most of the spectroscopic and physical methods employed by the chemist in structure determination are concerned only with the physics of molecules; mass spectroscopy deals with both the chemistry and the physics of molecules, particularly with gaseous ions. In conventional mass spectrometry the ions of interest are positively charged ions. The mass spectrometer has three functions:

1. To produce ions from the molecules under investigation.
2. To separate these ions according to their mass-to-charge ratio.
3. To measure the relative abundances of each ion.

Mass spectrometry is the latest method to join the array of spectroscopic methods at the disposal of the organic chemist. Interestingly, the demonstration of the basic principles of mass spectrometry preceded that of most of the other physical methods currently used for structure determination. As early as 1898, Wein showed that positive rays could be deflected by means of electric and magnetic fields; in 1912, J. J. Thompson recorded the first mass spectra of simple low molecular weight molecules. The earliest of the prototypes of today's instruments were the mass spectrometer of Dempster (1918) and the mass spectrograph of Aston (1918). By the early 1940's, instruments suitable for the examination of the spectra of organic molecules of moderate molecular weight were commercially available. The earliest extensive studies of organic systems were made by the petroleum industry on hydrocarbons. Although mass spectrometry was used for the quantitative analysis of hydrocarbon mixtures, there seemed to be little, if any, recognizable relationship between structure and spectra. This situation led to a dormant period in the utility of mass spectrometry for studies of organic molecules. In the late

1950's, Beynon, Biemann, and McLafferty clearly demonstrated the role of functional groups in directing fragmentation, and the power of mass spectrometry for organic structure determination began to develop. At that time the technology employed was essentially that available since 1940. Today, mass spectrometry has achieved status as one of the primary spectroscopic methods to which a chemist faced with a structural problem turns; great advantage is found in the extensive structural information which can be obtained from submilligram or even submicrogram quantities of material.

8.2 INSTRUMENTATION

A rather large variety of commercial mass spectrometers is currently available. There is considerable variation in the mechanics by which the various mass spectrometers accomplish their tasks of production, separation, and measurement of ions. It is entirely beyond the scope and purpose of this book to provide the reader with a detailed discussion of instrumentation. However, so that the reader may at least have general ideas of the principles by which most mass spectrometers operate, representative examples will be discussed briefly.

Throughout this text the term *low resolution* will refer to instruments capable of distinguishing only between ions of differing nominal mass, i.e., ions whose mass-to-charge ratios differ by one mass unit. *High resolution* will refer to spectrometers capable of distinguishing between ions whose masses differ in the third decimal place or less, for example, $C_2H_6^+$, CH_2O^+, and CH_4N^+ with masses (based on carbon 12.00000) of 30.0469, 30.0105, and 30.0344, respectively.

8.2.1 *Magnetic Focusing Instruments*

A schematic diagram of a representative single-focusing, low-resolution mass spectrometer is shown in Fig. 8.1. Sufficient sample is introduced into the ionization chamber to produce a pressure of 10^{-5} mm Hg. The vapor in the ionization chamber is bombarded with an electron beam of variable energy (usually 50 to 70 electron volts (ev), the latter value being most commonly employed). A small percentage of the molecules are ionized, by electron impact, into positively charged ions which subsequently form fragment ions. Negative ions are also formed to a small extent. The small repeller potential between the back wall of the ionization chamber and the first accelerator plate attracts the negative ions to the back wall and discharges them. At the same time, this potential pushes the positive ions toward the accelerating region. There they are accelerated by a potential of approximately 2000 volts between plates which have a slit in them. The

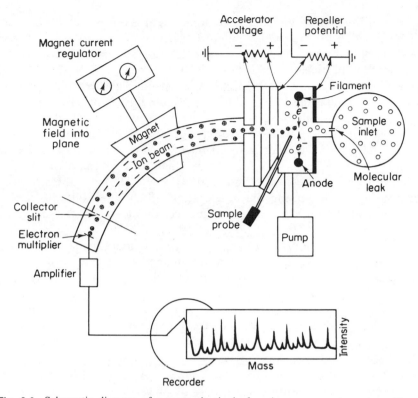

Fig. 8.1. Schematic diagram of a magnetic single focusing mass spectrometer with a 60° sector magnet.

ions are focused on an exit slit by means of subsidiary accelerating plates and slits. The positive ions are accelerated by this potential according to Eq. (8.1)

$$\tfrac{1}{2}mv^2 = eV \qquad (8.1)$$

where m is the mass of the ion, e the electronic charge, V the potential of the ion accelerating plate, and v the velocity of the particle. The accelerated ions then pass into the magnetic field H generated between the two poles of an electromagnet. In the magnetic field, the ions are deflected along a circular path according to Eq. (8.2)

$$r = \frac{mv}{eH} \qquad (8.2)$$

where r is the radius of the path. Elimination of v from the preceding two equations yields Eq. (8.3)

$$\frac{m}{e} = \frac{H^2 r^2}{2V} \qquad (8.3)$$

From this equation one can conclude that, at given values of H and V, only particles with a particular mass-to-charge ratio will arrive at the collector slit placed along the fixed path r. It can be further seen from the equation that particles with mass-to-charge ratios of $m/1$, $2m/2$, $3m/3$ would all follow the same path and be detected and recorded simultaneously. Fortunately, under the conditions usually employed in recording mass spectra of organic compounds, most of the particles are singly charged; ions of higher charge generally being produced in insignificant quantities. Equation (8.3) also indicates that a spectrum of ions could be obtained at the collector slit by variation of either the magnetic field strength or the accelerating voltage. In instruments with 180° magnetic arc path, the scanning is accomplished by decreasing the accelerating voltage at a suitable rate. In the 60° and 90° arc path instruments, the accelerating voltage is fixed and the magnetic field is varied.

The collector assembly consists of a series of collimating slits and an ion current detector. Modern electron multipliers are capable of detecting single ions. The current which is amplified and measured is directly related to the abundance of the ions at the mass being examined.

Two necessary features of an adequate recording system for mass spectrometry are (1) a fast response time and (2) an ability to record accurately peaks of widely varying intensity on the same spectrum. Although fast-response pen-and-ink recorders with manual or automatic attenuation are occasionally employed, the most practical solution is afforded by a multi-trace recording oscillograph. Ultraviolet light beams from an oscilloscope are played upon light-sensitive paper, simultaneously producing five tracings of intensity ratio $1:\frac{1}{3}, \frac{1}{10}, \frac{1}{30}$, and $\frac{1}{100}$ and permitting intensity variations up to 50,000 times. The recording system is very fast; typically, a complete spectrum may be recorded in a few seconds or less. The use of ultraviolet light and photosensitive paper makes any developing process unnecessary. A third recording system which is occasionally used is a *digitizer*, which prints the spectrum in the form of mass numbers and relative intensities. This system has the disadvantage that it is difficult for the chemist to grasp the important features of the spectrum at a glance from the examination of a tape of numbers.

The high-resolution, double-focusing instruments (Figs. 8.2 and 8.3) incorporate most of the principles and methods, with additional refinements, outlined in the foregoing discussion of low-resolution instruments. It will be seen from Fig. 8.3 that, at a constant magnetic field H, any spread in the magnitude of velocity (equivalent to the spread of V) will result in a spread in r for any given value of m/e. The peaks obtained in low-resolution mass spectrometry are indeed broadened by such a spread which is caused by the contributions of initial kinetic or thermal energy to the kinetic energy gained by the particle during acceleration in the ionization chamber.

A several-thousand-fold increase in the resolving power of an instrument can be achieved by elimination of this energy spread in the ion beam before

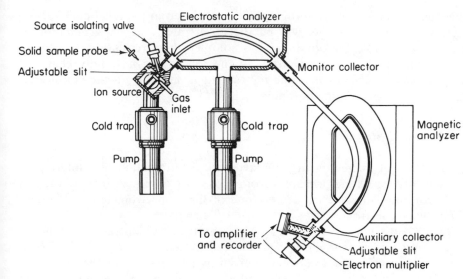

Fig. 8.2. Schematic diagram of a Nier-Johnson double focusing high resolution mass spectrometer showing the radial electric analyzer and the magnetic analyzer. (Courtesy Associated Electrical Industries, Ltd.)

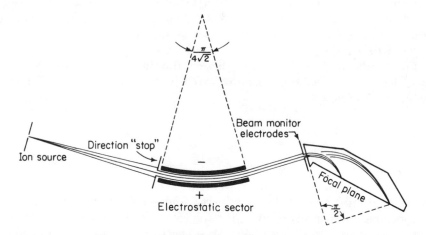

Fig. 8.3. Schematic diagram of a double focusing mass spectrometer showing Mattauch-Herzog geometry. (Courtesy Consolidated Electrodynamics Corporation.)

it enters the magnetic field. In the double-focusing instrument, ions are passed through a radial electrostatic field which sorts them into mono-energetic paths (velocity focusing) before they are passed into the magnetic field, where they are refocused according to their mass-to-charge ratio. According to the design of the instrument, the spectrum can be scanned by the conventional sweeping of the magnetic field or the accelerating potential, the ions being recorded as in the case of the low-resolution instrument as they come into focus on the collector (Fig. 8.2), or all ions may be simultaneously recorded on a photographic plate placed in the focal plane of the instrument (Fig. 8.3). As will be discussed later, this latter method is especially valuable for obtaining the complete high-resolution spectra of microsamples whose concentration in the ion source is rapidly changing and whose residence time in the instrument is short, as in the continuous monitoring of gas chromatography.

8.2.2 Other Instruments

Instruments of interest in organic structure determination but differing in fundamental operating principles from those just described include the ion cyclotron resonance spectrometers, time of flight spectrometers, and quadrapole mass spectrometers.

Briefly, the ion cyclotron resonance spectrometer operates as follows: When an ion of mass-to-charge ratio m/e is placed in a magnetic field H, it has a natural cyclotron frequency ω given by $\omega = He/m$. In the spectrometer, ions of differing m/e are detected by the absorption of energy from an oscillating electric field. The spectrum is obtained by sweeping the magnetic field so that the frequency ω of ions of differing mass are brought to the operating frequency (usually 77 to 770 kHz) of the instrument and energy is absorbed. An interesting feature of this type of instrument is that it may be set up for examination of negative ions by a simple reversal of electric and magnetic fields.

For details on the operation of these and other types of mass spectrometers, the reader is referred to more extensive books in the field of mass spectrometry (Sec. 8.9).

8.3   SAMPLE HANDLING

The sample must be introduced into the ionization chamber in the vapor state in sufficient quantity so that the vapor pressure exerted by the sample is in the range of 10^{-5} to 10^{-9} mm Hg. Most spectrometers have several sample introduction systems available; the method of choice is governed by the volatility of the sample.

For most organic compounds, direct insertion of the sample into the highly evacuated ionization chamber is advantageous. In a commonly employed system, the sample is placed in a small crucible on the end of a probe which is inserted into the ionization chamber through a vacuum lock system; if necessary to produce sufficient vapor pressure, the sample may be heated either by heating the entire ionization chamber or by heating elements associated with the probe. With probe heating, temperatures as high

as 1000° can be achieved; the method is limited only by the thermal stability
of the compound.

An extremely useful and versatile method for the direct introduction of
samples into the ionization chamber involves the direct hookup of a mass
spectrometer and a vapor phase chromatograph. Fortunately, the sensitivity
of mass spectrometer and gas chromatographic detectors are comparable;
the quantities of material usually eluted from a gas-liquid chromatography
(glc) column are much too small for analysis by infrared spectroscopy and
most other identification techniques. The eluent from a glc column can be
monitored by direct introduction into the ionization chamber. The dilution
of the sample by the carrier gas can be reduced by means of special sepa-
rators which pump off the carrier gas preferentially. Since the concentration
of any eluent is constantly and rapidly changing, this direct monitoring can
be achieved only with spectrometers having rapid recording systems. Photographic plate recordings of high-resolution spectra over a range from 28 to
800 m/e have been achieved from 1-μg samples having a residence time in
the ion source of only 1 second.

For gases and liquids or solids of high volatility, the *inlet system* is often
used for the introduction of the sample into the ion source. The inlet system
consists of a series of vacuum locks, a reservoir, a sensitive pressure-measur-
ing device, and a molecular leak into the ion source (Fig. 8.4). The volume
of the reservoir on commercial instruments varies from 1 to 5 liters; these
relatively large volumes require proportionately larger samples compared
with direct insertion. Sufficient sample is introduced into the reservoir to

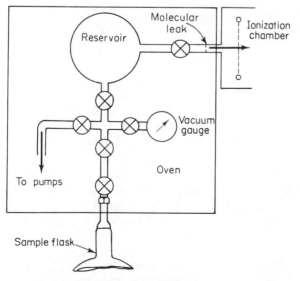

Fig. 8.4. Schematic diagram of an inlet system.

produce a pressure of 10^{-2} mm Hg. The entire system is in an oven which may be heated if necessary to produce sufficient pressure. The sample in the reservoir, under a pressure higher than that required in the ion source, is allowed to leak through a very small hole, producing a constant stream of the sample into the ionization region. The large volume of the reservoir minimizes depletion of the sample during the run; this is necessary if reproducible spectra are to be obtained at slow scan speeds.

Even with the variety of sample-handling techniques, mass spectrometry cannot be applied routinely to all compounds; the limiting factor is often thermal stability of the sample at the temperature necessary to produce the minimum pressure. It is often possible to convert thermally unstable compounds to more stable and/or more volatile derivatives, for example, acids to esters and alcohols to acetates or trimethyl silyl ethers.

8.4 PRODUCTION AND REACTION OF GASEOUS IONS

Ionization

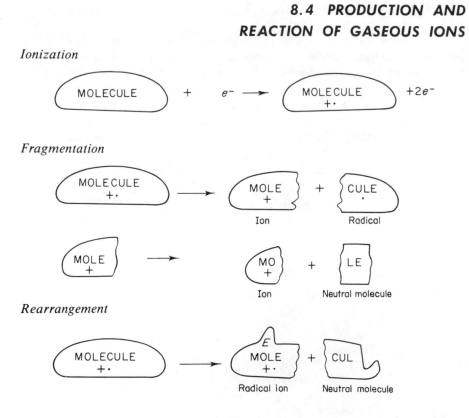

Fragmentation

Rearrangement

As implied in the introduction, mass spectrometry is largely concerned with chemical reactions. The initial chemical reaction is that between an electron and a molecule. A single electron is removed from the molecule by electron impact to produce a radical cation. For organic molecules, the electron beam energy required for this initial reaction is in the range of 9 to 15 electron volts (ev). The exact electron beam energy necessary to cause the ionization of a particular molecule is called its *ionization potential*; the electron beam energy necessary to cause a particular fragment to appear is the *appearance potential*. Mass spectra are usually obtained at an electron beam energy of 70 ev. Under these conditions considerable excess energy is imparted to the initially formed ion. Much of this excess energy is dissipated through fragmentations and rearrangements.

8.4.1 Determination of Molecular Weights and Formulas

The mass spectrum of an organic compound indicates that the compound is composed of several different molecular species of different combinations of naturally occurring isotopes. The masses that we should calculate for mass spectrometry should not be based on the chemical scale but should be the sum of the masses of the isotopes occurring in the species under consideration. For example, the molecular weight of methyl bromide is 94.95 on the chemical scale. This chemical scale weight is based on the weighted averages of the naturally occurring isotopes

$$
\begin{array}{ll}
C & 12.011 \\
H & 1.008 \\
Br & 79.909
\end{array}
$$

However, in the mass spectrum we, of course, do not see the average molecular species but rather the molecular species corresponding to each possible isotopic combination according to the following naturally occurring abundances:

$$
\begin{array}{lll}
^{12}C \;\; 98.89\% & ^{1}H \;\; 99.985\% & ^{79}Br \;\; 50.54\% \\
^{13}C \;\; 1.11\% & ^{2}H \;\; 0.015\% & ^{81}Br \;\; 49.46\%
\end{array}
$$

The low-resolution mass spectrum of methyl bromide in the region of the molecular ion is shown in Fig. 8.5.

In this text we shall refer to the radical cation produced by the removal of a single electron from a molecule composed of the lightest naturally occurring isotopes of the elements present in the compound as the *molecular ion*, symbolized by M. The terms M + 1, M + 2, etc., signify peaks at 1, 2, etc. nominal mass units higher than M.

If a molecular ion has sufficient stability to accord it a lifetime of approximately 10^{-5} seconds, it will be fully accelerated and recorded at its corresponding m/e value. Thus, for the large majority of compounds, mass

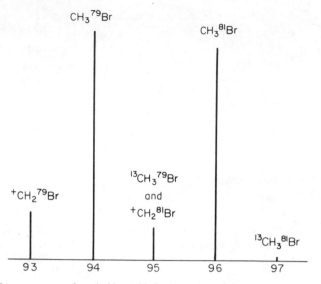

Fig. 8.5. Mass spectrum of methyl bromide in the region of the molecular ion. Small contributions by low abundance ^{2}H are not indicated. Contributions by species containing both ^{13}C and ^{2}H are negligible.

spectrometry provides an exact and unambiguous method for the determination of the molecular weight of a molecule. The molecular ion and related isotopic species correspond to the peaks of highest mass in the spectrum in the absence of collision processes (see below). However, in a number of cases, compounds give molecular ions of very low or negligible intensity, and care must be taken not to confuse peaks due to impurities or fragments with molecular ions. Reduction of the ionizing voltage leads to higher molecular ion intensities.

The stabilization of the positive charge in the molecular ion and the tendency toward fragmentation influence the intensity of the molecular ion peak. Biemann has suggested the following approximate order of decreasing stability of the molecular ion:

Aromatic compounds—conjugated olefins—alicyclic compounds—sulfides—straight chain hydrocarbons—mercaptans—ketones—amines—esters—ethers—carboxylic acids—branched hydrocarbons—alcohols.

Cyclic compounds naturally tend to give rise to correspondingly more intense molecular ion peaks than their acyclic analogs do, because in the cyclic structure, cleavage of one bond does not lead to the splitting off of a fragment of lower mass. It should be emphasized that one should always check to see that the fragmentation pattern exhibited in the rest of the spectrum can be accommodated by the suggested molecular ion and that the intensity of that peak is of the magnitude expected for the proposed structure. In some cases, even though a compound does not exhibit a molecular ion, it is possible to arrive at the molecular weight of the compound by the partial interpretation of the mass spectrum.   Another approach often used for determining molecular weights when compounds do not exhibit recognizable molecular ions is the formation of a derivative chosen to give a predicted mode of cleavage or a more stable molecular ion.

As an aid in determining whether a peak is due to a molecular ion or a fragment, the so-called *nitrogen rule*, which holds for all compounds containing carbon, hydrogen, oxygen, nitrogen, sulfur, and/or the halogens, as well as phosphorous, arsenic, and silicone, can be employed.   The rule states that a neutral molecule of even-numbered molecular weight must contain either no nitrogens or an even number of nitrogen atoms; an odd-numbered molecular weight compound must contain an odd number of nitrogen atoms. Simple fragments formed from even-numbered compounds will always have odd mass numbers, with the exception of rearrangement peaks which are usually formed by the elimination of a neutral molecule with the formation of a fragment of even mass number.

Since only a very small fraction of the molecules in the ionization chamber is actually ionized, those ions which are formed are present in a high population of un-ionized molecules. Ion-molecule collisions, which result in the abstraction of an atom or a group of atoms from the neutral molecule by the positive ion may occur resulting in a particle of mass higher than the expected molecular ion.   At the pressures usually employed in mass spectroscopy, the only significant reaction of this type is the abstraction of a hydrogen atom by the molecular ion resulting in a peak at M + 1. This M + 1 peak becomes particularly important in those cases (ethers, esters, amines, aminoesters, nitriles) in which the molecular ion is relatively unstable whereas the corresponding protonated species is quite stable. Since this M + 1 peak is formed in a bimolecular process, it can be recognized as such by the proportionality of its intensity to the square of the sample pressure. As the sample pressure is increased, the intensity of the M + 1 peaks will increase relative to the intensity of the other peaks in the spectrum. An M + 1 peak may also be recognized by its sensitivity to the repeller potential; an increase in the repeller potential will decrease the residence time of ions in the ionization chamber and thus decrease the likelihood of collision processes.

As indicated earlier in this chapter, with high-resolution mass spectrometry it is possible to measure masses to four decimal places. Table 8.1 gives the exact masses of the nuclides frequently found in organic compounds; masses of each atom differ from its mass number (nearest whole number) by its nuclear packing fraction.

It can be readily seen that each formulation of elements will have a unique mass associated with it. Measurement of the masses with sufficient accuracy allows one to determine the molecular formula.   An excellent example of this approach is found in the work of Djerassi on the alkaloid vobtusine. For this alkaloid, earlier work based on microanalysis and other

**Table 8.1. Nuclides of Importance in High-resolution
Mass Spectrometry of Organic Compounds**

Nuclide	Mass
H^1	1.0078252
C^{12}	12.0000000
C^{13}	13.003354
N^{14}	14.003074
O^{16}	15.99491415
F^{19}	18.998405
Si^{28}	27.976927
P^{31}	30.973763
S^{32}	31.972074
Cl^{35}	34.968855
Br^{79}	78.918348
I^{127}	126.904352

classical data had suggested possible molecular formulas of $C_{45}H_{54}N_4O_8$ (778), $C_{42}H_{48}N_4O_6$ (704), or $C_{42}H_{50}N_4O_7$ (722). Low-resolution mass spectrometry showed the molecular weight to be 718, which is inconsistent with all of the foregoing, but is consistent with the formulations of $C_{43}H_{50}N_4O_6$ (718.3730) or $C_{42}H_{46}N_4O_7$ (718.3366). A high-resolution mass spectrum indicated a molecular weight of 718.3743, clearly showing that the formula $C_{43}H_{50}N_4O_6$ was the correct one.

An extensive table of elemental composition vs. mass for use in high-resolution mass spectrometry has been constructed by McLafferty (ref. 8). A sampling from this table is shown in Table 8.2. From high-resolution spectra, one obtains the elemental composition, not only of the molecular ion, but of each ion that is examined. The spectrum is then interpreted directly according to the elemental formula of the molecule and its fragments.

**Table 8.2. Mass vs. Elemental
Composition at Nominal Mass 43**

Mass	Elemental composition
43.0058	CHNO
43.0184	C_2H_3O
43.0269	CH_3N_2
43.0421	C_2H_5N
43.0547	C_3H_7

From low-resolution spectra, it is sometimes possible to determine a unique molecular formula, or at least to severely limit the number of possibilities by consideration of the relative intensities of the various isotope peaks. In this discussion we shall restrict our considerations to compounds containing carbon, hydrogen, oxygen, nitrogen, sulfur, fluorine, chlorine, bromine, and iodine. The principal stable heavier isotopes of each of these and the relative abundances as percentages of the isotope of lower mass are listed in Table 8.3.

Table 8.3. Principal Stable Heavier Isotopes

Isotope	Percent of isotope of lowest mass
^{13}C	1.11
^{2}H	0.015
^{18}O	0.20
^{15}N	0.37
^{33}S	0.78
^{34}S	4.4
^{37}Cl	32.5
^{81}Br	98.0
F	single isotopic species
I	single isotopic species

To determine the molecular formula, one measures the intensity of the molecular ion M and expresses the intensity of the M + 1 peak and the M + 2 peak as percentages of the molecular ion peak. For any possible molecular formula, the observed values of the M + 1 and M + 2 peaks may be compared with the theoretically calculated intensities, or reference may be made to the table constructed by Beynon (ref. 6), available, in part, in the book by Silverstein and Bassler (ref. 23). An example of the use of this method follows:

From a mass spectrum we obtain the following data:

m/e	Percent
100(M)	(100)
101(M + 1)	5.64
102(M + 2)	0.60

We can readily eliminate from further consideration formulas containing sulfur, chlorine, or bromine from the relatively small contribution of the M + 2 peak (these elements are usually easy to detect because of their high contribution to the M + 2 peak). Iodine need not be considered because its mass alone is 127. The relatively large contribution of the M + 1 peak suggests that fluorine is also not present. If we look in the Beynon table of C, H, N, O compound formulas of mass 100, we find the following data:

100	M + 1	M + 2
$C_2H_2N_3O_2$	3.42	0.45
$C_2H_4N_4O$	3.79	0.26
$C_3H_2NO_3$	3.77	0.65
$C_3H_4N_2O_2$	4.15	0.47
$C_3H_6N_3O$	4.52	0.28
$C_3H_8N_4$	4.90	0.10
$C_4H_4O_3$	4.50	0.68
$C_4H_6NO_2$	4.88	0.50
$C_4H_8N_2O$	5.25	0.31
$C_4H_{10}N_3$	5.63	0.13
$C_5H_8O_2$	5.61	0.53
$C_5H_{10}NO$	5.98	0.35
$C_5H_{12}N_2$	6.36	0.17

100	*M + 1*	*M + 2*
$C_6H_{12}O$	6.72	0.39
$C_6H_{14}N$	7.09	0.22
C_7H_2N	7.98	0.28
C_7H_{16}	7.82	0.26
C_8H_4	8.71	0.33

On the basis of the nitrogen rule, we need not consider further any formulas containing an odd number of nitrogens. From the M + 1 and M + 2 values, the formula $C_5H_8O_2$ is found to be the best fit. The measured isotope peaks are usually slightly higher than the calculated ones because of small contributions from bimolecular collisions, impurities, M − 1 peaks, background, etc. Moreover, it should be emphasized that the method is limited to compounds which produce relatively intense molecular ions and to compounds of low-to-moderate molecular weight. Intense molecular ion peaks are necessary if the isotope peaks are to be of measurable intensity. In compounds that contain large numbers of atoms, the isotopic distribution functions are too complex for accurate analysis, and the distinction between possible formulae is not possible owing to experimental uncertainties in the peak intensities.

8.4.2 Fragmentation

The mass spectrum of a compound provides the structural chemist with two types of information: the molecular weight of the compound and the mass of the various fragments produced from the molecular ion. To obtain structural information, the chemist attempts on paper to reassemble these fragments in a way not unlike the assembling of a jigsaw puzzle. Hypothetical molecules containing these fragments are constructed, and the fragmentation patterns predicted for these models are checked against the spectrum obtained.

Fragmentation is a chemical process that results in bond breaking; the energetic considerations that are applicable to classical chemical reactions are also applicable to these fragmentation processes, but the situation is complicated by the distribution function of the excited-state energy residing in the molecular ion. The fragmentation patterns are best interpreted based on the known chemistry of carbonium ions in solution. The rate of production, and hence intensity, observed for any fragment ion appears to correlate with the stability of that ion and with the nature of the leaving group (a neutral molecule or a radical). Thus the rate at which the process $[A − B]^{\ddagger} \longrightarrow A^+ + B^{\cdot}$ or $[A − B]^{\ddagger} \longrightarrow A^{\ddagger} + B$ occurs will depend not only on the stability of the positive ion fragment A^+ or $A^{\ddagger}$ but also on the stability of the radical $B^{\cdot}$ or molecule B.

There are two major differences between conventional ion chemistry and chemistry in the mass spectrometer. First, we are dealing with particles in excited states, and the energies involved are considerably higher than those from typical chemical reactions. Second, at the very low operating pressures ($\approx$ low concentration) of the mass spectrometer, we are dealing with uni-molecular reactions, and energy is not dissipated to any appreciable extent by collisions with other molecules of the compound or solvent as might occur in solution chemistry.

Most of the important types of fragmentation are summarized in general form below.

1. Simple carbon-carbon bond cleavages

    (a) $[R-\overset{|}{\underset{|}{C}}-]^{\ddagger} \longrightarrow R^{+} + \cdot\overset{|}{\underset{|}{C}}-$

    (b) $R-\overset{|}{\underset{|}{C}}-\overset{|}{\underset{|}{C}}^{+} \longrightarrow R^{+} + \overset{\diagup}{\diagdown}C{=}C\overset{\diagup}{\diagdown}$

    Relative importance of R^{+} increases in the order

$$CH_3 < R'CH_2 < R_2'CH < R_3'C < CH_2{-}CH{=}CH_2 < CH_2{-}\langle\bigcirc\rangle$$

    (c) $[R-\overset{\overset{\textstyle O}{\|}}{C}-R]^{\ddagger} \longrightarrow R\cdot + R\overset{+}{C}{\equiv}O$

    (d) $R-\overset{+}{C}{\equiv}O \longrightarrow R^{+} + C{=}O$

2. Cleavages involving heteroatoms

$$X = \text{halogen, } OR', SR', NR_2' \ (R' = H, \text{Alkyl, Aryl})$$

    (e) $[-\overset{|}{\underset{|}{C}}-X]^{\ddagger} \longrightarrow -\overset{|}{\underset{|}{C}}^{+} + X\cdot$

    f) $[R-\overset{|}{\underset{|}{C}}-X]^{\ddagger} \longrightarrow R\cdot + \overset{\diagdown}{\diagup}\overset{+}{C}{\cdots}X$

    Heaviest substituent tends to be lost to greatest extent, e.g., $CH_3CH_2\cdot > CH_3\cdot$, etc.

3. Concerted cleavages

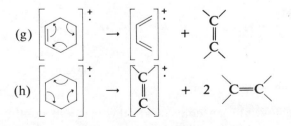

In many of the examples it will be noted that the indicated fragmentation pattern is not only facilitated by the production of a stable carbonium ion, but also that a stable neutral molecule results. It should be emphasized that, in the cleavage of any bond or bonds in an ion or an ion-radical, the positive charge may reside with either fragment.

$$n[AB]^+ \longrightarrow m(A^+ + B) + (n - m)(A + B^+)$$

The relative proportions are governed by the relative stabilities of the ions, radicals, and molecules formed. Thus in the spectrum of propiophenone, the following ions appear with the relative intensities as indicated.

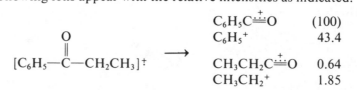

$$
\begin{array}{ll}
C_6H_5\overset{+}{C}\!\!=\!\!O & (100) \\
C_6H_5{}^+ & 43.4 \\
CH_3CH_2\overset{+}{C}\!\!=\!\!O & 0.64 \\
CH_3CH_2{}^+ & 1.85
\end{array}
$$

In the formation of a molecular ion by electron bombardment, removal of an electron from a nonbonded electron pair on a heteroatom appears easier than removal of a pi electron which, in turn, is easier to remove than a sigma electron. It is therefore possible and often desirable to depict fragmentation processes with the positive charge and free valence localized on a specific atom, e.g., $CH_3CH_2 \mid \overset{+\cdot}{O} \mid CH_2CH_3$. In diagrams in this text, we shall occasionally choose to show charges and radicals localized, but more often we shall represent ions and radical-ions by enclosing the structural formulas in brackets, e.g., $[CH_3CH_2{-}O{-}CH_2CH_3]^{\ddagger}$.

As an aid in the visualization of fragmentation and rearrangement processes, arrows are often used to symbolize bond breaking and bond making [Eq. (8.4)]. Since in conventional chemical notation such arrows

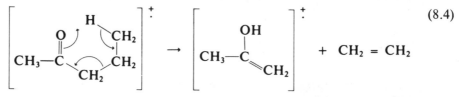

(8.4)

usually indicate two-electron shifts, other authors have suggested different representations, e.g.,

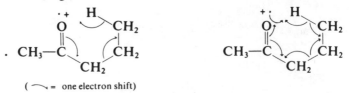

($\frown$ = one electron shift)

In the absence of a well-established convention, this text will continue to employ the method illustrated in Eq. (8.4). The student is cautioned to be well aware that the arrows in such a representation do not have physical significance; they indicate neither the direction of electron flow nor the numbers of electrons moving. It is simply a method of "electron book-keeping."

Often in the description of mass spectra, it is convenient to refer to a particular fragment ion in terms of the molecular ion less the mass or elemental composition of the lost neutral fragment(s). The $C_6H_5CO^+$ ion at mass 105 in the spectrum of propiophenone can be characterized as the $M - 29$ or the $M - C_2H_5$ ion.

8.4.3 Rearrangements

In the mass spectrum of almost all compounds, there are fragments whose presence cannot be accounted for by the simple cleavage of bonds in the parent compound. These fragments are the result of rearrangement processes. Such rearrangements may be expected to play a major role in fragmentation when the process is energetically and sterically favored; i.e., when the process results in the formation of more stable entities and the transition state required for the reaction is sterically accessible.

The most important of these rearrangement processes involves the migration of a hydrogen atom from one part of the molecule to another. Such rearrangements typically involve the migration of a specific hydrogen which is sterically accessible to a carbonium or radical site in the ionized molecule. Recall that in the ionization reaction electrons are most easily removed from π electron systems or nonbonded electrons on heteroatoms. Perhaps the most common rearrangement is the McLafferty rearrangement which involves a π electron system and a six-membered transition state. This type of rearrangement is illustrated in general terms as follows:

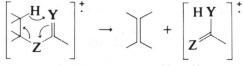

Olefins, ketones, esters, amides, acids, etc.

Other common rearrangements involving hydrogen migration include:

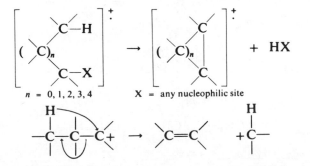

The second common type of rearrangement process is that involving 1,2 shifts. As in ordinary chemical reactions, the driving force for this type of rearrangement is the formation of a more stable species.

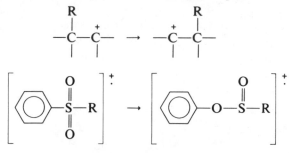

One can readily envision how 1,2 shifts might ultimately lead to fragments that seemingly have little to do with the initial structure of the compound; e.g., aryl sulfones lose carbon monoxide during fragmentation. Perhaps the most extensively studied rearrangement of this type is the formation of tropilium ion from toluene. In this case it appears that ring expansion

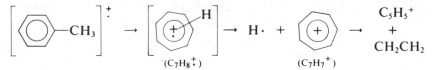

occurs in the parent ion before loss of a hydrogen atom. It has been demonstrated that in the intermediate $C_7H_8^+$ all hydrogens are equivalent. The spectra of thirteen nonaromatic isomers of toluene have been studied. In all cases, the spectra are remarkably similar and, in every example, the most abundant ion is mass 91 ($C_7H_7^+$), apparently formed from a common $C_7H_8^+$ intermediate.

Molecules which cannot undergo simple cleavage to form obviously stable particles or which contain no nucleophilic centers situated as to direct rearrangement often fragment in ways which defy facile explanation. Such is the case of many hydrocarbons, for example,

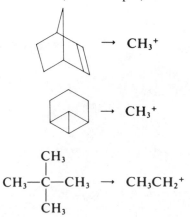

8.4.4 Metastable Peaks

$$m_1^+ \longrightarrow m_2^+ + \text{neutral fragment}$$

If the rate of decomposition of an ion m_1 formed in the ionization chamber is very fast, almost all such ions will decompose before reaching the acceleration region, and only the fragments m_2 will be accelerated, deflected, and detected in any significant quantity. If the original ion (m_1) is very stable, it will be observed as an intense peak relative to any daughter ions (m_2). Ions which decompose at an intermediate rate should show considerable intensity for both the primary (m_1) and the daughter (m_2) ions. With this intermediate rate of decomposition, some of the primary ions will decompose into fragments while traveling through the accelerating region of the instrument. Such ions will at first be accelerated as mass m_1, decompose with loss of some kinetic energy to the neutral fragment, and then continue to be accelerated and deflected as mass m_2. Such an ion will be recorded as a broad peak of low intensity (generally 1% or less) at mass m^*, as follows:

$$m^* = \left[\frac{(m_2)^2}{m_1}\right] \tag{8.5}$$

The metastable peaks (m^*) are usually found at 0.1 to 0.4 mass units higher than calculated from this equation. An example of the characteristic record of a metastable peak is shown in Fig. 8.6.

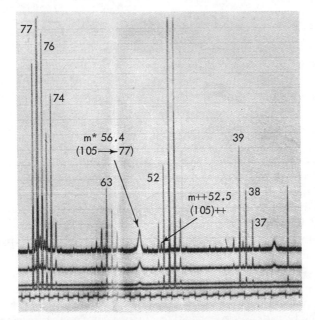

Fig. 8.6. A portion of a mass spectrum traced by a four-element galvanometer illustrating normal peaks, metastable peaks, and peaks due to doubly charged ions.

Metastable peaks are very important in the deduction of fragmentation mechanisms, for they indicate that the fragment of mass m_2 is formed in a one-step process from mass m_1. For example, in the mass spectrum of phthalic anhydride the following principal fragmentation pattern is indicated:

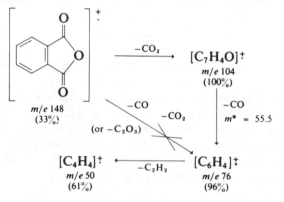

The metastable peak at 55.5 (calc'd. 55.5) tells us that the ion at m/e 76 (m_2)(benzyne?) is formed from the m/e 104 (m_1) ion and not by concerted loss of carbon monoxide and carbon dioxide from the molecular ion.

8.5   ISOTOPIC LABELING

Using isotopic labeling at strategically located points in a molecule and then following this label through to the products is a well-established method of studying chemical and biological reaction mechanisms. For the detection of radioisotopes, various counting techniques are employed; for the detection of deuterium and sometimes carbon-13, nmr techniques can be used (deuterium may also be measured by combustion of the sample to water and measurement of the density of the water). However, mass spectrometry is the only generally applicable method for the detection of stable isotopes. The mass spectrometric technique has the advantages of examining the intact molecule and often pinpointing the location of the label.

Isotopic labeling is also a very useful method of establishing reaction pathways which occur in the mass spectrometer. The method has special pertinence to low-resolution spectrometry which cannot distinguish between ions of the same nominal mass as illustrated by the following: The mass spectrum of cyclopentanone exhibits a prominent peak at mass 28 which could be attributed to either carbon monoxide or ethylene. Since this fragment retains two deuterium atoms (producing a peak at m/e 30) in the spectrum of the tetradeuterio derivative, the peak is due to ethylene.

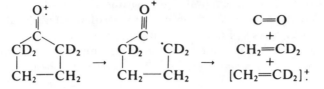

8.6   HANDLING AND REPORTING OF DATA

Because of space limitations in journals and books, just as in other forms of spectroscopy, it is impractical to present reproductions of the original mass spectra. Likewise, as is characteristic of other forms of spectroscopy, the methods of reporting data are considerably varied.

Data are frequently presented in tabular form, in which the mass numbers and the intensities of the peaks are listed. Two methods are commonly used to express intensity. The simplest and most frequently used method, though perhaps not the best, expresses the intensities in relation to the most intense peak in the spectrum. This peak, known as the *base peak* is arbitrarily assigned a value of 100. The second method relates the intensity of any given peak with the total intensity of the spectrum, i.e., the sum of the intensities of all the peaks, or the *total ionization* (Σ). The intensity of a peak is then expressed as percent of Σ, and it indicates the extent to which the molecular ions decompose into this fragment. It is frequently not practical to measure the peaks from $m/e = 1$ to the molecular weight; instead, they are measured over a more restricted range. In this case the designation Σ_m is used, where the value of m indicates the lowest mass in the range; e.g., Σ_{30} signifies that the intensities have been summed from mass 30 to the highest mass in the spectrum.

Chemists, especially organic chemists, immediately associate the term spectrum with a picture with lines on it. Mass spectral data are often presented in the form of a bar graph, the ordinate indicating the relative abundances (intensity) either as percent of the base peak or percent Σ, or both, and the abscissa indicating the mass-to-charge ratios. This method of presentation has the advantage that the overall characteristics of the spectrum are available at a glance.

The most extensive use of the large amount of data made available on a complete high-resolution mass spectrum involves the *element mapping* technique initiated by Biemann and co-workers. With this method, exact mass measurements are made on all, or almost all, of the ions produced, and possible elemental formulas are calculated for each ion. With large molecules this can involve hundreds of ions. The spectrum is then interpreted directly according to the elemental composition of ions responsible for each peak in the spectrum. The handling of the vast quantities of data necessary for the complete analysis of the spectrum is made practical only by the use of computer techniques. For high resolution instruments employing the Nier-Johnson geometry with scanning (Fig. 8.2), the data can be digitized directly on magnetic tape or punched cards which may then be analyzed by means of a computer. The Mattauch-Herzog instruments (Fig. 8.3) record the entire spectrum simultaneously on a photographic plate. The line intensity and position data from the plate are then digitized by means of an automatic or semiautomatic comparator. The calculated line positions are converted to exact masses by comparison with reference lines of a standard spectrum (usually perfluorokerosene) recorded simultaneously. Although the former method has the advantage of not requiring the intervention of an expensive and somewhat time consuming comparator, it has the disadvantage of requiring a relatively long time to record the complete spectrum. Thus it is not as suitable for use with direct hookup with a gas-liquid chromatograph.

From the digitized data, the computer calculates possible elemental compositions for each ion, sorts these into columns according to the content of heteroatoms, and prints out in the form of an element map. Figure 8.7 is the element map obtained from 3-octanone. In the first column, the nominal masses of the various ions are given. The second column indicates ions containing only carbon and hydrogen; the number of each element present is listed, followed by the difference in found and calculated mass expressed in millimass units, and then followed by 1 to 10 asterisks indicating on a log scale the approximate intensity of the particular ion. The third column gives the same information for ions containing carbon, hydrogen, and oxygen; and so on. In the element map the heteroatom composition increases from left to right, and the mass increases from top to bottom. The molecular ion is found in the entry at the bottom right, in this case, mass 128 and composition $C_8H_{16}O$. For an example of the interpretation of the information, we shall examine the information given at nominal mass 56. We note that there are two entries on this line, indicating two ions of the same nominal mass: the $C_4H_8^+$ ion which appears in the spectrum 1 millimass unit higher than calculated and with a relative intensity of 3 on a log scale, and the $C_3H_4O^+$ ion which appears in the spectrum 1 millimass unit lower than calculated with a relative intensity of 4.

The element map indicates that the largest saturated alkyl fragment (C_nH_{2n+1}) in the molecule is a pentyl group. The oxygen-containing fragments indicate acyl ions $(C_nH_{2n-1}O)$ corresponding to $n = 2$, 3, and 6 predominate. Acyl ions of $n = 3$ and 6 are indicative of a ketone with 8 carbons with the carbonyl oxygen at $C - 3$. Furthermore, the rearrangement ion corresponding to C_4H_8O implies that the chain is unbranched at the 4 position. At this point we can write the following structure:

$$C_4H_9-CH_2-CO-CH_2CH_3$$

A somewhat unexpected peak appears at mass 81 with composition C_6H_9, which probably arises from loss of water from the $C_6H_{11}O$ acyl ion. This example illustrates the utility of the element map for the interpretation of structural information and, furthermore, in uncovering unexpected fragmentation processes.

	CH	CHO	CHO$_2$	CHO$_3$
40	3/4 0**			
41	3/5 0*****			
42	3/6 0****			
43	3/7 0******	2/3 0*****		
44		2/4 0**		
45		2/5 0**		
50	4/2 0**			
51	4/3 0***			
52	4/4 0*			
53	4/5 0***			
54	4/6 0**			
55	4/7 1*****	3/3 1****		
56	4/8 1***	3/4 —1****		
57	4/9 1****	3/5 1******		
58		3/6 0***		
59		3/7 0**		
62	5/2 1*			
63	5/3 0*			
65	5/5 0*			
67	5/7 0**			
68	5/8 0**			
69	5/9 0****	4/5 0****		
70	5/10 0***	4/6 0***		
71	5/11 0******	4/7 0****		
72		4/8 0******		
73		4/9 —2*****		
75	6/3 1*			
79	6/7 0*			
80	6/8 0*			
81	6/9 0***			
83		5/7 0**		
84		5/8 0*		
85		5/9 1****		
86		5/10 0***		
97		6/9 0**		
98		6/10 0*		
99		6/11 0*****		
114		7/14 0*		
128		8/16 0*****		
	CH	CHO	CHO$_2$	CHO$_3$

Fig. 8.7. Element map of 3-octanone.

The important fragmentation patterns for a number of compounds representative of the most common chemical classes are described in this section. For a more comprehensive treatment, the general references cited at the end of this chapter should be consulted. *Mass Spectrometry of Organic Compounds,* by Budzikiewicz, Djerassi, and Williams is a very thorough and suitable treatment for chemists interested in organic structure determination. In many cases the structural formulas shown in this section for mass spectral fragments should be considered only as probable representations of the observed elemental composition of the fragments.

8.7.1 Hydrocarbons

8.7.1a Saturated hydrocarbons

The utility of mass spectra of hydrocarbons is greatly enhanced by the availability of a large number of reference spectra. Normal hydrocarbons exhibit clusters of peaks, 14 mass units (CH_2) apart, of decreasing intensity with increasing fragment weight (Fig. 8.8). The molecular ion peak is almost always present albeit of low intensity. Branching causes a prefer-

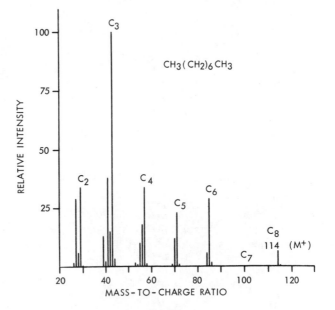

Fig. 8.8

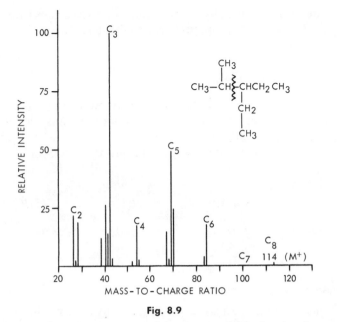

Fig. 8.9

ential cleavage at the point of the branch, resulting in the formation of a secondary carbonium ion (Fig. 8.9).   Low-intensity ions from random rearrangements are common.

Saturated alicyclic hydrocarbons give rise to complex spectra.   The

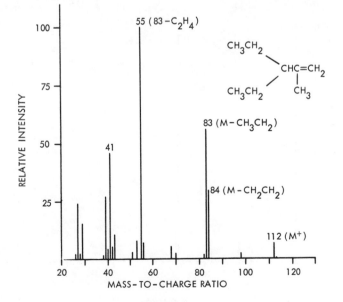

Fig. 8.10

molecular ion peak is usually rather intense. Characteristic peaks are usually formed from the loss of ethylene and side chains.

8.7.1b Unsaturated hydrocarbons

The molecular ion peak, apparently formed by removal of a π electron, is usually distinct. The prominent peaks from monoolefins have the general formula C_nH_{2n-1} (allyl carbonium ions), but fragments of the composition C_nH_{2n} formed in McLafferty rearrangements are also common.

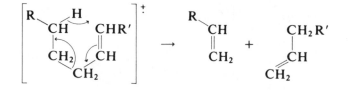

8.7.1c Aromatic hydrocarbons

Hydrocarbons containing an aromatic ring usually give readily interpretable mass spectra which exhibit strong molecular ion peaks. The molecular ion of benzene $[C_6H_6]^{\ddagger}$ accounts for 43% of the total ion current at 70 ev in the mass spectrum of benzene. The most characteristic cleavage of the alkyl benzenes occurs at the bonds beta to the aromatic ring. In the usual case,

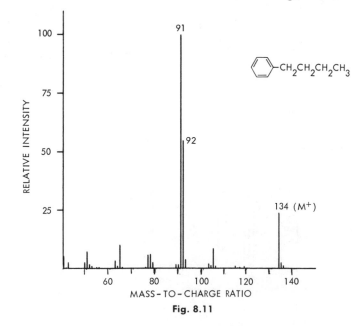

Fig. 8.11

the heaviest substituents are preferentially lost, resulting in the formation
of tropyllium or substituted tropyllium ions.

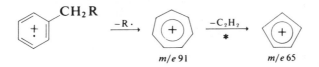

If side chains of propyl or larger are present, a McLafferty rearrangement may also occur.

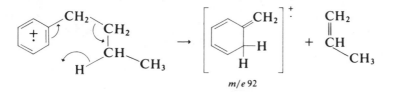

8.7.2 Halides

Iodine and fluorine are monoisotopic, whereas chlorine and bromine occur
naturally as mixtures of two principal isotopes. Hence a molecular ion or
fragment ion containing a chlorine will show a +2 isotope peak amounting
to about a third of the intensity, whereas a bromine-containing ion will be
accompanied by a +2 peak of almost equal intensity. For organic halides,
the abundance of the molecular ion in a given series of compounds increases
in the order $F < Cl < Br < I$. The intensity of the molecular ion peak
decreases with increasing size of the alkyl group and α branching. The
following fragmentation processes are listed in approximately decreasing
order of importance.

1. $R{-}X^{\ddagger} \longrightarrow R^+ + X\cdot$. Most important for X = I or Br.
2. $R{-}CH_2CH_2X^{\ddagger} \longrightarrow [RCH{=}CH_2]^{\ddagger} + HX$. More important for
   X = Cl and F than for I or Br.
3. $R{-}CH_2X^{\ddagger} \longrightarrow R\cdot + CH_2{=}X^+$. Heaviest group preferentially
   lost.
4.

$$R{-}CH_2 \underset{\underset{CH_2}{CH_2}}{\overset{X^{\ddagger}}{\underset{|}{|}}}CH_2 \longrightarrow \underset{\underset{CH_2}{CH_2}}{\overset{CH_2{-}X^{\ddagger}}{\underset{|}{|}}}CH_2 + R\cdot$$

X = Br or Cl

The molecular ion peak of aromatic halides is usually abundant. An
M $-$ X peak is usually observed (Fig. 8.15).

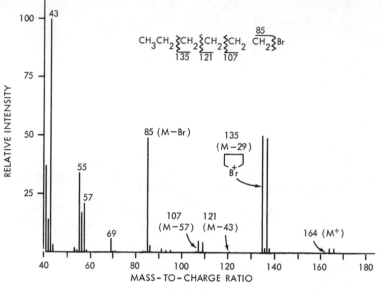

Fig. 8.12

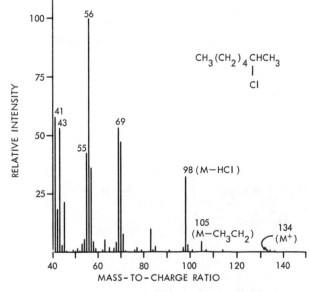

Fig. 8.13

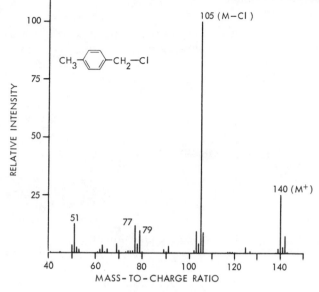

Fig. 8.14

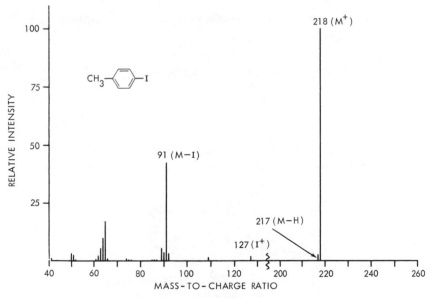

Fig. 8.15

8.7.3 Alcohols and Phenols

The molecular ion peak of primary and secondary alcohols is usually weak; that of tertiary alcohols is usually not detectable. The most important general fragmentation process involves cleavage of the bond beta to the oxygen atom. The largest group is lost most readily.

$$R_2-\underset{\underset{R_3}{|}}{\overset{\overset{R_1}{|}}{C}}-\overset{+\cdot}{O}H \longrightarrow R_i + \underset{R_3}{\overset{R_2}{\diagdown}}C\overset{+}{=}OH$$

Loss of water may occur by a cyclic mechanism

but it also may occur by a thermal process on the hot inlet surfaces.

$$R-CH_2-CH_2CHR \xrightarrow{\ \ \Delta\ \ } RCH_2-CH=CHR$$
$$\underset{OH}{|}$$

A concurrent elimination of water and an olefin is usually observed with alcohols of four or longer carbon chains.

$$\xrightarrow[-C_2H_4]{-H_2O} [CH_2=CHR]^{\ddagger}$$

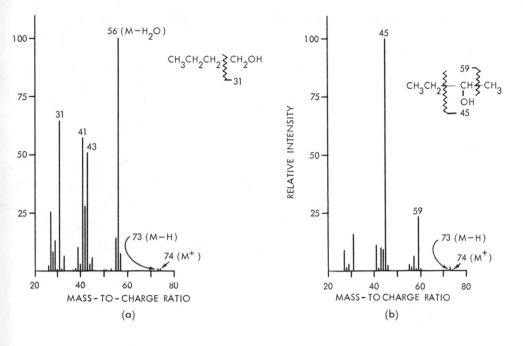

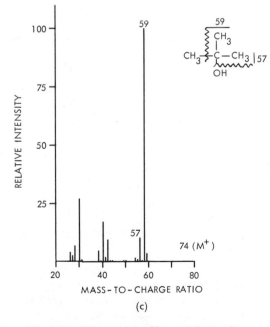

Fig. 8.16. Mass spectra of isomeric butanols.

The fragmentation of a typical cyclic alcohol is illustrated below.

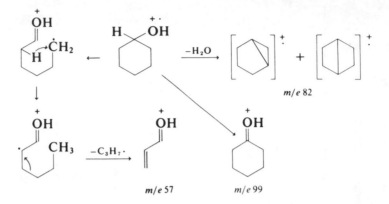

m/e 82

m/e 57   *m/e* 99

Benzylic alcohols typically exhibit intense molecular ion peaks.

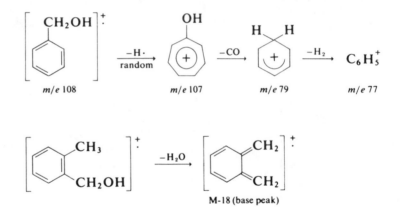

m/e 108   *m/e* 107   *m/e* 79   *m/e* 77

M-18 (base peak)

Phenols also exhibit intense molecular ion peaks; the M-CO peak is characteristic.

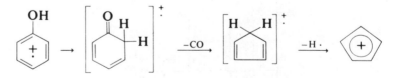

Trimethyl silyl ethers of alcohols are widely employed in mass spectrometry and gas liquid chromatography because of their higher volatility compared with the parent alcohol. Although the molecular ion peak of these

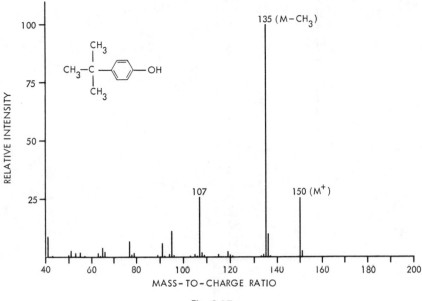

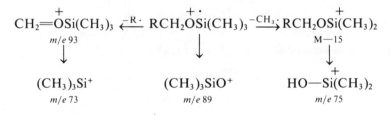

Fig. 8.17

ethers is often of low intensity, the molecular weight can usually be inferred from the strong M − CH_3 peak.

$$CH_2 \!\!=\!\! \overset{+}{O}Si(CH_3)_3 \xleftarrow{-R\cdot} R\overset{+\cdot}{CH_2OSi(CH_3)_3} \xrightarrow{-CH_3} R\overset{+}{CH_2OSi(CH_3)_2}$$
$$\qquad m/e\ 93 \qquad\qquad\qquad\qquad\qquad\qquad\qquad M\!-\!15$$
$$\downarrow \qquad\qquad\qquad\qquad \downarrow \qquad\qquad\qquad\qquad \downarrow$$
$$(CH_3)_3Si^+ \qquad\qquad (CH_3)_3SiO^+ \qquad\qquad HO\!\!-\!\!\overset{+}{Si}(CH_3)_2$$
$$m/e\ 73 \qquad\qquad\quad m/e\ 89 \qquad\qquad\qquad m/e\ 75$$

8.7.4 Ethers and Acetals

The molecular ion peak of ethers is weak but can usually be observed. M + 1 peaks frequently occur at higher pressures. The ethers undergo fragmentations similar to those of alcohols. The two most important pathways are:

1. Cleavage of a bond beta to the oxygen

$$R\!\!-\!\!CH_2\overset{+\cdot}{O}R \longrightarrow CH_2\!\!=\!\!\overset{+}{O}R + R\cdot$$

In the case of branching at the α-carbon, the largest R group is lost preferentially.

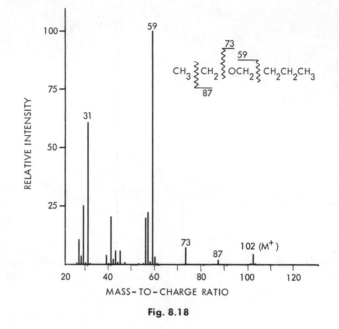

Fig. 8.18

2. Cleavage of a C—O bond

$$R\overset{+\,\cdot}{-}O{-}R \longrightarrow RO\cdot + R^+$$

Acetals and ketals behave in a similar fashion.

Important fragmentations of aromatic ethers are illustrated below.

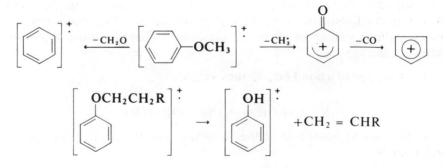

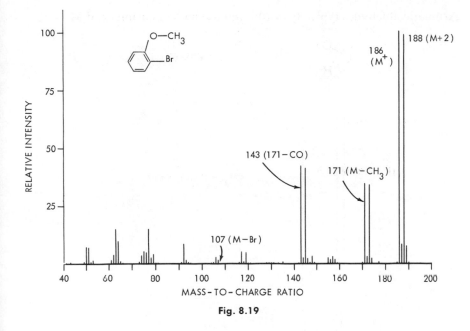

Fig. 8.19

8.7.5 Aldehydes

Aliphatic aldehydes exhibit weak molecular ion peaks.

α Cleavage

$$[R—CHO]^{+} \xrightarrow{-H\cdot} R—C\equiv O^{+}$$

The M − 1 peak is diagnostic of aldehydes.

$$[R—CHO]^{+} \xrightarrow{-R\cdot} HC\equiv O^{+}$$
$$\text{m/e 29}$$

The peak at m/e 29 from higher aldehydes may also be due to $C_2H_5^{+}$.

β Cleavage

$$[R—CH_2—CHO]^{+} \longrightarrow R^{+} + CH_2=CH—O\cdot$$
$$\text{M—43} \qquad\qquad\qquad 43$$

McLafferty rearrangements

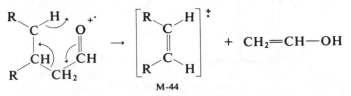

Aromatic aldehydes typically exhibit intense molecular ions and M − 1.

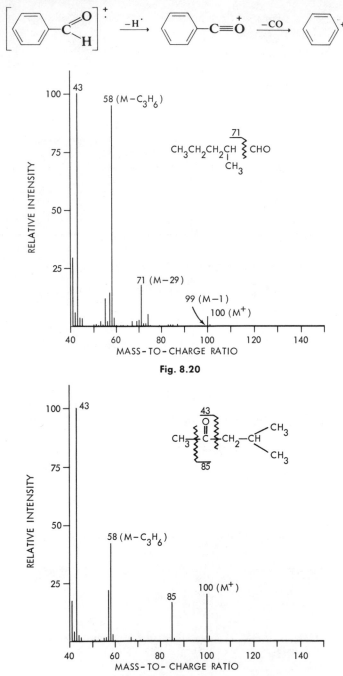

Fig. 8.20

Fig. 8.21

8.7.6 Ketones

The molecular ion of aliphatic ketones is intense. Fragmentation pathways are similar to those of aldehydes. The peak arising from α cleavage with loss of the larger alkyl group is more intense than that from loss of the smaller one. Thus the base peak for most methyl ketones is m/e 43 (CH_3CO^+); intense peaks may also occur at m/e 43 with propyl or iso-propyl ketones.

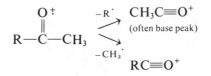

With ketones containing γ-hydrogens, McLafferty rearrangements are prevalent.

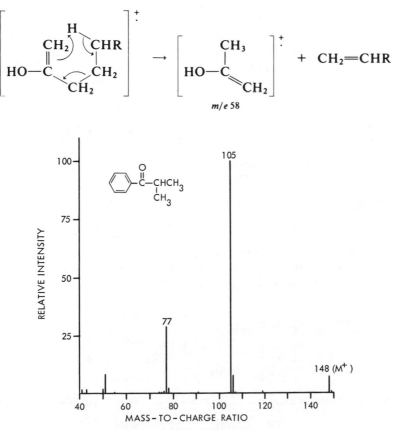

Fig. 8.22

Important fragmentations of cyclic and aromatic ketones are illustrated below.

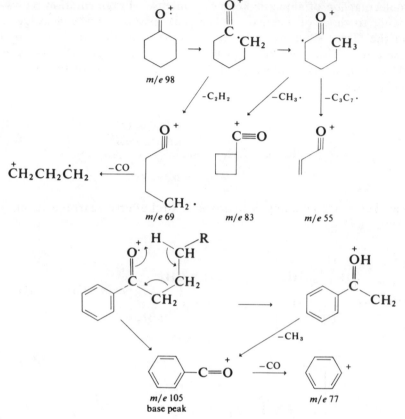

m/e 98

$CH_2CH_2\dot{C}H_2$ $\xleftarrow{-CO}$

m/e 69   *m/e* 83   *m/e* 55

m/e 105
base peak

m/e 77

8.7.7 Acids

Acids are more frequently and often more conveniently examined as their methyl esters. The free acids exhibit weak but discernible molecular ion peaks. Normal aliphatic acids in which a γ-hydrogen is available for transfer have their base peak at *m/e* 60.

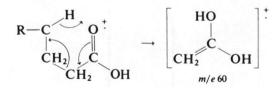

m/e 60

Both aliphatic and aromatic acids exhibit a peak due to [COOH] at *m/e* 45.

$$[R-COOH]^{+} \xrightarrow{-R\cdot} {}^{+}COOH$$
m/e 45

In the spectra of aromatic acids, the M − OH peak is prominent.

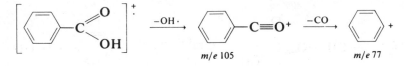

m/e 105   m/e 77

8.7.8 Esters and Lactones

8.7.8a Methyl esters

The molecular ion is usually discernible. Fission of bonds adjacent to the carbonyl group may occur to yield four ions.

$$\left[\begin{matrix} & O \\ & \| \\ R+C+OCH_3 \end{matrix}\right]^{\ddot{+}} \longrightarrow RC\equiv O^+ \; > \; ^+O\equiv C-OCH_3 \; > \; ^+OCH_3 \; > \; R^+$$

The McLafferty rearrangement is a most important process in the longer chain methyl esters. The base peak in the mass spectrum of methyl esters of C_6 to C_{26} fatty acids occurs at m/e 74.

$$RCH_2CH_2CH_2COOCH_3 \longrightarrow \left[\begin{matrix} OH \\ | \\ CH_2=C \\ \qquad \diagdown OCH_3 \end{matrix}\right]^{\ddot{+}} + \; RCH=CH_2$$

m/e 74

8.7.8b Higher esters

The higher esters undergo fragmentations similar to those of the methyl esters, but the overall spectrum is complicated by additional fragmentation involving the alkoxy group. Such esters usually exhibit a peak corresponding to a protonated acid which may arise by the following pathway.

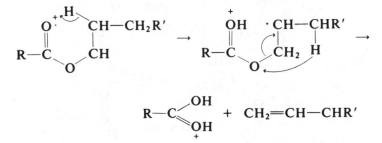

Rearrangement ions occur at m/e 60 and 61, corresponding to acetic acid and protonated acetic acid in the spectrum of ethyl or higher esters of butyric or higher acids.

Benzyl esters fragment with the loss of ketene.

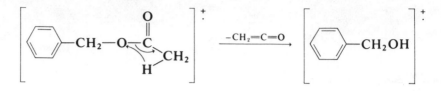

The loss of the alkoxy group is a very important fragmentation process for alkyl benzoates.

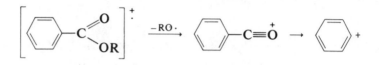

It is typical that ortho groups exert a marked influence on fragmentation processes.

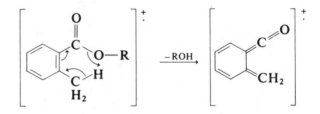

Some of the important ions recorded in the spectrum of γ-valerolactone are illustrated below.

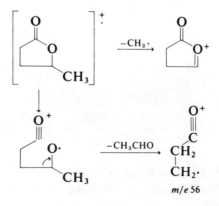

m/e 56

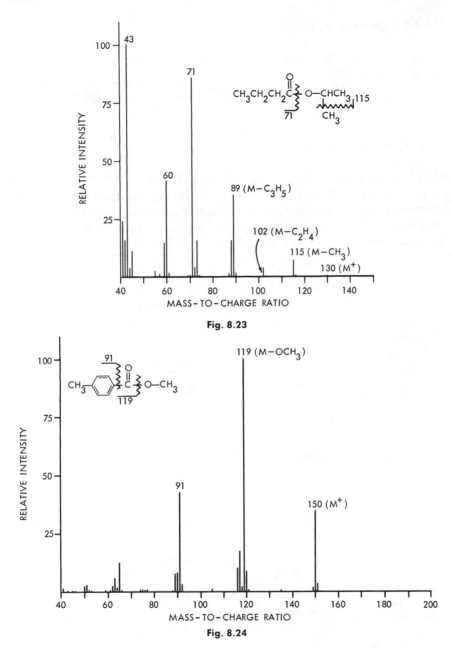

Fig. 8.23

Fig. 8.24

8.7.9 Amines

Amines and other nitrogen compounds containing an odd number of
nitrogen atoms have an odd-numbered molecular weight.

8.7.9a Aliphatic amines

The molecular ion peak is weak to absent. The most intense peak results from β cleavage.

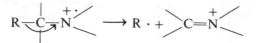

Again the largest alkyl group is preferentially lost. The base peak for all primary amines unbranched at the α-carbon is at m/e 30 $[CH_2{=}\overset{+}{N}H_2]$. The presence of this peak is good but not conclusive evidence for a primary amine since this fragment may arise from sequential fragmentation from secondary and tertiary amines. Cyclic fragments apparently occur during the fragmentation of longer chain amines.

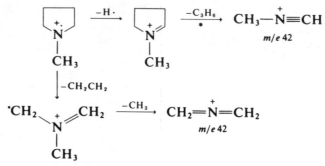

Cyclic amines give rise to strong molecular ion peaks; primary fragmentation occurs at α-carbons.

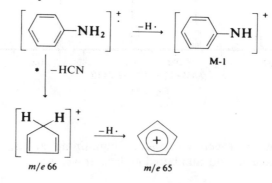

Aromatic amines give very intense molecular ion peaks accompanied by a moderate M − 1 peak.

$$\left[\bigcirc\!\!-NH_2\right]^{\overset{+}{\cdot}} \xrightarrow{-H\cdot} \left[\bigcirc\!\!-NH\right]^{+}$$

$$M\text{-}1$$

$$\downarrow^{\bullet} -HCN$$

$$\left[\begin{array}{c}H \quad H\\ \square\end{array}\right]^{\overset{+}{\cdot}} \xrightarrow{-H\cdot} \bigpentagon{\oplus}$$

$$m/e\ 66 \qquad\qquad m/e\ 65$$

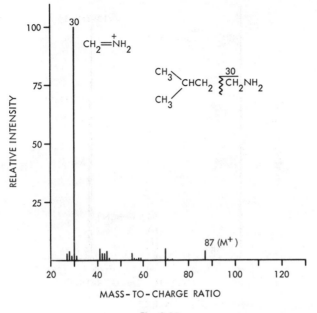

Fig. 8.25

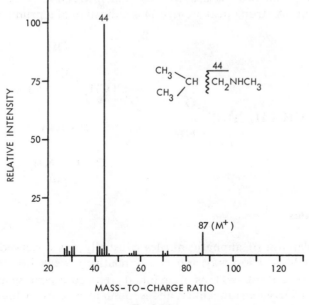

Fig. 8.26

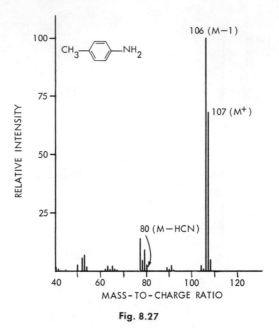

Fig. 8.27

8.7.10 Amides

The molecular ion peak is usually observed. McLafferty rearrangements are important. A strong peak at m/e 44 is indicative of a primary amide.

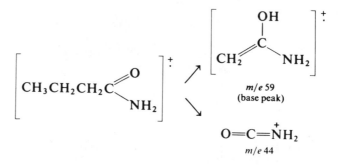

8.7.11 Nitriles

The molecular ion of aliphatic nitriles is usually not observed. At higher pressures an M + 1 ion may appear. A relatively weak but useful M − 1 ion due to R—CH=C=N⁺ is often found. The base peak generally results from a McLafferty rearrangement and is found at m/e 41 in the spectrum of C_4 to C_9 straight chain nitriles.

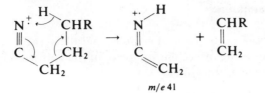

$m/e\,41$

8.7.12 Nitro Compounds

The molecular ion peak of aliphatic nitro compounds is seldom observed. The spectra of such compounds are composed mainly of hydrocarbon ions; with observable peaks containing nitrogen occurring at m/e 30 (NO) and 46 (NO_2). On the other hand, nitrobenzene exhibits an intense molecular ion. A rearrangement reaction also occurs to give $C_6H_5O^+$.

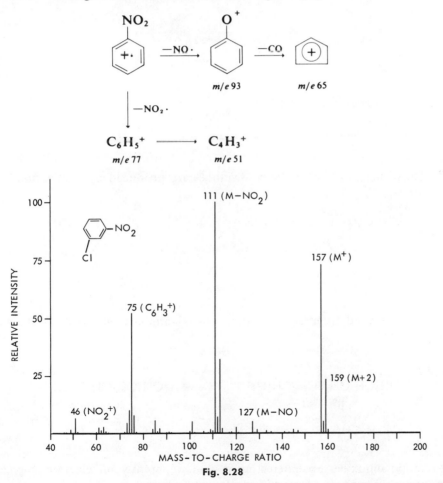

Fig. 8.28

8.7.13 Sulfur Compounds

The fragmentation of thiols and sulfides parallels those of alcohols and ethers. In each case the molecular ion peak of the sulfur compounds is more intense. Cyclic ions of the type

have been postulated in a number of cases.

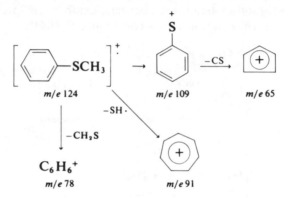

The principal fragments from disulfides are produced by elimination of olefins.

$$[CH_3CH_2-S-S-CH_2CH_3]^+ \xrightarrow{-CH_2CH_2} [CH_3CH_2SSH]^+$$
$$m/e\ 122 \qquad\qquad m/e\ 94$$

$$\downarrow -CH_2CH_2$$

$$[HSSH]^+$$
$$\cdot m/e\ 66$$

The principal fragmentation of a typical aliphatic sulfoxide is shown below.

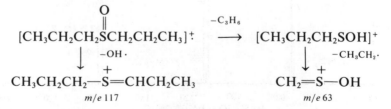

Aromatic sulfoxides are interesting in that apparently an electron impact induced aryl migration from sulfur to oxygen occurs.

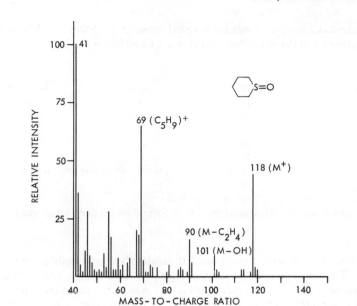

Fig. 8.29

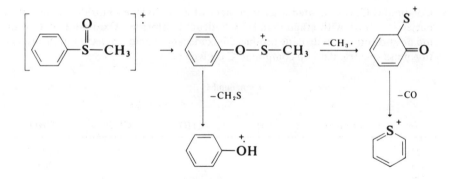

Similar migrations occur with aromatic sulfones.

8.8   MASS SPECTRAL PROBLEMS

Identify each of the following unknowns on the basis of the mass spectral and other structural data given. Suggest plausible explanations for the mass spectral patterns observed.

1. Compound **A** exhibits an intense infrared band at 1715 cm^{-1}. The important
ions appearing in the mass spectrum of **A** are listed below:

m/e	% base peak	m/e	% base peak
41	44	85	66
42	8	100	12
43	10	113	6
55	10	142	12
57	100		
58	58	metastable peak at 38.3	
71	6		

Compound **A** reacts with semicarbazide to form a single derivative, mp 89 to 90°.

2. The infrared spectrum of compound **B** displays strong bands at 1690 and 826
cm^{-1}. The nmr spectrum consists of a pair of AB doublets near $\delta\,7.60$ (4) and
a singlet at $\delta\,2.45$ (3). The peaks at highest m/e in the mass spectrum of **B** are
found at 198 (26), 199 (4), 200 (25), 201 (3), and the base peak is found at 183
(100).

3. Compound **C** was isolated after treatment of a methylene chloride extract of a
bacterial culture with ethanol and dry hydrogen chloride. Oxidation of **C** with
iodine and sodium hydroxide produced hexadecanoic acid. Selected high resolution mass spectral data on compound **C** is given below.[1]

Compound C

m/e	Intensity	CH	CHO	CHO$_2$	CHO$_3$
327	2				20/39
326	5				20/38
239	15		16/31		
196	2	14/28			
143	23				7/11
131	25				6/11
130	100				6/10
115	6				5/7
102	5				4/6
97	11	7/13	6/9	5/5	
88	20			4/9	
71	17	5/11	4/7		

---

[1] The high resolution mass spectral data for this problem were kindly supplied by Dr. Wal-
ter J. McMurray, Yale University.

4. Compound **D**, mp 115 to 116°, gave the mass spectrum reproduced in Fig. 8.30.

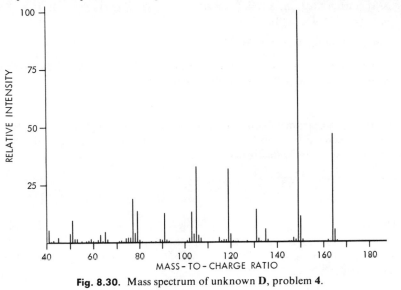

Fig. 8.30. Mass spectrum of unknown **D**, problem **4**.

5. The mass spectrum of compound **E** is shown in Fig. 8.31. At low operating pressures the following data are obtained:

m/e	Relative intensity
114	12.0
115	0.81
116	0.07

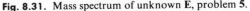

Fig. 8.31. Mass spectrum of unknown **E**, problem **5**.

6. The infrared spectrum of compound **F** is devoid of absorption in the 6 μ region; mass spectrum (70e) m/e (rel. intensity) 27 (26), 29 (39), 31 (18), 45 (100), 47 (16), 61 (6), 73 (44), 75 (8), 89 (3), 103 (15), 117 (2).

   Anal.   Found: C, 61.10; H, 11.9; N, 0.00.

   Compound **F**, bp 102°, upon treatment with 2,4-dinitrophenylhydrazine reagent forms a yellow precipitate, mp 166 to 168°.

7. The infrared spectrum of compound **G** contains a carbonyl band at 1830 cm^{-1}. The nmr spectrum consists of a singlet at $\delta 7.34$.   Important ions in the mass spectrum of **G** are indicated below.

m/e	% base peak
63	48
64	100
92	56.5
136	75.5
137	6.1
138	0.8

Metastable peaks are found at 44.5 and 62.6.

8. Extraction of the leaves of *Nepeta cataria Linné*, otherwise known as catnip, with methylene chloride followed by chromatography of the crude extract over alumina employing carbon-tetrachloride-acetone as the eluent resulted in the isolation of an oil, compound **H**, λ_{max}^{EtOH} 240 mμ.

Mass Spectral Data

m/e	% base peak	m/e	% base peak
27	43	53	12
28	7	55	100
29	46	70.3	--
36.8 (m^*)	--	83	96
39	40	84	5
41	12	98	51
43	92	99	3.5
51.1 (m^*)	--	100	0.2

9. Compound **J**, n_D^{20} 1.4800, bp 77° 10 mm, gave the mass spectrum shown in Fig. 8.32.

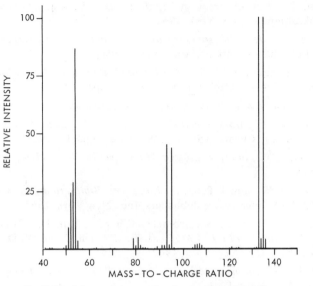

Fig. 8.32. Mass spectrum of unknown **J**, problem **9**.

10. Compound **K**, which has an exceptionally disagreeable odor, yields the following mass spectral data:

m/e	% base peak	m/e	% base peak	m/e	% base peak
27	24	57	38	83	46
29	32	58	2.9	84	51
39	19	59	3.0	89	11
41	76	61	18	98	4
42	47	68	18	112	23
43	74	69	59	146	37
47	29	70	70	147	3.9
55	74	71	15	148	1.8
56	100	82	11		

8.9 GENERAL REFERENCES ON MASS SPECTROSCOPY

1. Budzikiewicz, H., C. Djerassi, and D. H. Williams, *Mass Spectrometry of Organic Compounds*, Holden-Day, Inc., San Francisco, 1967. This is the most complete general reference written for the organic chemist interested in employing mass spectrometry in structure determination.

2. Budzikiewicz, H., C. Djerassi, and D. H. Williams, *Interpretation of Mass Spectra of Organic Compounds*, Holden-Day, Inc., San Francisco, 1964.

3. Budzikiewicz, H., C. Djerassi, and D. H. Williams, *Structure Elucidation of Natural Products by Mass Spectrometry*, Vols. I and II, Holden-Day, Inc., San Francisco, 1964.

4. Biemann, K., *Mass Spectrometry, Applications to Organic Chemistry*, McGraw-Hill Book Company, New York, 1962.

5. Beynon, J. H., *Mass Spectrometry and Its Application to Organic Chemistry*, Elsevier Publishing Company, Amsterdam, 1960.

6. Beynon, J. H., and A. E. Williams, *Mass and Abundance Tables for Use in Mass Spectrometry*, Elsevier Publishing Company, Amsterdam, 1963.

7. Reed, R. I., *Ion Production by Electron Impact*, Academic Press, New York, 1962.

8. McLafferty, F. W., *Mass Spectral Correlations*, Advances in Chemistry Series No. 40. American Chemical Society, Washington, D. C., 1963.

9. McLafferty, F. W., *Interpretation of Mass Spectra*, W. A. Benjamin, Inc., New York, 1966.

10. McLafferty, F. W., and J. Penzelik, *Index and Bibliography of Mass Spectrometry*, 1963–1965, Interscience Publishers, Inc., New York, 1967.

11. McLafferty, F. W., Mass Spectrometry, Chap. 2, Vol. II, in *Determination of Organic Structure by Physical Methods*, F. C. Nachod and W. D. Phillips, eds., Academic Press, Inc., New York, 1962.

12. Reed, R. I., *Application of Mass Spectrometry to Organic Chemistry*, Academic Press, Inc., New York, 1966.

13. Reed, R. I., *Ion Production by Electron Impact*, Academic Press, Inc., New York, 1962.

14. Kiser, R. W., *Introduction to Mass Spectroscopy and Its Applications*, Prentice-Hall, Inc., Englewood Cliffs, N. J., 1965, Chapter 5.

15. McDowell, C. A., ed., *Mass Spectrometry*, McGraw-Hill Book Company, New York, 1963.

16. "Catalog of Mass Spectra Data," American Petroleum Institute Res. Project No. 44, Carnegie Institute of Technology, Pittsburgh, Pa., and the Manufacturing Chemists Association Research Project, Agricultural and Mechanical College of Texas, College Station, Texas.

17. *Index of Mass Spectral Data*, A.S.T.M. Special Technical Publication No. 356, American Society for Testing and Materials, Philadelphia, 1963.

18. Cornu, A., and R. Massot, *Compilation of Mass Spectral Data*, Heydon and Sons Ltd., London, 1966.

19. Lederberg, J., *Tables and an Algorithm for Calculating Functional Groups of Organic Molecules in High Resolution Mass Spectrometry*, NASA Sci. Techn. Aerosp. Rep., N64-21426 (1964).

20. Waldron, J. D., ed., *Advances in Mass Spectrometry*, Pergamon Press, New York, 1959.

21. Elliott, R. M., ed., *Advances in Mass Spectrometry*, Pergamon Press, New York, 1963.

22. Mead, W. L., ed., *Advances in Mass Spectrometry*, Elsevier Publishing Company, Amsterdam, 1966.

23. Silverstein, R. M., and G. C. Bassler, *Spectrometric Identification of Organic Compounds*, 2nd ed., John Wiley & Sons, Inc., New York, 1967.

Part III

IDENTIFICATION OF
ORGANIC COMPOUNDS

9

Characterization

of an Unknown Compound

In Part III we shall be concerned with the classification of a compound with respect to the functional groups present and the ultimate unambiguous identification of the compound. Such a task is most efficiently accomplished through a systematic integration of physical and chemical methods. Just as the various forms of spectroscopy furnish us information in a synergistic manner, the combination of physical data with chemical observations also acts in a synergistic manner. Through the careful and systematic selection of appropriate physical and chemical methods, the task of reaching our goal—the final identification of an organic structure—may be greatly simplified; and the consumption of both time and material may be greatly reduced.

Of the various forms of spectroscopy, the state of the art, the cost and availability of the instrument, and the ease and rapidity of operation place infrared spectroscopy as the primary spectroscopic method to which the organic chemist turns when faced with a structural problem. Typically, as soon as there is some reasonable assurance of the purity of a compound, an infrared spectrum should be obtained. Chemists often find it convenient to examine the infrared spectrum of a mixture, not only as an aid in deciding on purification methods, but also as a method of checking to see that no artifacts appear during the purification process and to provide assurance that all components of a mixture are isolated during a separation process (Sec. 1.6).

Even upon cursory examination of the infrared spectrum of a compound, one can often classify the material with respect to functional groups present. Equally important, and often more so, one can readily establish the absence of many functional groupings, thus immediately eliminating many wet tests which may be laborious and costly in materials. For example, the casual observation of the absence of infrared bands in the

3600 to 3100 cm^{-1} (2.8 to 3.2 μ) region and of strong bands in the 1850 to 1650 cm^{-1} (5.4 to 6.1 μ) region readily eliminates from further consideration all N—H compounds (primary and secondary amines and amides, etc.), all O—H compounds (alcohols, phenols, carboxylic acids, oximes, etc.), and all C=O compounds (acids, esters, ketones, aldehydes, etc.). The same information gained from this quick observation (information which should be accessible in not more than 20 minutes from the time the sample is obtained) might also be obtained in several days by the indiscriminant use of more than 50% of the commonly used wet classification tests.

Compounds will be classified according to functional group; the most useful spectroscopic and chemical methods for the identification of the functional group will be outlined. Finally, the crucial physical and chemical methods that allow structure assignment will be discussed and the procedures outlined in some detail. The final chapter in this section provides instructions for making a thorough literature search for a compound, its reactions, and properties.

9.1   NOTEBOOKS AND REPORTS

The authors recommend the use of a bound notebook with numbered pages. In keeping a laboratory notebook, whether for class or research purposes, there is one cardinal rule: when one makes an observation, it should be written down immediately. Neatness and order, though important, are secondary. Chemists should never get into the habit of recording experimental observations on loose sheets of paper to be transcribed later into a bound notebook. Loose pages tend to get lost and one's immediate impressions are often tempered with time. One poor recall can be very costly in terms of time, materials, and reputation. The laboratory notebook should be kept at your side in the laboratory at all times. Observations should be recorded when they are made, and plenty of space should be allowed in your notebook for comments, additions of later information, and computations. Adequate references for the procedures used and data cited should be included. For each unknown identified, a summary page should be made, briefly outlining the arguments and literature citations employed in the structural identification.

Chemists should get into the habit of coding all samples, spectra, analyses, etc. with their initials and notebook page numbers. For example, the code RAF-2-147-D appearing on an infrared spectrum weeks, months, or years after the spectrum was originally run indicates that the spectrum is that of compound D described on page 147 of laboratory notebook 2 of chemist RAF. This system will allow the student or anyone else to immediately look up the history and source of the compound.

Step 1. Gross Examination

A. Physical state. The physical state of the unknown should be indicated. Additional description, such as "amorphous powder," "short needles," and "viscous liquid," often proves useful.

B. Color. Many organic compounds possess a definite color owing to the presence of chromophoric groups. Among the most common of the simple chromophoric groups are nitro, nitroso, diazo, azo, and quinone. Compounds with extensive conjugation are likely to be colored. The color of many samples is due to impurities frequently produced by air oxidation. Freshly purified aromatic amines and phenols are usually colorless; on storage, small amounts of these compounds are oxidized to highly colored quinone impurities. Stable, colorless liquids or white crystalline solids are not likely to contain the usual chromophoric groups or functional groups that are easily oxidized.

C. Odor. With a good nose, educated by experience, a chemist can often make a tentative identification of a common chemical, or he can make an intelligent guess as to functional groupings present. Alcohols, phenols, amines, aldehydes, ketones, and esters all have odors more or less characteristic of the general group. Mercaptans, low molecular weight amines, and isonitriles (particularly toxic) have characteristic odors that one seldom mistakes once experience has been gained. The student will soon learn to recognize the commonly used solvents by their odors. Many organic compounds are exceedingly toxic and will produce at least temporary discomfort upon inhalation. The reader is cautioned about the indiscriminate "whiffing" of compounds or mixtures about which little is known. To note the odor of an unknown substance, hold the tube pointed away from the face in one hand and gently wave the vapors from the tube toward the nose with the other hand.

Step 2. Determination of Purity and Physical Constants

One should be assured that the physical constants used in a structure identification are obtained with pure material. The necessity for purification or fractionation may be indicated by melting point and/or boiling point ranges, behavior on thin-layer or gas-liquid chromatography, or any inhomogeneity or discoloration. For all liquid samples, the refractive index should be determined.

Step 3. Classification by Functional Group

A. Determination of acidity or basicity and solubility behavior. The acidity or basicity of a compound, as determined with indicators or potentiometer or by solubility in acids or alkali, as well as general solubility behavior (Secs. 10.1–10.2), is useful, not only in the determination of possible chemical classes to which the compound may belong, but also to serve as a guide in the choice of solvents and procedures used for spectroscopic analyses.

B. Classification of functional groupings by spectroscopic and chemical means. As indicated earlier, in this text we shall use spectroscopy with significant emphasis on infrared as the primary method of determining the major functional groups present. Supplemental chemical tests will be used to confirm or clarify the assignments made by spectroscopic means.

C. Elemental analysis. Qualitative elemental analysis (Sec. 11.1) by the sodium fusion method is not a procedure which the chemist finds necessary to apply to every unknown compound with which he is faced. The presence or, more likely, the absence of many heteroatoms may be known from the source and/or history of the sample. The presence or absence of many heteroatoms may also be inferred from spectroscopic data. High-resolution mass spectrometry will provide exact elemental compositions. Evidence of the presence or absence of sulfur, halogens, and nitrogen can usually be obtained from low-resolution mass spectrometry. Infrared spectroscopy, by virtue of detection of various functional groups, can establish the presence or absence of elements. The presence of an amino, nitrile, or nitro group in the infrared spectrum provides obligatory evidence for the presence of nitrogen compounds. The presence of sulfur may be inferred from mercaptan, sulfone, sulfoxide, etc. bands in the infrared spectrum. The basicity of a compound may be sufficient to establish the presence of nitrogen. A simple Beilstein test is usually reliable for establishing the presence or total absence of halogen within a compound. Finally, quantitative microanalytical data for C, H, and N may often be available. Such data, together with the known presence or absence of oxygen as inferred by chemical or spectroscopic means, may also be sufficient to eliminate or establish the presence of other elements within the molecule.

Step 4. Final identification

At this point, based on the foregoing chemical and physical data, classification of the unknown compound as belonging to a specific functional group class should be possible. A comparison between the physical data obtained for an unknown compound with information about known compounds listed in the literature should be made, and an initial list of possibilities which melt or boil within 5° of the value observed for the unknown should be prepared. This initial list is usually obtained by consulting tables of compounds arranged in order of melting points or boiling points and according to the functional groups present. The most extensive of the tables—which also gives data on derivatives—is the *Tables for Identification of Organic Compounds*, published by the Chemical Rubber Publishing Co. as a supplement to the *Handbook of Chemistry and Physics*. It should be apparent that no one compendium is available that provides complete coverage of the literature on physical constants useful in identification work (refer to the more extensive discussion of literature searching provided in Sec. 13).

The initial list of possibilities may be immediately condensed to include only those compounds which conform to the data on elemental composition, solubility, chemical behavior, and other physical and spectral properties. In most cases, a tentative list of possibilities will not contain more than three to five compounds. A more extensive literature search is then made to ascertain additional properties of the compounds listed as possibilities. The final proof of structure of an unknown is accomplished through the preparation of selected derivatives and/or the collection, comparison, and evaluation of other physical and chemical data, such as neutralization equivalent, nuclear magnetic resonance, ultraviolet, and mass spectra, optical rotation or optical rotatory dispersion, and dipole moments.

The classical criteria for the conclusive identification of a compound are that (1) the compound has the expected chemical and spectral properties, and the physical constants match those given in the literature; (2) the appropriate derivatives prepared from the unknown melt within 1 to 2° of the melting points given in the literature. The actual criteria necessary to establish the identity of the compound beyond reasonable doubt is, of course, subject to considerable variation, depending on the kind and content of the information available to the experimenter and the complexity of the compound in question. In many cases, one derivative may be sufficient, and in some cases, it may not be necessary to prepare derivatives at all. Indeed, more rigorous and detailed proof of structure may be often obtained from a combination of carefully obtained and interpreted spectral and other physical data (e.g., molecular weights by osmometry or other methods), chemical reactivity (e.g., kinetics, pK_a's, etc.), and microanalytical data (e.g., found percentages of C, H, and other elements agreeing with the theoretical within 0.3%).

The task of the rigorous establishment of the identity of two compounds is considerably simplified if a known sample is available. In such a case, identity may be established by one or more of the following methods.

A. Mixture melting point. Two samples having the same melting point are identical if, upon admixture of the two samples, the melting point observed is not depressed relative to that of pure samples (see Sec. 2.2.2). This procedure, wherever possible, should be applied not only to the original unknown but also to derivatives thereof.

B. Spectral comparison. In cases of molecules of moderate complexities the superimposability of the infrared spectrum of the unknown with that of the known establishes the identity of the two. However, different molecules of simple structure, such as undecane and dodecane, may have remarkably similar spectra, whereas in exceedingly complicated molecules, such as steroidal glycosides, peptides, and many antibiotics, the resolution obtained in the infrared region is usually not sufficient to allow identification with certainty by means of spectral comparison. The method applies equally well to the use of spectra recorded in the literature; in either event, care must be taken to be sure that the spectra are recorded under identical conditions.

The comparison of spectra obtained by other spectroscopic methods also provides helpful criteria for the establishment of identity. Since nmr spectra run the gamut from the exceedingly simple to the extraordinarily complex, direct comparison may provide anything from very tenuous to very conclusive information. Mass spectral data may be expected to match very closely only when the spectra are obtained under exactly identical conditions (usually run one after the other on the same instrument). It should be noted that the mass spectral fragmentation pattern is often not sensitive to differences in spacial orientation of atoms or groups; the mass spectra of geometrical isomers are often almost indistinguishable. The superimposability of ultraviolet spectra only serves to indicate that compounds have the same chromophore.

C. Chromatographic comparison. Good but not necessarily rigorous criteria of the identity of two compounds can be established by comparative gas-liquid, thin-layer, and/or paper chromatography (Sec. 1.5). The standard procedure involves chromatographing the two compounds separately and as a mixture. Safer conclusions can be drawn if the comparisons are made under several conditions and by more than one method.

It should be apparent to the reader that it is impossible to set down an exacting set of conditions which must be met each time a compound is said to be conclusively identified. There are numerous methods by which identity may be established; some are more rigorous than others, some are more appropriate in one case than in another. In any event, there is no substitute for common sense and the careful and thoughtful planning and execution of crucial experiments, and a thorough evaluation of data.

9.3 SELECTION OF DERIVATIVES

Ideally, the derivatives should be easily and quickly made from readily available reagents and should be easily purified. The best derivatives meet the following standards:

1. A crystalline solid melting between 50 and 250°. Solids melting below 50° are often difficult to crystallize and recrystallize. Accurate determinations of melting points above 250° are difficult.

2. The derivative should have a melting point and other physical properties that are considerably different from the original compound.

3. Most importantly, the derivative chosen should be one whose melting point will single out one of the compounds from the list of possibilities. The melting points of the derivatives to be compared should differ from each other by a minimum of 5°.

In this text, for each functional group class, detailed procedures will be found only for the most generally suitable derivatives for which extensive data are readily available. In most cases, compilations of the melting points of the derivatives chosen will be found in *Tables for Identification of Organic Compounds*.

9.4 LABORATORY SAFETY

The first thing a chemist should do when beginning work in a new laboratory is to learn the locations and methods of use of the emergency facilities:

exits	fire blankets
eye wash facilities	gas masks
fire extinguishers	first aid supplies
safety showers	

Of equal importance is knowing how to secure help fast when it is needed. Emergency phone numbers should be posted. Is there a dime available for the public phone?

9.4.1 Eye Protection

1. Wear safety glasses at all times in the laboratory. If you normally wear prescription lenses, these will usually suffice for safety purposes during the small-scale routine laboratory work described in this text. Many persons find that the general-purpose safety glasses (especially plastic glasses) soon tire the eyes because of poor fit and poor lenses. The chemist who will be spending long hours in the laboratory is advised to purchase custom frames fitted with safety lenses from his local optometrist. Those who wear glasses should consider having hardened safety lenses in their next glasses.

Contact lenses are *not* eye protection devices; in fact, in an accident, they may increase the degree of injury to the eye. It is recommended that contact lenses not be worn in the laboratory or that full eye protection be used in conjunction with them at all times.

2. Never look directly into the mouth of a flask containing a reaction mixture.

3. Avoid measuring of acids, caustics, or other hazardous materials at eye level. Place a graduate cylinder on the bench and add liquids a little at a time.

9.4.2 Fire

1. Use flames only when absolutely necessary. Promptly extinguish any flame not being used.

2. Learn the location and use of fire extinguishers. For wood, paper, or textile fires, almost any kind of extinguisher is suitable. For grease or oil fires, avoid the use of water extinguishers—they simply spread the burning material. For fires involving electrical equipment, use carbon dioxide or dry chemical extinguishers. For fires involving active metals or metal hydrides, use dry chemical extinguishers or sand.

To put out a fire, first cool the area immediately surrounding the fire with the extinguishers to prevent the spread of the flames; then extinguish the *base* of the blaze. Remember to aim the extinguisher at the base of the fire and not up into the flames.

3. When clothing is afire, the victim should not run any distance. This merely fans the flames. Smother the fire by wrapping the victim in a fire blanket. Use a coat or roll the victim on the floor if a fire blanket is not readily available, or douse the flames under the emergency shower.

9.4.3 Handling of Chemicals

1. Be cautious at all times when handling chemicals, especially those about which you know little.

2. Handle all chemicals that produce corrosive, toxic, or obnoxious vapors in an exhaust hood.

3. Avoid direct contact with organic chemicals. Use plastic or rubber gloves when handling hazardous materials.

4. A lab coat or apron can protect you as well as your clothing.

5. Handle compressed gas cylinders with care. Always move them with a cart and strap them down when they are in place.

6. Never pour large quantities of volatile solvents into the sink.

7. Use special precautions with sealed glass vials of hazardous materials. Never subject a sealed glass container to severe thermal shock; e.g., do not place a commercial vial of methyl amine, etc., into a Dry Ice bath.

8. Never use your mouth to pipette dangerous liquids. It is always best to use a rubber safety bulb for pipetting.

9. When working with any potentially dangerous reaction, use adequate safety devices—safety shields, gloves, goggles, etc.

9.5 FIRST AID

Although severe injuries seldom occur in the chemistry laboratory, it is wise for all chemists to be familiar with such important first aid techniques as stoppage of severe bleeding, artificial respiration, and shock prevention. Why not consult a first aid manual today?

9.5.1 Treatment of Chemical Injuries
to Eyes

The most important part of the treatment of a chemical injury to the eye is that done by the victim himself in the first few seconds. Get to the eye wash fountain or any source of water immediately, and wash the injured eye thoroughly with water for 15 minutes. Thorough and long washing is particularly important in the case of alkaline materials. If appreciable injury has been suffered, a physician should be consulted at once.

9.5.2 Burns from Fire and Chemicals

Chemical burns of all types should be immediately and thoroughly washed with water. Alcohol may prove more effective in removing certain organic substances from the skin.

For simple thermal burns, ice cold water is a most effective first aid measure. If cold water or a simple ice pack is applied until the pain subsides, healing is usually more rapid.

For extensive burns, place the cleanest available cloth material over the burned area to exclude air. Have the victim lie down, and call a physician and/or an ambulance immediately. Keep the head lower than the rest of the body, if possible, to prevent shock. Do not apply ointments to severe burns.

9.5.3 Cuts and Wounds

The most common minor laboratory accident involves cuts on the hand. Such cuts can usually be treated by applying an antiseptic and a bandage. If the cut is deep and possibly contains imbedded glass, a physician should be consulted.

For severe wounds—don't waste time. Use pressure directly over the wound to stop bleeding. Use a clean cloth over the wound and press with your hand or both hands. If you do not have a pad or bandage, close the wound with your hand or fingers. Raise the bleeding part higher than the rest of the body unless broken bones are involved. Never use a tourniquet except for a severely mangled arm or leg. Keep the victim lying down to prevent shock. Secure professional help immediately.

10

Classification by Solubility

and Acid-Base Properties

10.1 DETERMINATION OF
SOLUBILITY CHARACTERISTICS

The solubility characteristics of a compound may be very useful in providing structural information. Fairly elaborate solubility and indicator determination schemes have been presented in previous texts on this subject. The authors of this text contend that such elaborate schemes are not required because of the present availability of spectral methods of functional group detection. In addition, the solubility test results obtained with rather complex molecules may not lend themselves to unambiguous interpretation. In view of these comments, a rather limited series of solubility tests are being recommended in this text. The solvents recommended for use in these tests are water, 5% hydrochloric acid, 5% sodium hydroxide, and 5% sodium hydrogen carbonate.

The effect of structure on the acidity or basicity of organic molecules will not be discussed in this text owing to a necessary limitation of space. For discussions concerning the inductive, resonance, and steric effects on the acidity and basicity of organic compounds, the student is referred to the modern organic chemistry texts.

The quantities of compound and solvent used in the solubility tests are critical if reliable data are to be derived. A compound will be considered "soluble" if the solubility of that compound is greater than 3 parts of compound per 100 parts of solvent at room temperature (25°). The experimentalist should use as little of the unknown compound as is necessary, generally 10 mg in 0.33 ml of solvent. Employing such small quantities of material requires a fairly accurate determination of the weight of the material to be used; however, it is not necessary to weigh accurately the amount of compound each time. A 10-mg portion of the compound should

307

be accurately weighed out. Subsequent 10-mg portions of the compound may then be estimated visually.

A summary of the solubility characteristics of some of the most important classes of compounds is found in Table 10.1.

Table 10.1. General Solubility Guidelines

Soluble in Water and Ether

Monofunctional alcohols, aldehydes, ketones, acids, esters, amines, amides, and nitriles containing five carbons or less.

Soluble in Water; Insoluble in Ether

Amine salts, acids salts, polyfunctional compounds such as polyhydroxy alcohols, carbohydrates, polybasic acids, amino acids, etc.

Insoluble in Water; Soluble in NaOH and NaHCO₃

High molecular weight acids and negatively substituted phenols.

Insoluble in Water and NaHCO₃; Soluble in NaOH

Phenols, primary and secondary sulfonamides, primary and secondary aliphatic nitro compounds, imides, and thiophenols.

Insoluble in Water; Soluble in Dilute HCl

Amines except diaryl and triaryl amines, hydrazines, and some tertiary amides.

Insoluble in Water, NaOH, and Dilute HCl, but Contain Sulfur or Nitrogen

Tertiary nitro compounds, tertiary sulfonamides, amides, azo compounds, nitriles, nitrates, sulfates, sulfones, sulfides, etc.

Insoluble in Water, NaOH, and HCl; Soluble in H₂SO₄

Alcohols, aldehydes, ketones, esters, ethers (except diaryl ethers), alkenes, alkynes, and polyalkylbenzenes.

Insoluble in Water, NaOH, HCl, and H₂SO₄

Aromatic and aliphatic hydrocarbons and their halogen derivatives, diaryl ethers, perfluoro-alcohols, -esters, -ketones, etc.

10.1.1 Solubility in Water

Water is a highly polar solvent possessing a high dielectric constant and is capable of acting as a hydrogen bond donor or acceptor. As a result, molecules possessing highly polar functional groups capable of entering into hydrogen bonding with water (hydrophilic groups) display a greater solubility in water than do molecules without such functional groups. Irrespective of the presence of a highly polar functional group in a molecule, a limiting factor on its solubility in water is the amount of hydrocarbon structure (*lipophilic* portion of the molecule) associated with the functional group. For example, methanol, ethanol, *n*-propanol, and *n*-butanol all possess solubilities in water at room temperature which would allow classification as "water soluble." However, n-pentanol would be classified as "insoluble." The solubility within a series of compounds is also dependent on the extent and position of chain branching, the solubility in water increasing as the branching increases. For example, all of the isomeric pentanols except 3-methyl-1-butanol and n-pentanol are soluble in water.

Similar solubility trends are observed with aldehydes, amides, amines, carboxylic acids, and ketones, although minor variations in the cutoff solubility point may occur in the various classes of compounds. The upper limit of water solubility for most compounds containing a single hydrophilic group occurs at about five carbon atoms. Most difunctional and polyfunctional derivatives are soluble in water as are most salts of organic acids and bases.

Distinction between monofunctional and polyfunctional compounds and salts can usually be made by testing their solubility in diethyl ether, a relatively nonpolar solvent. Glycols, diamines, diacids, or other similar di- and polyfunctional derivatives, as well as salts, are not soluble in diethyl ether.

Further solubility tests of water-soluble compounds in 5% hydrochloric acid, 5% sodium hydroxide, or 5% sodium hydrogen carbonate are meaningless in that their solubility in these solvents is primarily dependent on the solubility in water and not on the pH of the solvent. It is still possible, however, to detect the presence of acidic and basic functional groups in a water soluble compound. The pH of the aqueous solution of the compound, derived from the solubility test, is determined by placing a drop of the solution on a piece of pHydrion paper. If the compound contains an acidic functional group, the solution will be acidic (with phenols and enols, the solution will be only very weakly acidic, and a control test should be run on an aqueous solution of a phenol for comparison); and for compounds containing a basic functional group, the solution will be basic. It is usually possible to distinguish between a strong and a weak acid by the addition of one drop of dilute sodium hydrogen carbonate solution. Carbon dioxide bubbles will appear if a strong acid is present (pK_a less than 7). The pK of the acidic or basic functional group can be determined readily if it is desired (Sec. 10.2).

10.1.2 Solubility in 5 Percent Hydrochloric Acid

Compounds containing basic functional groups, e.g., amines (except triarylamines), hydrazines, hydroxylamines, aromatic nitrogen heterocyclics, but *not* amides (although some N,N-dialkyl amides are soluble), are generally soluble in 5% hydrochloric acid. Occasionally, an organic base will form an insoluble hydrochloride salt as rapidly as the free base dissolves, thus giving the appearance of an insoluble compound. One should observe the sample carefully as the solubility test is run.

10.1.3 Solubility in 5 Percent
Sodium Hydroxide

Compounds containing acidic functional groups with pK_a's of less than approximately 12 will dissolve in 5% sodium hydroxide. Compounds falling into this category include sulfonic acids (and other oxysulfur acids), carboxylic acids, β-dicarbonyl compounds, β-cyanocarbonyl compounds, β-dicyano compounds, nitroalkanes, sulfonamides, enols, phenols, and aromatic thiols. Certain precautions should be observed since some compounds may undergo reaction with sodium hydroxide, for example, reactive esters and acid halides, yielding reaction products which may be soluble or insoluble in the 5% sodium hydroxide solution. Certain active methylene compounds (β-dicarbonyl compounds, etc.) may undergo facile condensation reactions, producing insoluble products. If any reaction appears to occur during the course of the solubility test, it should be noted. Long chain carboxylic acids (C_{12} and longer) do not form readily soluble sodium salts but tend to form a "soapy" foam.

10.1.4 Solubility in 5 Percent
Sodium Hydrogen Carbonate

The first ionization constant of carbonic acid (H_2CO_3) is approximately 10^{-7}, which is less than the ionization constant of strong acids (carboxylic acids) but is greater than the ionization constant of weak acids (phenols). Therefore, only acids with pK_a's less than 6 will be soluble in 5% sodium hydrogen carbonate. This category includes sulfonic (and other oxysulfur acids) and carboxylic acids and highly electronegatively substituted phenols (for example, 2,4-dinitrophenol and 2,4,6-trinitrophenol).

10.1.5 Additional Solubility Classifications

Two additional solubility tests may be used to further classify compounds, although these tests are not generally necessary in that more valuable and definitive structural information can be derived from the various spectra of the compound.

Compounds containing sulfur and/or nitrogen but which are not soluble in water, 5% hydrochloric acid, or 5% sodium hydroxide are almost always sufficiently strong bases to be protonated and dissolved in concentrated sulfuric acid. For such compounds, the sulfuric acid solubility test provides no additional information. Compounds falling into this solubility category include most amides, di- and triarylamines, tertiary nitro, nitroso, azo, azoxy and related compounds, sulfides, sulfones, disulfides, etc.

Compounds not soluble in water, 5% hydrochloric acid, or 5% sodium hydroxide and which do not contain nitrogen or sulfur may be further

classified based on their solubility in concentrated sulfuric acid and 85%
phosphoric acid. Compounds insoluble in concentrated sulfuric acid include
hydrocarbons, unreactive olefins and aromatics, halides, diaryl ethers, and
many perfluoro compounds.

Compounds soluble in concentrated sulfuric acid include compounds
which contain very weakly basic functional groups. This solubility class
includes alcohols, aldehydes, ketones, esters, aliphatic ethers, reactive olefins
and aromatics, and acetylenes. Dissolution in concentrated sulfuric acid is
often accompanied by reaction, as indicated by color changes or decomposition, and should be noted. This solubility class may be subdivided further
on the basis of solubility in 85% phosphoric acid. The foregoing compounds
which are soluble in sulfuric acid and contain *5 to 9 carbon atoms* generally
will be soluble in 85% phosphoric acid.

10.2 DETERMINATION
OF IONIZATION CONSTANTS

The ionization constant of an acidic or basic functional group provides important information concerning the type of functional group present and the
type of substituents near the functional group. Ionization constant data are
particularly useful in the determination of structures of complex natural
products, for example, alkaloids and proteins. The use of ionization constant data in structure work requires the availability of ionization constant
data for simple model systems which contain all the essential structural features in the immediate vicinity of the acidic or basic functional group. The
original literature abounds with ionization constant data for a great variety
of systems, and no compilation of data will be presented in this text.

There are two relatively simple methods for the determination of ionization constants; one is a titrametric method requiring the use of a pH meter,
and the other is spectrophotometric, requiring the use of an ultraviolet-
visible spectrophotometer.

In the case of amines, especially liquid amines, it is usually more convenient and more accurate to make these determinations on a nonhygro-
scopic crystalline salt of the amine by titration with base rather than to
attempt a direct titration of the organic base with an acid.

10.2.1 Titrametric Method

The titrametric method of ionization constant determination involves the
titration of a known quantity of the acid or base with a base or acid, respectively, of known concentration. The pH, actually the hydrogen ion concen-

tration, is determined at several degrees of neutralization, and the ionization constant is calculated by means of Eqs. (10.1) and (10.2) for each point (method a) where [H-anion] and [cation-OH] represent the concentration of undissociated acid and base, respectively.

$$K_A = \frac{[H^+][anion]}{[H\text{-}anion]} \tag{10.1}$$

$$K_B = \frac{[cation][OH^-]}{[cation\text{-}OH]} \tag{10.2}$$

The ionization constant should remain constant throughout the neutralization curve. (Actually, the ionization constants calculated from Eqs. (10.1) and (10.2) will vary throughout the titration curve owing to changes in total ion concentration (ionic strength) and, in cases of relatively weak acids and bases, hydrolysis of the resulting ions. More elaborate calculations incorporating Debye-Hückel approximations and hydrolysis constants may be employed if very accurate ionization constants are desired.)

Another simple titrametric procedure involves the determination of the entire neutralization curve, pH vs. volume of titrant (method b). The pH at the half-neutralization point is taken as the value of the pK. This method should be used only with acids having pK_a's less than 4, and bases having pK_b's less than 4, to avoid errors introduced owing to hydrolysis of the ions in solution. This method requires the determination of the entire neutralization curve in order to calculate accurately the half-neutralization point.

10.2.1a Procedure: Determination of pK

A 50 to 100 mg accurately weighed portion of the acid or base is dissolved in 50 ml of distilled water. The electrodes of a pH meter, previously standardized carefully with a buffer having a pH in the expected pK region of the acid or base, are immersed in the solution, and quantities of carefully standardized 0.05 N acid or base are added from a burette. After each addition of titrant, the solution is allowed to mix thoroughly (the use of a magnetic stirrer is recommended) until equilibrium is attained, and the pH of the solution is recorded.

If one employs method a to calculate the ionization constant, the concentration of all species in solution must be calculated (the student should note that the total volume of the solution changes during the titration, and appropriate corrections in the calculations must be made). Utilization of method b does not require the exact calculation of the concentrations of the various species in solution.

If the quantity of sample is limited, as little as 5 mg of sample may be used. Corresponding reductions in the amount of solvent and in the concentration of the titrant may be required. Many times the solubility of the compound will not permit the determination of the ionization constant in water.   Mixed or nonaqueous solvents may be used, for example. water-ethanol, water-dioxane, or pure ethanol or dioxane; however, pK values determined in these solvents will differ substantially from the values determined in pure water.   Ionization constant comparisons should be made only when the same solvent system is employed.

10.2.2 Spectrophotometric Method

The electronic absorption spectra of acids and their conjugate bases, and similarly, bases and their conjugate acids, generally differ in the wavelength of maximum absorption and/or in extinction coefficient. The extent of the difference depends on the type of system present in the molecule.   The greater this difference, the more accurately one can determine the ionization constant.   The spectroscopic method is generally applicable only to conjugated systems which display absorption in the ultraviolet and visible regions.   The principal advantage of the spectroscopic method is the small amount of sample required, the final solutions being 10^{-3} to 10^{-5} molar.

To determine the ionization constant by spectroscopic methods, the absorption spectra of the protonated and unprotonated species must be available.   This is accomplished by recording the ultraviolet or visible spectrum of the compound in a sufficiently acidic solution such that the compound is completely protonated, and in a sufficiently basic solution such that only the unprotonated form is present.   These spectra provide the extinction coefficients and wavelengths required in the calculations.   The spectrum of the compound is then recorded in buffer solutions in which the compound will be present in both the protonated and unprotonated forms.   The concentrations of the protonated and unprotonated forms are then calculated (see Sec. 3.3 for the details of calculating extinction coefficients), and the ionization constant is calculated using Eqs. (10.1) and (10.2), in which the hydrogen-ion, or hydroxide-ion, concentration is calculated from the pH of the buffer used as solvent.   The ionization constant should be determined with the use of at least two different buffers.   A plot of the concentration of the protonated, or unprotonated, form vs. buffer pH will resemble a typical neutralization curve.

10.3   DETERMINATION OF NEUTRALIZATION AND SAPONIFICATION EQUIVALENTS

The neutralization equivalent is defined as the equivalent weight of an acid, or base, as determined by titration with base, or acid.   The neutralization equivalent may be used to determine the empirical formula of a molecule, or the number of acidic or basic functional groups contained in the molecule if the molecular weight is known.

The neutralization equivalent is calculated from Eq. (10.3). For procedural details, see Sec. 12.10.

$$\text{N.E.} = \frac{\text{weight of sample}}{\text{volume of titrant} \times \text{normality}} \tag{10.3}$$

The saponification equivalent is the equivalent weight of an ester or amide based on the ester and amide functional groups. The saponification equivalent is used in the same way as the neutralization equivalent to determine the empirical formula of the ester or amide. For procedural details, see Sec. 12.13.

11

Qualitative and

Quantitative Elemental Analyses

11.1   QUALITATIVE ELEMENTAL ANALYSIS

Before proceeding with chemical and spectral functional group analysis, it is helpful to determine what elements are present in a molecule and in what ratio these elements are present. The former data are derived via qualitative tests which can be readily carried out by the reader, whereas the latter data are usually obtained by trained analysts in laboratories specifically set up for such quantitative analysis. In addition to quantitative analysis of the elements present, these laboratories are usually capable of performing quantitative functional group analyses, for example, methyl, methoxyl, acetoxyl, and deuterium. The following paragraphs of this section outline the handling and interpretation of such analytical data.

The presence of carbon in organic compounds, or salts of organic compounds, will be assumed, as will the presence of hydrogen except in perhalogenated compounds; no specific tests for the presence of carbon and hydrogen will be given. (The presence of hydrogen is readily indicated by infrared and hydrogen nuclear magnetic resonance spectroscopy, as discussed in detail in Chapters 4 and 5.) Heteroatoms most commonly found in organic compounds include the halogens, oxygen, sulfur, and nitrogen. The detection of oxygen is relatively difficult by qualitative analytical means, and we shall rely on infrared spectral analysis and on solubility data to indicate its presence.

The presence of metallic elements, for example, lithium, sodium, and potassium (usually as the salts of acidic materials), can be indicated by an ignition test. A few milligrams of the substance are placed on a clean stainless steel spatula and carefully burned in a colorless burner flame. The presence of a noncombustible residue after complete ignition indicates the presence of a metallic element. The color of the flame during ignition may give some indication of the type of metal present. The ignition residue is dissolved in two drops of concentrated hydrochloric acid, and a platinum wire is dipped into the solution and placed in a colorless burner flame. The following colors are produced on ignition: blue by lead, green by copper and boron, carmine by lithium, scarlet by strontium, reddish-yellow by calcium, violet by potassium, cesium, and rubidium, and yellow by sodium. Precautions must be taken to avoid contamination of the sample by sodium because the intense yellow flame produced by sodium generally masks all other elements. If only trace amounts of sodium are present, the initial yellow flame will soon disappear, leaving the final colored flame of the elements present in greater amounts.

The qualitative tests for most of the elements are based on reactions involving an anion of the element, for example, the halide, sulfide, and cyanide; hence, a reductive decomposition of the compound is required. This is generally accomplished by means of fusion with metallic sodium.

$$C, H, O, N, S, X, \xrightarrow[\Delta]{Na} NaX, NaCN, Na_2S, NaCNS$$

A few of the elements are more readily detected as the oxyanions, formed in an oxidative decomposition of the compound by fusion with sodium peroxide. Such examples include phosphorous, sulfur, silicon, and boron. The reductive and oxidative procedures will be discussed separately.

11.1.1 Sodium Fusion

To accomplish the decomposition of organic compounds with sodium, the compound is fused with metallic sodium. Since, in general, the sample sizes employed are quite small, 1 to 10 mg, and the detection tests are quite sensitive, the student must take certain precautions to avoid contamination of the sample during and after the fusion. All glassware must be thoroughly cleaned and rinsed with distilled water, or preferably deionized water, and the solvents and reagents must be of analytical quality. *Safety glasses should be worn during these tests!*

11.1.1a Procedure: Sodium fusion

(a) If the organic sample to be analyzed is quite volatile, a 1 to 10-mg sample of the material is placed in a 4-in. test tube and 30 to 50 mg of sodium, a piece about half the size of a pea, is cautiously added to the test tube. (The mouth of the test tube should be pointed away from the experimenter and other people in the laboratory to prevent their being splashed with material from the tube should a violent reaction ensue on the addition of the sodium.) The test tube is then gently heated until decomposition and charring of the sample occurs. When it appears that all the volatile material has been decomposed, the test tube is strongly heated until the residue becomes red. The test tube is allowed to cool to room temperature, and a few drops of methanol are added to decompose the excess sodium. If no gas evolves on the addition of the methanol, an excess of sodium was not present, and there is a distinct possibility of incomplete conversion of the elements to their anions. The fusion should be repeated with a larger quantity of sodium. (Some procedures call for the use of soft-glass test tubes, which are plunged into cold distilled water when red hot, causing the tube to fracture and allowing the salts to dissolve in the water. This procedure may prove dangerous if an excess of sodium remains in the hot tube!)

The contents from the tube are boiled with 1.5 to 2.0 ml of distilled water, diluted to 10 ml with distilled water, and the mixture is then filtered or centrifuged. The decomposition of the organic material usually leads to the extensive formation of carbon, which may prove very difficult to remove. Occasionally, filtration through a filtering aid (e.g., Celite) which has been thoroughly washed with distilled water, will remove the finely divided carbon particles. The resulting solution should be clear and nearly colorless. If the solution is highly colored, the entire fusion process should be repeated because the color may interfere with the detection tests. This final solution will be referred to as the working solution in the discussions of the following elemental tests.

(b) An alternative procedure for carrying out a sodium fusion involves the slow addition of the organic material to a portion of molten sodium contained in the test tube. This procedure should be used only with relatively nonvolatile compounds to avoid loss of sample by volatilization. After addition of the sample, the tube and its contents are heated until red and then allowed to cool; a few drops of methanol are then added to decompose the excess sodium. Proceed as in (a) above.

Although the sodium fusion procedure for the decomposition of organic compounds is the simplest and most useful, several other reductive decompositions have been employed. These procedures involve the use of magnesium, magnesium and an alkali carbonate, zinc and an alkali carbonate, or soda lime. In the decompositions utilizing magnesium and soda lime, the organic nitrogen is converted eventually into ammonia. Organic nitrogen is converted to cyanide in the zinc and alkali carbonate decomposition. The decomposition mixtures are dissolved in water, clarified, and used in the following elemental detection tests.

11.1.2 Detection of Halides

The presence of chlorine, bromine, and iodine can be readily detected by the precipitation of the corresponding silver halides on treatment with silver ion. Fluorine cannot be detected in this test because silver fluoride is soluble.

11.1.2a Procedure: Detection of halides

The presence of nitrogen and sulfur interfere in this test; sulfide and cyanide must be removed before treatment of the solution with silver ion. This is accomplished by acidification of the working solution followed by gentle heating to boil off the hydrogen cyanide and hydrogen sulfide formed on acidification. *This procedure must be carried out in a hood.* A great excess of sulfuric acid should be avoided since silver sulfate may precipitate from solutions containing a high concentration of sulfate ions. Take 0.5 ml of the working solution and carefully acidify by the dropwise addition of 10% sulfuric acid. (The pH of the solution may be checked readily by immersing a clean stirring rod into the acidified solution and then applying a piece of pHydrion paper to the liquid adhering to the stirring rod.) The solution is gently boiled, over a microburner, for about 3 to 5 minutes. One or two drops of aqueous silver nitrate solution are added; the formation of a substantial quantity of precipitate indicates the presence of chloride, bromide, or iodide. A white precipitate soluble in ammonium hydroxide indicates the presence of chloride; a pale yellow precipitate slightly soluble in ammonium hydroxide indicates bromide; and a yellow precipitate insoluble in ammonium hydroxide indicates iodide. Fluoride does not form a precipitate with silver ion. The formation of a slight cloudiness in the solution is not indicative of a positive test. It is recommended that a halide analysis on a known compound be carried out for comparison purposes.

Tests for all of the halogens must be carried out if the initial silver nitrate test is positive. This prevents missing one or more of the halogens if more than one halogen was originally present in the unknown. The chemical distinction between chloride, bromide, or iodide is based on the ease of oxidation of iodide over bromide, and bromide over chloride, and the ability of bromine, but not iodine, to add to an active site of unsaturation.

11.1.2b   Procedure: Distinction between chloride, bromide, and iodide

A 0.5-ml portion of the working solution is placed in a small test tube and acidified with 10% sulfuric acid. It is then gently boiled in the hood to remove any hydrogen cyanide and hydrogen sulfide, as described before. To this solution are added 4 to 5 drops of 0.1 N potassium permanganate. The contents of the test tube are shaken for approximately 1 minute, 10 to 20 mg oxalic acid is added to discharge the color of the excess permanganate, and 0.5 ml of carbon disulfide (carbon disulfide is extremely flammable and toxic, and due precautions should be taken) is added to the tube. The contents of the tube are again shaken and the two phases allowed to separate. The presence of a color in the carbon disulfide layer indicates the presence of bromine or iodine; a purple color is formed if iodine is present, and a red-brown color if bromine or both bromine and iodine are present. The formation of no color in the carbon disulfide layer indicates the presence of chlorine, although further confirmation is required.

Distinction between bromine and/or iodine is readily made by the addition of 1 to 2 drops of allyl alcohol. If only iodine is present, the color of the carbon disulfide layer will remain, whereas with only bromine, the color will be immediately discharged. If both bromine and iodine are present, the color of the carbon disulfide layer will turn from red-brown to purple.

To check for the presence of chloride ion, the aqueous layer is removed from the test tube by means of a capillary pipette and placed in another clean test tube. Several drops of nitric acid (6 N) are added, and the contents of the tube are gently boiled for 2 to 3 minutes (the nitric acid oxidizes any bromide or iodide, but not chloride, to the corresponding free halogen which is vaporized during the boiling process). The solution is then cooled and 2 drops of 0.1 M silver nitrate solution are added. The production of a white precipitate (but not a light cloudiness) indicates the presence of chloride ion.

Detection of fluorine: Owing to the solubility of silver fluoride in aqueous solutions, the presence of fluorine cannot be detected by the use of silver nitrate. The simplest procedure for the detection of fluoride involves the reaction of fluoride ion with the red zirconium alizarin complex to produce free alizarin (yellow in color) and the colorless zirconium hexafluoride complex.

11.1.2c Procedure: Detection of fluoride

A 0.5-ml portion of the working solution is acidified with 3 to 4 drops of concentrated hydrochloric acid. To this solution are added 2 to 3 drops of the zirconium alizarin complex solution, prepared by mixing equal volumes of a 1% ethanol solution of alizarin and a 2% solution of zirconium chloride, or nitrate, in 5% hydrochloric acid. The change in color from red to yellow indicates the presence of fluoride ion.

Beilstein test for halogens: The presence of chlorine, bromine, or iodine in organic compounds can be detected by the Beilstein test. The test depends on the production of a volatile copper halide when an organic halide is strongly heated with copper oxide. The test is extremely sensitive, and a positive test should always be confirmed by other methods.

11.1.2d Procedure: Beilstein test

A small loop in the end of a copper wire is heated to redness (an oxide film is formed) in a Bunsen flame until the flame is no longer colored. After the loop has cooled, it is dipped into a little of the compound to be tested and then reheated in the non-luminous Bunsen flame. A blue-green flame produced by volatile copper halides constitutes a positive test for chlorine, bromine, or iodine (copper fluoride is not volatile).

Very volatile compounds may evaporate before proper decomposition occurs, causing the test to fail. Certain compounds such as quinoline, urea, and pyridine derivatives give misleading blue-green flames owing to the formation of volatile copper cyanide.

Mass spectrometric detection of halides: All the halogens except fluorine exist as a mixture of isotopes. The mass spectrum of a halogen-containing compound will display several peaks in the parent mass region, which will be separated by the difference in mass of the halogen isotopes and will be in an intensity ratio equal to the natural abundance ratio (see sec. 8.4.1). Mass spectrometry is also valuable in indicating the number of halogen atoms that occur in a single molecule.

11.1.3 Detection of Nitrogen

The detection of nitrogen is based on the conversion of the nitrogen present in a given molecule to the cyanide ion which is usually detected as the cyano complex formed from ferrous ammonium sulfate (Prussian blue).

11.1.3a Procedure: Detection of nitrogen

(a) A 0.5-ml portion of the working solution is adjusted to pH 13 (as determined by testing with pHydrion paper) by the addition of 6 N sodium hydroxide. Two drops each of a saturated ferrous ammonium sulfate solution and 30% potassium fluoride solution are added, and the resulting solution is boiled gently for 30 sec. The solution is immediately acidified by the addition of 30% sulfuric acid until the colloidal iron hydroxide dissolves. An excess of acid must be avoided as it may interfere with the test. The production of a blue color indicates the presence of nitrogen in the original compound.

The formation of a very faint blue color, or a pale green color, is indicative of a poor fusion, resulting in an incomplete conversion of the nitrogen in the original molecule to cyanide ion. This is particularly true of compounds in which the nitrogen is in a high state of oxidation, for example, nitro compounds. The fusion should be repeated with the use of a greater excess of sodium and more drastic heating. It is recommended that, in such cases, a control fusion be carried out on a known nitro derivative. It should be pointed out here that the functional groups containing nitrogen in a high oxidation state are quite readily detected by infrared spectroscopy (Chapter 4), and it is not necessary to be concerned about poor fusion results.

The presence of sulfide interferes in the foregoing test for nitrogen. If sulfur is present in the original molecule, the iron sulfide formed when the ferrous ammonium sulfate is added must be removed by centrifugation and decantation of the supernatant liquid. The test for nitrogen is then continued.

(b) An alternative procedure for the detection of cyanide involves acidification of 1 ml of working solution with acetic acid, followed by the addition of two drops of a freshly prepared 1% solution of benzidine in 50% acetic acid and one drop of 1% copper sulfate solution. The formation of a blue color or precipitate indicates the presence of nitrogen. The presence of iodide ion results in the formation of a green precipitate; the other halides do not interfere with this test.

11.1.4 Detection of Sulfur

11.1.4a Procedure: Detection of sulfide

A 1-ml portion of the working solution is acidified with acetic acid, and 2 to 3 drops of dilute lead acetate solution are added. The formation of a black precipitate indicates the presence of sulfur.

11.1.5 Detection of Other Metallic and Nonmetallic Elements

If the ignition test reveals the presence of metallic elements, decomposition of the material in the presence of ammonium nitrate is carried out, and the detection of the elements is carried out by the general inorganic qualitative analysis scheme.

11.1.5a Procedure: Fusion with ammonium nitrate

Approximately 200 to 250 mg of compound is mixed with 500 mg of ammonium nitrate in a clean crucible, and the mixture is strongly heated for 15 to 20 min. The residue is allowed to cool and is then dissolved in concentrated sulfuric acid or aqua regia. The excess acid is evaporated (*in a fume hood*!), and the residue is ignited again for a short period of time. If carbonaceous material still remains, a few drops of concentrated nitric acid are added and ignition is repeated. The resulting residue is analyzed by the usual inorganic qualitative analysis scheme. Mercury is not detected in this procedure since it distills off as free mercury. Mercury can be detected by boiling 10 to 20 mg of the compound with 10 ml of concentrated potassium chlorate solution until the reaction mixture is colorless. To test the solution for mercury, a piece of clean copper wire is immersed into the solution for 10 min. The formation of a mercury amalgam, as evidenced by the color of the wire, indicates the presence of mercury.

11.1.6 Oxidative Decomposition

Oxidative decomposition of organic compounds is generally accomplished by fusion of the compound with sodium peroxide. This procedure requires the use of a small bomb. The qualitative tests employed are those outlined earlier and found in inorganic qualitative analysis texts.

Swift and Nieman have developed procedures for the complete analysis of organic compounds based on sodium peroxide fusion.[1]

11.2   QUANTITATIVE ELEMENTAL ANALYSIS

Data on the quantitative elemental composition of an unknown are particularly useful in establishing structure. Traditionally, microanalytical data have been accepted as important criteria for proof of structure and purity. In most primary chemical journals, microanalytical data are given for all new compounds reported in a paper; the data are usually presented in the following form:

> *Anal.*   Calcd for $C_{10}H_{14}N_2O_4$: C, 53.09; H, 6.24; N, 12.38.
> Found: C, 53.20; H, 6.35; N, 12.44.

(In writing a formula, after C and H, other elements are listed in alphabetical order.) Such quantitative elemental analyses are not generally performed by the individual chemist; they are usually made by trained technicians in laboratories (often outside independent laboratories) specifically designed for this purpose.

The most common, and hence least expensive, analytical data are those for carbon-hydrogen and nitrogen. In these analyses, the sample is quantitatively combusted to yield carbon dioxide, water, and nitrogen. In the classical technique, the water is collected by adsorption on calcium chloride, the carbon dioxide is collected on Ascarite (sodium hydroxide on asbestos), and the nitrogen is determined by volume. In this method, a sample of about 5 mg is required for a carbon-hydrogen determination, and an additional 5 mg for nitrogen and each other determination.

The simultaneous microdetermination of carbon, hydrogen, and nitrogen may be achieved by automatic gas chromatographic methods. With such

---

[1] *A System for the Ultimate Analysis of Chemical Warfare Agents*, published by the Chemical Warfare Service, U.S.A., and briefly described in Analytical Chemistry, **26**, 538 (1954).

instruments, a sample of about 1 mg is converted to water, carbon dioxide, and nitrogen, which are recorded as a three-peak chromatogram. The composition can be determined by peak height to an accuracy of about 0.3%.

To derive the empirical formula of a compound which contains n different elements, elemental analyses of $n - 1$ of the elements present must be determined. The percentage of the remaining element may be determined as the difference between the sum of the percentages of the other elements and 100%. (The percentage of oxygen is usually determined by difference.) Certain errors may be introduced by this method. The acceptable experimental error involved in the determination of the percentage of each element is ±0.3%. If the analysis of a majority of the elements is high, or low, then the analysis of the element obtained by difference may be anomalously low, or high. With molecules of a high degree of complexity, such errors may result in an inability to distinguish between two or more empirical formulas unless they are very accurately determined. It is therefore recommended that, in cases when complex structures are thought to be involved, the quantitative analysis of each element contained in the compound be determined. Multiple analyses are also recommended in such cases, using an average value for each element in the calculations of the empirical formula.

For the conversion of microanalytical data to an empirical formula, the first step is to divide the percentage of each element by its atomic weight and next divide the resulting numbers by the smallest one to determine the atomic ratios.

Element	Percentage	Atomic weight		Atomic ratio
C	$67.38 \div 12.01 = 5.61$			10
H	$7.92 \div 1.01 = 7.84$	$\times \dfrac{1}{0.56} =$		14
N	$15.72 \div 14.01 = 1.12$			2
O	$8.98 \div 16.00 = 0.56$			1

The empirical formula for the above compound is $C_{10}H_{14}N_2O$. Conversion of an empirical formula to a molecular formula requires molecular weight data (see discussion on the determination of molecular weights, Sec. 2.8). Sometimes, it may be known from other data that an unknown contains only one chlorine, two nitrogens, etc.; in such cases, the molecular formula may be determined directly from analytical data.

A compilation of calculated analytical data for a great number of compounds containing C, H, N, O, and S in various combinations has been published in book form and is very useful in handling analytical data.[2]

It should be obvious from the foregoing paragraphs that great care must be exercised in the preparation of a sample for analysis. The sample must be of very high purity. Recrystallized solids should be heated at low temperatures, below the melting or decomposition points, and under vacuum to

---

[2]George H. Stout, *Composition Tables for Compounds Containing C, H, N, O, S.* W. A. Benjamin, Inc., New York (1963).

remove solvent molecules. Hygroscopic compounds must be protected from moisture. The presence of as little as 1 mole percent of water, which is generally a very low weight percent, may be sufficient to produce a bad analysis. The analyst should always be informed about the hygroscopic nature of a compound. He should also be warned of compounds of explosive or very toxic nature.

11.2.1 Interpretation of Empirical Formula Data

Once the chemist has determined the molecular formula of a compound, valuable information can be derived from the molecular formula with respect to gross features of the molecule. In particular, the molecular formula data can be used to calculate the total number of rings and/or double and triple bonds (sites of unsaturation) in the unknown molecule.

The number of sites of unsaturation in a molecule is conveniently calculated by means of Eq. (11.1)

$$N = \frac{\sum_i n_i(v_i - 2) + 2}{2} \tag{11.1}$$

where N is the number of sites of unsaturation, n_i is the number of atoms of element i, and v_i is the absolute value of the valence of element i. The following examples will illustrate the use of Eq. (11.1).

EXAMPLE 1. Elemental analysis indicates an empirical formula of C_6H_6. Using v_i of 4 for carbon and 1 for hydrogen, we have

$$N = \frac{6(4 - 2) + 6(1 - 2) + 2}{2} = \frac{8}{2} = 4$$

sites of unsaturation.

EXAMPLE 2. Elemental analysis indicates an empirical formula of C_7H_6ClNO. Using v_i of 4 for carbon, 1 for hydrogen, 1 for chlorine, 3 for nitrogen, and 2 for oxygen, we have

$$N = \frac{7(4 - 2) + 6(1 - 2) + 1(1 - 2) + 1(3 - 2) + 1(2 - 2) + 2}{2} = 5$$

sites of unsaturation.

EXAMPLE 3. Elemental analysis indicates an empirical formula of $C_{19}H_{18}BrP$. The physical and chemical properties indicate that the bromide is ionic and that we must be dealing with a pentavalent phosphorus. Thus, using v_i of 4 for carbon, 1 for hydrogen and bromine, and 5 for phosphorus, we have

$$N = \frac{19(4 - 2) + 18(1 - 2) + 1(1 - 2) + 1(5 - 2) + 2}{2} = 12$$

sites of unsaturation.

In these calculations, no distinction can be made among the various types of sites of unsaturation. The experimentalist must rely on spectral and chemical tests to indicate what types of functional groups, and hence their worth in sites of unsaturation, are present in any given molecule. A single ring, C=C, C=O, C=N, N=O, or any other doubly bonded system is considered as a single site of unsaturation; C≡C and C≡N are considered as two sites of unsaturation. The phosphorus-oxygen and sulfur-oxygen coordinate covalent bonds ($\equiv\overset{+}{P}$—O^- and $>\overset{+}{S}$—O^-) are incorporated as single bonds and thus are *not* sites of unsaturation. An aromatic ring, containing a combination of rings and double bonds, represents four sites of unsaturation.

11.2.2 Deuterium Analysis

Of increasing importance is the availability of quantitative deuterium content analysis in the presence of hydrogen. Three procedures are currently available: combustion analysis, nuclear magnetic resonance, and mass spectrometric methods.

Combustion analysis for deuterium involves the conversion of all hydrogen and deuterium into a mixture of light and heavy water. The water from the combustion of the compound is collected, and the density of the sample is determined by comparison with mixtures of light and heavy water of known composition in a density gradient solvent system. The result is recorded as the mole percent of deuterium based on the total hydrogen content.

11.2.3 Quantitative Functional Group Analysis

Most analytical laboratories are capable of carrying out quantitative functional group analyses. (See Table 5 for examples of typical available functional group analyses in addition to elemental analysis.) Since most of these analyses involve a chemical degradation, the yields of the fragments to be analyzed will not correspond to 100% of the functional group contained in the starting molecule. The net result is that less than integral functional group analytical results will be obtained, and a correction factor must be applied. This correction factor is generally $1/0.75$ to $1/0.85$ (to be multiplied times the observed results).

Table 5. **Available Analysis for Elements and Functional Groups***

Elements	*Functional groups*
Carbon-hydrogen	Acetyl
Nitrogen	Alkoxyl
Phosphorus	Benzoyl
Oxygen	C-Methyl (Kuhn-Roth)
Sulfur	N-Methyl
Fluorine	Amino, primary (van Slyke)
Chlorine	Hydroxyl
Bromine	Peroxide oxygen
Iodine	Bromine or iodine number
Metallic elements of	Microhydrogenation
groups Ia, IIa, IIIa,	Active hydrogen (Zeisel)
IVa and a few selected	Ash
transition metals	Lactone titration
	Molecular weight (Rast and osmometric)
	Neutralization equivalent
	pK of acids and bases
	Saponification equivalent
	Volatile acid

*Selected from price lists of a variety of analytical companies.

11.2.4 Active Hydrogen Determination

Active hydrogen may be classified in two categories: The first consists of the relatively acidic hydrogens bonded to oxygen or sulfur (e.g., alcohols and acids), and the second type consists of the weakly acidic hydrogens bonded to a carbon alpha to a strong electron withdrawing group (e.g., a carbonyl or cyano group). The first category of active hydrogen is determined readily by treatment of the compound with methyl magnesium iodide (Zerevitinov determination), the methyl Grignard reagent reacting to give one mole of methane for each mole of active hydrogen. A carefully weighed portion of the sample is dissolved in a high boiling anhydrous ether, e.g., dimethoxyethane, and placed in a flash attached to a gas measuring apparatus (an inverted, filled burette whose opening is maintained below the surface of water, or other neutral liquid, in a beaker and connected to the flask) and equipped with a pressure equilibrating addition funnel. The quantity of material used will depend on the expected number of acidic hydrogens and on the volume of the gas measuring burette. The volume of the solvent should be kept to a minimum. The Grignard reagent is prepared from methyl iodide and excess magnesium in the solvent used, and it is put in the addition funnel. The funnel is stoppered, and the methyl Grignard reagent is slowly added to the compound. The volume of the displaced gas is measured, corrected to standard pressure and temperature and for liquid vapor pressure, and is converted into moles of methane produced. The ratio of the number of moles of methane produced to the number of moles of unknown employed is the number of acidic hydrogens. In a similar test the volume of hydrogen produced from the reaction of active hydrogens with lithium aluminum hydride is measured.

The second category of active hydrogen bonded to carbon is determined by deuterium exchange experiments. A small amount of compound (100 mg) is dissolved in deuterium oxide, or an ethanol-OD deuterium oxide mixture, in the presence of a small amount of basic catalyst, usually sodium carbonate or occasionally very dilute sodium deuteroxide. The solution is allowed to stand for several hours, and the compound is recovered by extraction or evaporation. The process is repeated to ensure maximum isotope concentration in the sample. The sample is carefully purified for deuterium analysis by combustion, nuclear magnetic resonance, or mass spectroscopy (the sample should be kept away from sources of replaceable protium). Much smaller samples may be used when analysis is available by mass spectrometry. Most hydrogens alpha to an activating group capable of stabilizing anions will undergo exchange under the foregoing conditions.

$$ -\overset{|}{\underset{H}{C}}-\overset{/}{C}=X \xrightarrow{\text{base}} -\overset{|}{\underset{\ominus}{C}}-\overset{/}{C}=X \longleftrightarrow \; >C=\overset{/}{C}-X^- \xrightarrow{D^+} -\overset{|}{\underset{D}{C}}-\overset{/}{C}=X $$

A further distinction among active hydrogens on methine, methylene, and methyl positions can be made. Activated methylene and methyl groups undergo condensation reactions with benzaldehyde to give benzylidene derivatives, as illustrated by the following equation:

$$ -\overset{\overset{X}{\|}}{C}-CH_3 + C_6H_5CHO \xrightarrow{\text{base}} -\overset{\overset{X}{\|}}{C}-CH=C\overset{\diagup H}{\diagdown C_6H_5} $$

The sample is dissolved in absolute ethanol in the presence of benzaldehyde and sodium ethoxide (prepared by the cautious addition of sodium metal to absolute ethanol) or sodium hydroxide. The benzylidene derivative may precipitate and be collected by filtration, or it may be recovered by extraction with ether after dilution of the reaction mixture with water. The sample is purified and its molecular formula determined by combustion analysis or molecular weight determination. The introduction of each benzylidene moiety will result in an increase in the molecular formula by C_7H_4. Activated methine positions do not undergo condensation reactions to produce benzylidene type derivatives.

Functional Group Classification

and Characterization

This text gives procedures using semimicro quantities of materials, typically in the 50 to 500-mg range. The experienced and careful chemist will have no trouble manipulating quantities on a much smaller scale. Valuable material may be conserved by running solubility and classification tests in capillary tubes. In most cases, infrared spectra of compounds typical of the functional class under consideration are reproduced in this chapter. All temperatures are given in °C.

12.1   HYDROCARBONS

12.1.1 Alkanes and Cycloalkanes

12.1.1a Classification

The following observations are indicative of saturated hydrocarbons:

1. *An exceptionally simple infrared spectrum.* In the infrared spectra of hydrocarbons, the strongest bands appear near 2900, 1460, and 1370 cm^{-1} (3.45, 6.85, and 7.30 μ) (Fig. 12.1). In simple cyclic hydrocarbons containing no methyl groups, the latter peak is absent. Other compounds which typically give rise to rather simple infrared spectra include aliphatic sulfides and disulfides, aliphatic halides, and symmetrically substituted olefins and acetylenes.

2. *Negative iodine charge-transfer test.* When iodine is dissolved in compounds containing π electrons or nonbonded electron pairs, a brown solution results. The brown color is due to the formation of a complex between

the iodine and the π or nonbonded electrons. These complexes are called charge-transfer or π complexes. Solutions of iodine in nonparticipating solvents are violet.

$$\bigcirc\kern-1.6em\big\|\cdots I_2 \qquad R_2O \cdots I_2$$

12.1.1b Procedure: Iodine charge-transfer test

On a spot-test plate place a very small crystal of iodine; add one or more drops of the unknown liquid. Saturated hydrocarbons and their fluorinated and chlorinated derivatives, and arenes and halogenated derivatives give violet solutions; all other compounds give brown solutions. The test should be employed only with liquid unknowns. For color comparisons, it is strongly recommended that knowns be run concurrently with the unknowns.

3. *Insoluble in cold concentrated sulfuric acid.* Compounds that are unsaturated or possess a functional group containing nitrogen or oxygen are soluble in cold concentrated sulfuric acid. Saturated hydrocarbons, their halogen derivatives, simple aromatic hydrocarbons, and their halogen derivatives are insoluble.

4. *Negative test for halogens.* On the basis of the chemical, physical, and spectroscopic behavior, the most likely compounds to be confused with saturated hydrocarbons are the alkyl halides. If an unknown is suspected of being a hydrocarbon, the absence of halides should be verified by the Beilstein or sodium fusion test. The presence or absence of halogens may also be indicated by quantitative or qualitative measurements of the density of the unknown.

5. *No resonance appears below δ2 in the nmr spectrum.* In most functionalized compounds, resonance of one or more hydrogens is usually found downfield from δ2; exceptions include compounds in which the functional group is located on a tertiary carbon and certain alkyl metal derivatives.

6. *No absorption above 185 mμ in the ultraviolet spectrum.*

12.1.1c Characterization

There are no suitable general chemical methods for the conversion of saturated hydrocarbons to useful derivatives; they are typically inert or undergo nondiscriminate reactions which produce inseparable mixtures. Saturated hydrocarbons are, therefore, best identified through physical constants and spectroscopic methods.

The most useful physical constants are the boiling point and refractive index. Specific gravity measurements are sometimes employed. Standard pycnometers or gravitometers require careful temperature control and large amounts of material; however, accurate determinations of specific gravity

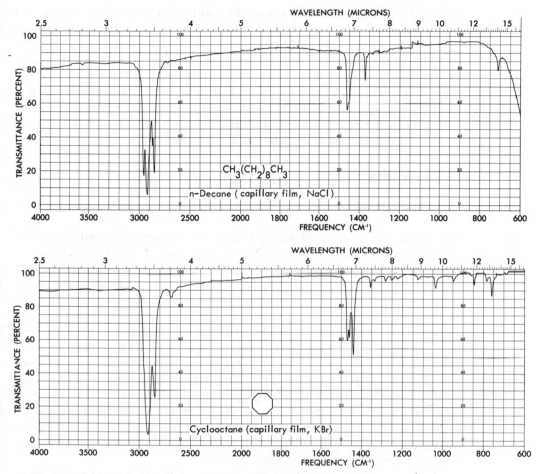

Fig. 12.1. Typical infrared spectra of alkanes. Note the absence of a 1370 cm^{-1} band for methyl in the spectrum of cyclooctane.

may be made on very small amounts of material by means of a capillary technique (Sec. 2.5). Retention times in gas-liquid chromatography are often very helpful for the tentative identification of a hydrocarbon.

Of the spectroscopic methods, mass spectroscopy is perhaps the most useful for the final identification of the saturated hydrocarbons. The utility of the method is greatly enhanced by the large number of reference spectra available. For the larger saturated hydrocarbons, nmr is of limited utility because of the small differences in chemical shift of the various types of hydrogens. In some cases, integration of the nmr spectrum may provide information on the ratio of methyl to other types of hydrogens in the molecule. In those cases in which the ring bears hydrogens, nmr is the best way to detect the presence of cyclopropanes because of their unique high-field resonance. Careful examination of the infrared spectrum can provide information about certain structural features of an unknown hydrocarbon, e.g., geminal methyl groups, but typically cannot conclusively lead one to the assignment of structure. There is even some danger in assigning structures to hydrocarbons based upon comparison of infrared spectra because of the small differences in the relatively simple spectra of compounds with closely related structure, e.g., 4-methyl and 5-methylnonane. Comparative gas-liquid chromatography is used in distinguishing hydrocarbons once some idea of the structure or molecular weight is available.

12.1.2 Alkenes

12.1.2a Classification

The infrared spectra of alkenes are generally considerably more complex than those of the saturated hydrocarbons (*cf.* Figs. 12.1–12.2). The C=C absorption appears in the 1680 to 1620 cm^{-1} (5.95 to 6.17 μ) region; unfortunately, this band varies from very intense to entirely absent in completely symmetrical olefins. However, unsaturation may also be detected by the vinyl hydrogen band just above 3000 cm^{-1} (3.33 μ) (if they are not hidden by the stronger, main band below 3000 cm^{-1}) or by the C—H out-of-plane bending bands in the 1000 to 700 cm^{-1} (10.0 to 14.3 μ) region. In many cases, the latter may be intense and very diagnostic. The presence of vinyl and allylic hydrogens may often be inferred from nmr spectra by the characteristic bands near $\delta 5$ to $\delta 7$ and near $\delta 2$, respectively. Unconjugated olefins show only simple end absorption in the ultraviolet spectra recorded to 200 mμ. The characteristic maxima and extinction coefficients at higher wavelengths for conjugated systems make ultraviolet spectroscopy quite suitable for dealing with conjugated systems.

As noted in Sec. 12.1.1, alkenes may be distinguished from saturated hydrocarbons by the positive iodine charge-transfer test and by their solubility in concentrated sulfuric acid.

Compounds suspected of being alkenes or alkynes should be tested with *bromine in carbon tetrachloride*. The majority of alkenes and alkynes add bromine quite rapidly; ethylenes or acetylenes substituted with electronegative groups add bromine slowly or not at all.

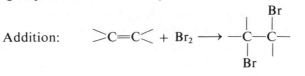

Substitution reactions accompanied by the liberation of hydrogen bromide gas occur with phenols, amines, enols, aldehydes, ketones, and

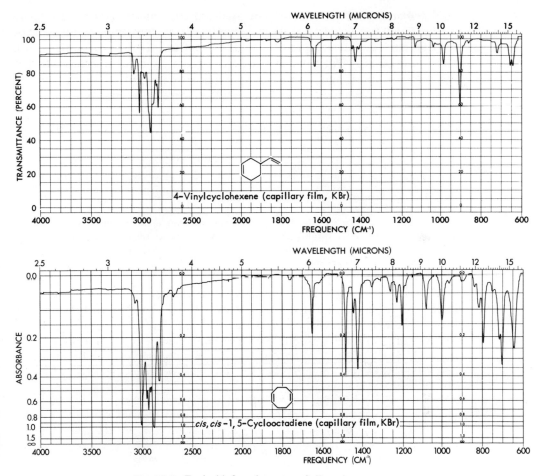

Fig. 12.2. Typical infrared spectra of alkenes.

other compounds which contain active methylene groups. With amines, the first mole of hydrogen bromide is not evolved but reacts with the amine to form the amine hydrobromide.

12.1.2b Procedure: Bromine in carbon tetrachloride

Dissolve 50 mg or two drops of the unknown in 1 ml carbon tetrachloride, and add dropwise a 2% solution of bromine in the solvent until the bromine color persists. If more than two drops of the bromine solution are required to cause the color to remain for 1 min, an addition or substitution reaction has occurred. Substitution reactions are indicated by the evolution of hydrogen bromide, which is insoluble in carbon tetrachloride. The evolved hydrogen bromide may be detected by blowing breath across the top of the tube and noting the fog that is produced, or better, by denoting the acidic reaction on dampened pH paper held across the mouth of the tube.

A *permanganate test* may also be used to detect unsaturation. The positive test given by alkenes and alkynes is indicated by the disappearance of the purple permanganate ion and the appearance of the sparingly soluble brown manganese oxide. Positive tests are also given by phenols, aryl amines, aldehydes, primary and secondary alcohols (reaction often slow), and all organic sulfur compounds in which the sulfur is in a reduced state.

$$3 \text{>C=C<} + 2KMnO_4 + 4H_2O \longrightarrow 3 \underset{HO}{-\overset{|}{C}}\underset{OH}{-\overset{|}{C}-} + 2MnO_2 + 2KOH$$

12.1.2c Procedure: Permanganate test

Dissolve 25 mg or one drop of the unknown in 2 ml of water or reagent grade acetone in a small test tube. Add dropwise with vigorous shaking a 1% aqueous solution of potassium permanganate. If more than one drop of the permanganate is reduced, the test is positive.

Tetranitromethane forms colored charge-transfer or π complexes with olefins and other unsaturated compounds. Many times, unsaturation may be detected with this reagent when other methods fail. The reagent is toxic, explosive, and expensive, and it should be employed in as small a quantity as practical. The substitution pattern of olefins may be determined by measurement of the absorption maxima of the complexes in the visible region.[1]

$$\underset{/\ \ \backslash}{\overset{\backslash\ \ /}{\underset{C}{\overset{C}{\|}}}} + C(NO_2)_4 \longrightarrow \underset{/\ \ \backslash}{\overset{\backslash\ \ /}{\underset{C}{\overset{C}{\|}}}} \cdots C(NO_2)_4$$

Yellow-to-red complex

12.1.2d Procedure: Tetranitromethane test

Add a small amount of the compound to be tested to a 25% solution of *tetranitromethane* in chloroform. (The reagent solution can be made up in advance and stored in the refrigerator.) The color increases from yellow to dark red with increasing unsaturation. Simple olefins and benzene give a yellow coloration; naphthalene gives an orange color. The test is positive with cyclopropane, feeble with α, β-unsaturated carbonyl compounds and other electrophilic olefins, and negative with allylic alcohols and alkynes.

12.1.2e Characterization

Although alkenes undergo numerous reactions, mainly involving addition to or cleavage of the carbon-carbon double bond, relatively few of these re-

---

[1] E. Heilbronner, *Helv. Chim. Acta*, **36**, 1121 (1953).

actions have wide general applicability for the preparation of suitable derivatives. Like saturated hydrocarbons, many of the simple alkenes may be identified solely by reference to physical and/or spectral properties.

Upon oxidation, alkenes yield aldehydes, ketones, and/or acids. The product obtained depends upon the reagent and the conditions as well as on the degree of substitution about the double bond. *Ozonization*[2] is the classical method for determining positions of double bonds in alkenes. The ozonides which are formed in these reactions are usually not isolated since they are often unstable and may decompose with explosive violence. Hydrolysis of ozonides with water, and reduction with zinc, yields aldehydes and ketones. When the ozonides are worked up in the presence of acidic hydrogen peroxide, the aldehydic components are oxidized to carboxylic acids. Ozonization reactions have the advantage that they are relatively simple to run and work up. A nice technique for the ozonolysis of double bonds is to add the ozone to a solution of the olefin in methylene chloride: pyridine (2:1) at $-70°$. The reaction mixture is then allowed to slowly warm to room temperature, poured into water, and the aldehydes and/or ketones isolated by ether extraction. The pyridine acts as a solvent and reducing agent. Ozone may be generated from oxygen by a commercially available apparatus[2] or may be purchased diluted with Freon-13 in stainless steel cylinders.[3]

$$R_2C{=}CHR' \xrightarrow{O_3} R_2C\underset{O}{\overset{O-O}{\diagup\diagdown}}CHR' -
\begin{cases}
\xrightarrow[H_2O]{Zn} R_2C{=}O + R'CHO \\[2ex]
\xrightarrow[H^+]{H_2O_2} R_2C{=}O + R'COOH
\end{cases}$$

An aqueous solution of excess *periodate ion with catalytic amount of permanganate ion*[4] is capable of the oxidative cleavage of carbon-carbon double bonds. The best conditions for carrying out an oxidative cleavage of olefins with the reagent are the use of a large excess of periodate and 5 to 10 mole percent of permanganate in an aqueous solution at pH 7 to 10. If the olefin is insoluble in water, *t*-butanol, dioxane, or pyridine may be used as a co-solvent up to 25%. The reaction is illustrated by the following equations:

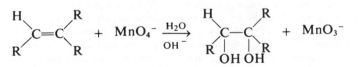

^{2}"Organic Ozone Reactions and Techniques" and "Basic Manual of Applications and Laboratory Ozonization Techniques," The Welsbach Corp., Ozone Process Division, Westmoreland and Stokley Sts., Philadelphia, Pa. 19129.

^{3}The Matheson Company, P.O. Box 85, East Rutherford, N.J.

^{4}R. U. Lemieux and E. von Rudloff, *Can. J. Chem.*, **33**, 1701, 1710 (1955).

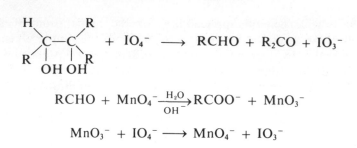

12.1.1f Procedure: Periodate-permanganate oxidation

Dissolve 4 g of sodium metaperiodate in 75 ml of water, and add 0.2 g of potassium permanganate. Stir and warm until the salts are dissolved. Add 25 ml of *t*-butanol and enough sodium carbonate to produce a pH of 8 to 9. Add 1 mmole of olefin, and stir the mixture overnight. Add dilute sulfuric acid until the solution is acid, and add sufficient sodium metasulfite to convert all of the periodate, iodate, and iodine into iodide. Add sufficient dilute sodium hydroxide to make the colorless solution basic. Remove the *t*-butanol under reduced pressure. Acidify the solution and obtain the acid(s) by thorough extraction with ether. (If volatile ketones are produced it may be necessary to modify the workup.)

A useful modification of this reaction involves substituting osmium tetroxide for permanganate.[5] In this case, the oxidation is stopped at the aldehyde stage. The expensive and poisonous osmium tetroxide is used only in catalytic amounts, so the major objections to this reagent are minimized. Aqueous dioxane is used as a solvent, or the reaction is run in a two-phase system of water and ether.

Olefins may be directly oxidized to carboxylic acid by shaking or refluxing with *alkaline permanganate*. Under these conditions, small amounts of other carboxylic acids often arise owing to overoxidation. This may lead to difficulties in purification and in the interpretation of the results.

$$RCH{=}CHR \xrightarrow[\text{(2) } H_3O^+]{\text{(1) } KMnO_4, OH^-} 2RCOOH$$

12.1.2g Procedure: Alkaline permanganate oxidation

The compound (500 mg) is added to 50 ml of water containing 2 g of potassium permanganate; 1 ml of 5% sodium hydroxide solution is added, and the mixture is shaken at intervals for 10 to 15 min, or until the purple color of the permanganate disappears. For less reactive compounds, the mixture is heated under reflux for 1.5 to 3 hr, and then allowed to cool. The mixture is carefully acidified with sulfuric acid. Any excess manganese dioxide may be removed by the addition of small amounts of sodium hydrogen sulfite. The acid is collected by filtration or by extraction with chloroform, ether, or methylene chloride.

[5]R. Pappo, D. S. Allen, Jr., R. U. Lemieux, and W. S. Johnson, *J. Org. Chem.*, **21**, 478 (1956).

Unsaturated hydrocarbons may also be characterized by the addition of reagents to the double bond. Addition of bromine is a particularly facile reaction. In some cases, the dibromide formed is a suitable solid derivative and may be characterized and identified by means of physical constants and spectral data. The bromide, if a liquid, may be converted to a 2,4-dinitro-phenylthioether by reaction with 2,4-dinitrothiophenol. The dibromide is a particularly useful derivative for the identification of gaseous or low-boiling olefins which may distill out of reaction mixtures. In this case, the derivative may be prepared by passing the gas, aided by a nitrogen stream, directly into a carbon tetrachloride solution containing bromine. Any excess bromine may be removed from the carbon tetrachloride by shaking with aqueous sodium hydrogen sulfite.

The *Kharasch reagent (2,4-dinitrobenzenesulfenyl chloride)* readily adds to many olefins to yield crystalline chlorosulfides. With symmetrical alkenes, or styrene derivatives, a single adduct is formed. Other olefins often yield mixtures of isomeric sulfides. The reagent will also add to acetylenes and allenes. The reagent adds in a stereospecific *trans* manner; thus *cis-* and *trans*-2-butene give different products. Tetra-substituted olefins do not react.

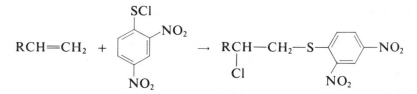

12.1.2h Procedure: Addition of 2,4-dinitrobenzenesulfenyl chloride

> CAUTION: The reagent, 2,4-dinitrobenzenesulfenyl chloride, is known to be explosive. Do not heat above 100°.
>
> Mix 200 mg of 2,4-dinitrobenzenesulfenyl chloride with an equal amount of the olefin in 5 ml of glacial acetic acid. Heat on a steam bath for 15 min or until the reaction is complete. The presence of unreacted reagent may be detected by the liberation of iodine on starch iodide paper.
>
> The mixture is cooled and the precipitate is collected by vacuum filtration. If crystals do not form, the reaction mixture is poured onto a small amount of crushed ice. The crude derivative may be recrystallized from ethanol.

12.1.2i Chemical conversion to other
       functional classes

When the saturated hydrocarbon is well characterized, identification of an olefin may be expedited by conversion to the saturated hydrocarbon. Addition of hydrogen across the double bond may be effected through catalytic hydrogenation, dissolving metal reductions, diimide reductions, or the hydroboration-protolysis reaction. The catalytic hydrogenation of very small amounts of material has often been used as a means of determining the amount of unsaturation in a molecule. Sensitive gas burettes or pressure gauges are used to quantitatively measure the amount of hydrogen uptake. In most catalytic hydrogenations, the hydrogen is supplied to the reaction vessel at atmospheric or higher pressures from a storage tank. The Brown-II Hydrogenator[6] employs hydrogen generated *in situ* by the action of acid on sodium borohydride. The apparatus consists of a reaction flask, a generator flask and a burette, which is used to add standard sodium borohydride solution to the generator through an automatic valve. The catalyst used with the apparatus can be prepared by the reduction of nickel or platinum metal salts with borohydride immediately prior to use. Diimide reductions may be performed by the generation of the diimide *in situ* by the oxidation of hydrazine with hydrogen peroxide under the influence of copper catalyst.[7]

$$NH_2NH_2 + H_2O_2 \longrightarrow H—N{=}N—H + 2H_2O$$

$$\text{\large$\diagdown$}C{=}C\text{\large$\diagup$} + H—N{=}N—H \longrightarrow \underset{\substack{| \\ H}}{-C}\underset{\substack{| \\ H}}{-C-} + N_2$$

In a reaction known as hydroboration,[8] diborane (B_2H_6) adds as BH_3 to alkenes to give organoboranes. Diborane and alkylboranes are very reactive, toxic substances (the lower molecular weight alkyl boranes are also highly flammable); but for most synthetic purposes, the diborane and the alkylboranes are generated and reacted *in situ*. Diborane is conveniently prepared by reaction with boron trifluoride etherate and sodium borohydride in glyme, or by reaction of the boron trifluoride with lithium borohydride in diethyl ether. The utility of the hydroboration reaction is illustrated by the following equations:

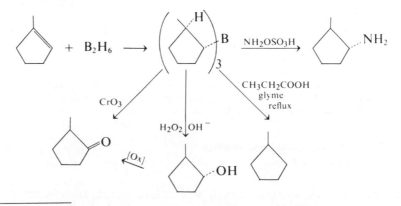

---

[6]H. C. Brown and C. A. Brown, *J. Am. Chem. Soc.*, **85**, 1005 (1963) and previous papers.

[7]E. J. Corey, W. L. Mock, and D. J. Pasto, *Tetrahedron Letters*, 347 (1961).

[8]H. C. Brown, *Hydroboration*, W. A. Benjamin, Inc., New York, 1962; G. Zweifel and H. C. Brown, *Org. Reactions*, **13**, 1 (1963).

Protonolysis of the adduct achieves a simple reduction of the alkene to the saturated hydrocarbon. The reaction is readily adapted to small-scale use; the hydrogen atoms are introduced stepwise in a *cis* manner. The protonolysis is usually carried out by the use of aqueous propionic acid. (The procedure may be readily adapted to introduce one or two deuterium atoms by employing borodeuteride or by reaction in deuterium oxide or both. The selective introduction of one or two deuterium atoms is a useful structural aid in conjunction with mass spectrometry or nmr spectroscopy.) The peroxide reaction achieves the anti-Markonikoff addition of water to the carbon-carbon double bond. The alcohols thus obtained may be converted to solid derivatives or may be further oxidized to aldehydes or ketones for identification purposes. Alkyl boranes may be directly oxidized to ketones. In certain cases, the conditions necessary to effect this direct oxidation may induce rearrangements. It is often advantageous to first prepare the alcohol and then oxidize to the ketone. This procedure—conversion of an olefin to a ketone—has utility in that it indicates the size of the ring in which the carbon-carbon double bond is located by the carbonyl frequency of the final product. Alkyl boranes also react with hydroxylamine-O-sulfonic acid to produce amines.

The Diels-Alder reaction occasionally may be found to be the most suitable method for preparing a derivative of an olefin or a diene.[9] Maleic anhydride and N-phenylmaleiimide are among the most suitable of the dienophiles. They are rather reactive and usually form solid derivatives which may be characterized by further chemical conversion if necessary. Diphenylisobenzofuran[10] is recommended as a highly reactive diene which may be used to trap unstable olefins and acetylenes.[11]

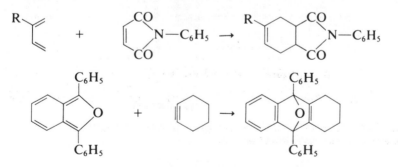

The addition of nitrosyl chloride, nitrogen trioxide, and nitrogen tetroxide have found special applications in the characterization of terpenes.[12]

[9]M. C. Kwitzel, *Org. Reactions*, **4**, 1(1948); H. L. Holmes, *Ibid*, **4**, 60 (1948).

[10]M. S. Newman, *J. Org. Chem.*, **26**, 2630 (1961).

[11]G. Wittig and R. Pohlke, *Chem. Ber*, **94**, 3276 (1961).

[12]J. L. Simonsen and L. N. Owen, The Terpenes, Vols. I and II, Cambridge University Press, Cambridge, 1947–1957.

12.1.3 Alkynes and Allenes

Alkynes and allenes, like alkenes, are soluble in concentrated sulfuric acid, add bromine in carbon tetrachloride, and are oxidized by cold aqueous permanganate.

12.1.3a Classification of alkynes

There are two general types of alkynes—terminal and nonterminal. The terminal alkynes are easily detected by their very characteristic infrared spectra and by their ease of formation of metal derivatives, e.g., mercury alkynides (see below). The terminal alkynes show a sharp characteristic $\equiv$C—H band at 3310 to 3200 cm^{-1} (3.02 to 3.12 μ) and a triple-bond stretch of medium intensity at 2140 to 2100 cm^{-1} (4.67 to 4.76 μ). In the nmr, the $\equiv$C—H occurs as a singlet near $\delta 2.8$ to 3.0. The nonterminal alkynes display triple-bond absorption at 2250 to 2150 cm^{-1} (4.44 to 4.65 μ) which may be absent in symmetrically substituted cases (Fig. 12.3).

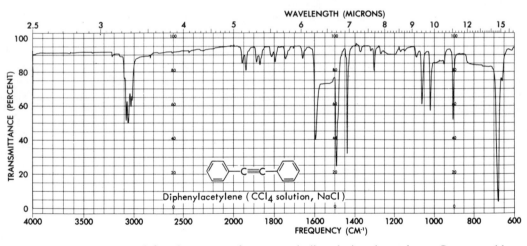

Fig. 12.3. Infrared spectrum of a symmetrically substituted acetylene. Compare with spectrum of phenylacetylene (Fig. 4.10).

12.1.3b Characterization of alkynes

Like other hydrocarbons, the characterization of alkynes is heavily dependent on physical properties. Only a few general methods have been developed for the preparation of solid derivatives.

Alkynes may be *hydrated to ketones* by the action of sulfuric acid and mercuric sulfate in dilute alcohol solution. Terminal acetylenes give methyl ketones; nonterminal, unsymmetrically substituted acetylenes often give mixtures of ketones.

$$ RC{\equiv}CR + H_2O \xrightarrow[H_2SO_4]{HgSO_4} RCOCH_2R $$

12.1.3c Procedure: Hydration of alkynes

A solution of 0.2 g of mercuric sulfate and 3 to 4 drops of concentrated sulfuric acid in 5 ml of 70% aqueous methanol is warmed to 60°. The alkyne (0.5 g) is added drop-wise, and the solution is maintained at 60° with stirring for 1 to 2 hr. The methanol is distilled off, and the residue is saturated with salt and extracted with ether. Appropriate spectral and physical properties of the ketone are determined, and a solid derivative is made.

Addition of 2,4-dinitrobenzenesulfenyl chloride proceeds readily to give colored, crystalline products.

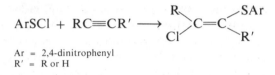

$$ ArSCl + RC{\equiv}CR' \longrightarrow $$

Ar = 2,4-dinitrophenyl
R' = R or H

12.1.3d Procedure: Addition of 2,4-dinitrobenzenesulfenyl chloride

Dissolve the reagent in ethylene chloride, cool to 0°, and add dropwise an excess of the alkyne. Keep the reaction mixture at 0° for 2 hr or longer. Remove the solvent and the excess alkyne by vacuum evaporation. Keep the residue cold until crystallization occurs. Recrystallize from ethanol using activated carbon, if necessary, to remove any highly colored impurities.

When a terminal alkyne is added to the *alkaline mercuric iodide* reagent, an immediate white or gray-white precipitate forms. This reaction may be used as a convenient test for terminal acetylenes, as well as a method for the preparation of a derivative. *Certain heavy metal acetylides are explosive when dry and therefore should be handled in small quantities only.*

$$ 2\,RC{\equiv}CH \xrightarrow[NaOH]{K_2HgI_4} (RC{\equiv}C)_2Hg $$

12.1.3e Procedure: Mercury derivatives

Reagent: Dissolve 6.6 g of mercuric chloride in a solution of 16.3 g of potassium iodide in 16.3 ml of water, and add 125 ml of 10% sodium hydroxide.

Cool an excess of the above reagent and add dropwise, with stirring, a solution of alkyne in 20 volumes of 95% ethanol. Stir for 2 to 3 min, filter, wash with 50% ethanol, and recrystallize from ethanol.

12.1.3f Allenes

Allenes are usually identified by spectral and physical characteristics. The best method of detection is by the presence of a strong stretching vibration, often observed as a doublet, in the 2200 to 1950 cm^{-1} (4.55 to 5.13 μ) region. Allenes absorb only below 200 mμ in the ultraviolet. In the nmr, terminal hydrogens are near δ4.4 and allenic hydrogens near δ4.8 with ^4J values of up to 7 Hz. Allenes are capable of exhibiting optical isomerism.

Derivation may be made by the addition of 2,4-dinitrobenzenesulfenyl chloride (see Sec. 12.1.3d); however, mixtures may be formed.

12.1.4 Aromatic Hydrocarbons

12.1.4a Classification

Benzene and a large number of its alkylated derivatives are liquids. Most compounds containing more than one aromatic ring, either fused or non-fused, are solids. Aromatic hydrocarbons burn with a characteristic sooty flame and give negative tests with bromine in carbon tetrachloride and alkaline permanganate (Sec. 12.1.2a). The simple aromatic hydrocarbons are insoluble in sulfuric acid but soluble in fuming sulfuric acid. With two or more alkyl substituents, the nucleus is sufficiently reactive to sulfonate and such compounds dissolve in concentrated sulfuric acid.

The infrared spectra of aromatic hydrocarbons exhibit the expected aromatic (and, in the case of alkylated nuclei, aliphatic) C—H stretching and bending deformations, plus characteristic ring absorption bands (Sec. 4.2.1b). The spectra show only bands of moderate intensity above 1000 cm^{-1} (10 μ) and generally only moderate to weak bands between 1000 and 800 cm^{-1} (10.0 and 12.5 μ).

In the nmr spectra, the presence, position, and, especially, intensity (integration) of the aryl hydrogens, plus the chemical shift and coupling pattern of any aliphatic protons, provide important and often definitive structural information.

In all alkyl benzenes, the substituted tropylium ion is an important and diagnostic fragment in the mass spectra.

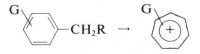

12.1.4b Characterization

The *nitro and polynitro derivatives* are generally useful in identification of aromatic hydrocarbons. In certain cases, it is advantageous to reduce

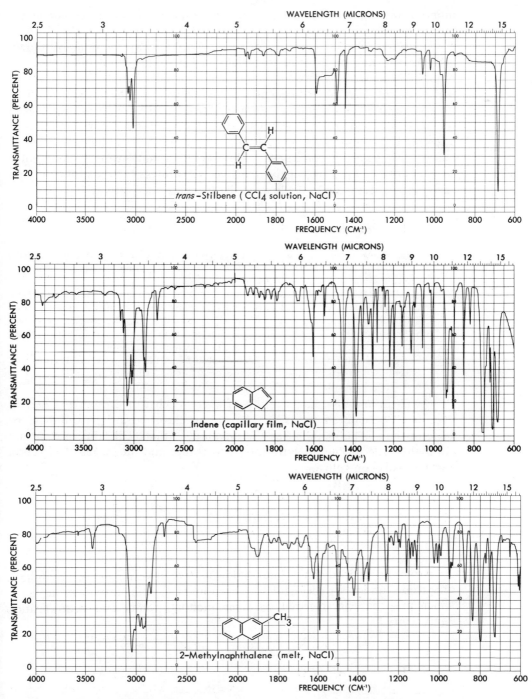

Fig. 12.4. Typical infrared spectra of aromatic hydrocarbons. Note the absence of a $\nu_{C=C}$ in the spectrum of *trans*-stilbene.

mononitro compounds to primary amines from which solid derivatives are made.

The nitration of aromatic compounds should always be done with great caution and on a small scale, especially when dealing with unknowns. A number of different procedures are used for nitration. The conditions which should be employed depend on the reactivity of the substrate and on the degree of nitration desired. It is very helpful to know something about the structure and, hence, reactivity of the compound before nitration is attempted. The following methods in procedure 12.1.4c are listed in order of increasingly vigorous conditions. Method (b) usually leads to mononitro compounds, except for highly reactive substrates such as phenol and acetanilide. Method (c) usually gives di- or trinitro derivatives.

12.1.4c Procedure: Nitration

(a) Reflux a mixture of 0.5 g of the compound, 2 ml of glacial acetic acid, and 0.5 ml of fuming nitric acid for 10 min. Pour into ice water. Filter the precipitate, wash with cold water, and recrystallize from aqueous ethanol.

(b) Add 0.5 to 1 g of the compound to 2 to 4 ml of concentrated sulfuric acid. An equal volume of concentrated nitric acid is added drop by drop, with shaking after each addition. Heat on a water bath at 60° for 5 to 10 min. Remove the tube from the bath and shake every few minutes. Cool and pour into ice water. Filter off the precipitate, wash with water, and recrystallize from aqueous ethanol.

(c) Follow the foregoing procedure, substituting fuming nitric acid instead of concentrated nitric acid. Heat the mixture for 10 min on a steam bath.

The clean and efficient nitration of aromatic hydrocarbons and derivatives can be achieved by use of nitronium fluoroborate in sulfolane or acetonitrile.[13] The method has the advantages that acid labile groups are not hydrolyzed during the nitration or work-up, and the stages of nitration may be easily controlled.

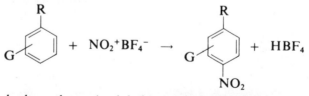

Aromatic hydrocarbons (and halogen derivatives) react with phthalic anhydride under Friedel-Crafts conditions to yield aroylbenzoic acids which may be characterized by their melting points and neutralization equivalents.

[13]S. J. Kuhn and G. A. Olah, *J. Am. Chem. Soc.*, **83**, 4564 (1961); reagent available from Ozark-Mahoning Co., 310 West 6th St., Tulsa, Okla. 74119.

12.1.4d Procedure: Aroylbenzoic acids

CAUTION: Carbon disulfide is highly inflammable.

In a small apparatus equipped with a reflux condenser and a hydrogen chloride trap, place 0.4 g of phthalic anhydride, 10 ml of carbon disulfide, 0.8 g of anhydrous aluminum chloride, and 0.4 g of the aromatic hydrocarbon. Heat the mixture on a water bath until no more hydrogen chloride is evolved, or about 30 min. Cool under tap water. Decant the carbon disulfide layer slowly. Add 5 ml of concentrated hydrochloric acid with cooling and then 5 ml of water to the residue. Shake or stir the mixture thoroughly. Cool if necessary to induce crystallization. Collect the solid, wash with cold water, and recrystallize from aqueous ethanol. If the product fails to crystallize, take the oil up in dilute ammonium hydroxide, treat with activated carbon, filter, cool, and neutralize with concentrated hydrochloric acid.

With chromic acid or alkaline permanganate, *alkyl benzenes are oxidized to aryl carboxylic acids.* This procedure is recommended when there is one side chain, in which case benzoic or a substituted benzoic acid is obtained, or when there are two ortho side chains, in which case *o*-phthalic acids are obtained (the melting point of a number of substituted *o*-phthalic acids and anhydrides have been recorded). Permanganate oxidation is usually the preferred procedure.

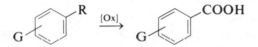

12.1.4e Procedure: Permanganate oxidation

In a small flask equipped for reflux, place 1.5 g of potassium permanganate, 25 ml of water, 6.5 ml of 6 *N* sodium hydroxide, 2 boiling chips, and 0.4 to 0.5 g of the alkyl benzene. Reflux gently for 1 hr, or until the purple color of the permanganate has disappeared. Cool the reaction mixture and acidify with dilute sulfuric acid. Heat to boiling; add a pinch of sodium hydrogen sulfite if necessary to destroy any brown manganese dioxide. Cool and filter the acid. Purify by recrystallization from water or aqueous ethanol, or by sublimation.

Polynitro aromatic compounds form stable charge-transfer complexes[14] with many electron-rich aromatic systems. The picrates of polynuclear aromatic hydrocarbons and aryl ethers are well-known examples (Sec. 12.4).

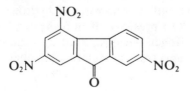

2,4,7-trinitro-9-fluorenone (TNF)

---

[14]L. N. Ferguson, *The Modern Structural Theory of Organic Chemistry*, Prentice-Hall, Inc., Englewood Cliffs, N.J., 1963.

The *complexes of 2,4,7-trinitro-9-fluorenone (TNF)* and aromatic hydro-carbons, or derivatives, are of considerable utility in identification work, especially with polynuclear systems.[15]

12.1.4f Procedure: TNF complexes

Prepare separate, nearly saturated hot solutions of TNF and the aromatic substance in absolute ethanol, ethanol-benzene, or glacial acetic acid. Mix the two hot solutions and heat for 1 min, cool, and recrystallize the complex from any of the above solvent systems.

12.2   HALIDES

12.2.1a Classification

Halogen substituents may be found in combination with all other functional groupings. When another functional group is present, chemical transformations for the purpose of making derivatives are usually carried out on that functional group rather than the halogen; e.g., esters are made from chloro-acids, carbonyl derivatives are made from chloroketones, and urethanes are made from bromoalcohols.

The presence or absence of halogen in an unknown can seldom be inferred from its infrared spectrum (Fig. 12.5). The carbon fluorine stretch-ing absorptions occur in the 1350 to 960 cm^{-1} (7.41 to 10.42μ) region, those for chlorine occur in the 850 to 500 cm^{-1} (11.7 to 20μ) region (often ob-scured by chlorinated solvents or aromatic substitution bands), while those for bromine and iodine occur below 667 cm^{-1} (16 μ) and are not observed in the normal infrared region.

The presence or absence of halogen may be determined by the Beilstein test, analysis of the filtrate from a sodium fusion, or by application of the silver nitrate or sodium iodide tests (see below). The very characteristic cracking patterns and/or isotope peaks observed in the mass spectra of halogenated compounds provide definitive evidence for the presence and types of halogen(s). The presence of halogen is sometimes suggested by the characteristic chemical shift in the nmr of hydrogens on the carbon atom bearing the halogen. The presence of fluorine can be demonstrated by nmr spectroscopy. In hydrogen spectra, hydrogen-fluorine coupling may be observed; in fluorine spectra, the relative number and kind of fluorines

---

[15]M. Orchin and E. O. Woolfolk, *J. Am. Chem. Soc.*, **68**, 1727 (1946); M. Orchin, L. Reggel, and E. O. Woolfolk, *Ibid.*, **69**, 1225 (1947); D. E. Laskowski and W. C. McCrone, *Anal. Chem.*, **30**, 1947 (1958).

present may be determined through integration and measurement of characteristic fluorine chemical shifts.

In addition to alkyl and aryl halides, halogen occurs in the following types of compounds:

1. Salts (amine hydrochlorides, sulfonium salts, etc.)
2. Acid halides (acyl halides, sulfonyl halides, etc.)
3. Haloamines and haloamides (N-bromosuccimide, Chloramine-T, etc.)
4. Miscellaneous compounds (iodobenzene dichlorides, alkyl hydrochlorites, halosilanes, alkyl mercuric halides, and other organometallic halides, etc.)

Alkyl and aryl halides (along with hydrocarbons, diaryl ethers, and most perfluoro compounds) are insoluble in concentrated sulfuric acid.

The following tests are useful in classifying the various types of alkyl and aryl halogen compounds. It is advisable to test the reactivity of any unknown halide toward each reagent.

12.2.1b Alcoholic silver nitrate

$$AgNO_3 + X^- \longrightarrow AgX + NO_3^-$$

The test depends on the rapid and quantitative reaction of silver nitrate with halide ion to produce insoluble silver halide (excepting the fluoride).

Silver nitrate in aqueous or ethanolic solution gives an immediate precipitate with compounds which contain ionic halide, or with compounds such as acid chlorides which react immediately with water or ethanol to produce halide ion. Many other halogen-containing substances react with silver nitrate to produce insoluble silver halide. The rate of such reactions is a measure of the reactivity of the substrate in $S_N 1$ reactions:

$$RX \longrightarrow R^+ + X^-$$

Silver and other heavy metal salts catalyze $S_N 1$ reactions of alkyl halides by complexation with the unshared electrons of the halide, making the leaving group a metal halide rather than halide ion.

$$R-\ddot{X}: \xrightarrow{Ag^+} R-\ddot{X}:---Ag^+ \xrightarrow{slow} R^+ + AgX$$

$$R^+ \longrightarrow \text{alcohols, ethers, olefins, etc.}$$

The rate of precipitation of silver halide depends on the leaving group, $I > Br > Cl$ (silver fluoride is soluble), and upon the structure of the alkyl group. Any structural factors which stabilize the electron-deficient carbonium ion (R^+) will accelerate the reaction. We thus find the expected order of reactivity

benzyl $\approx$ allyl > tertiary > secondary > primary > methyl > vinyl $\approx$ aryl

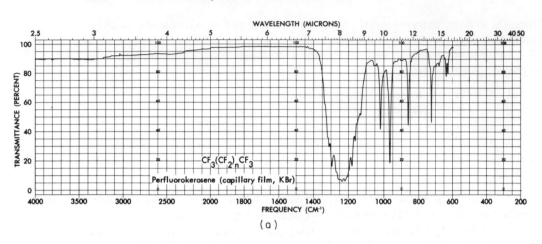

(a)

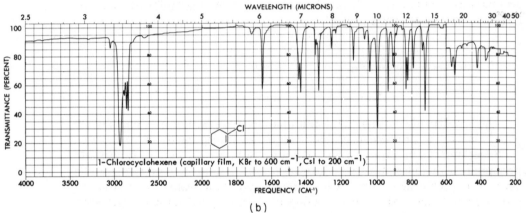

(b)

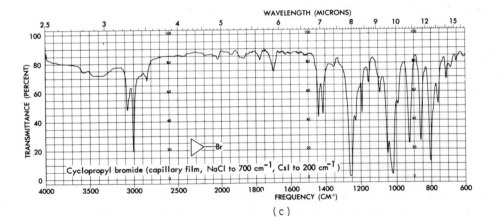

(c)

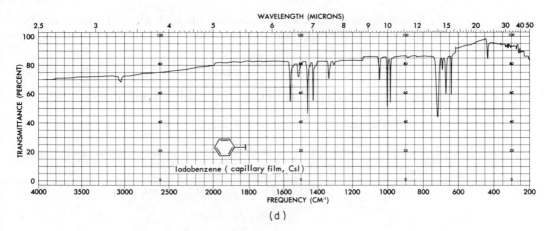

Fig. 12.5. Typical infrared spectra of halides.

The following equations illustrate other structural classes which are un-
usually reactive in S_N1 reactions

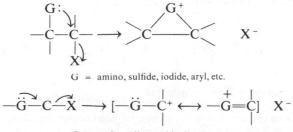

G = amino, sulfide, iodide, aryl, etc.

G = amino, alkoxy, thioalkoxy, etc.

Factors which tend to destabilize the incipient carbonium ion decelerate
the rate of S_N1 reactions. The following compounds are not reactive under
S_N1 conditions:

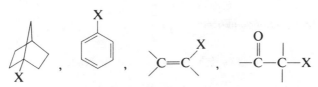

A summary in order of decreasing reactivity of silver nitrate with
various organic compounds containing halogen is as follows (X=I, Br,
or Cl):

1. Water-soluble compounds which give immediate precipitates with
   aqueous silver nitrate.
   (a) Salts.
   (b) Halogen compounds which are hydrolyzed immediately with
       water, such as low molecular weight acyl or sulfonyl halides.

2. Reactivity of water-insoluble compounds with alcoholic silver nitrate.
   (a) Immediate precipitation

RCOCl   RSO₂Cl $RO-\overset{|}{\underset{|}{C}}-X$

R—CH=CH—CH₂—X   R—CHBrCH₂Br

ArCH₂X   R₃CX   RI

   (b) Silver halide precipitates slowly or not at all at room temperature, but readily when warmed on the steam bath.

$$RCH_2Cl \quad R_2CHCl \quad RCHBr_2$$

   (c) Inert toward hot silver nitrate

$$ArX \quad RCH=CHX \quad HCCl_3 \quad -\overset{O}{\overset{\|}{C}}-CH_2X$$

12.2.1c Procedure: Silver nitrate test

Add 1 drop (or several drops of an ethanol solution) of the halogen compound to 2 ml of 2% ethanolic silver nitrate. If no reaction is observed after 5 min, heat the solution to boiling for several minutes. Note the color of any precipitate formed. Add two drops of 5% nitric acid. Certain organic acids give insoluble silver salts. Silver halides are insoluble in dilute nitric acid; the silver salts of organic acids are soluble.

For water-soluble compounds, the test should be run using aqueous silver nitrate.

12.2.1d Sodium iodide in acetone

$$NaI + RX \xrightarrow{\text{acetone}} RI + NaX$$

$$X = Br \text{ or } Cl$$

The rate of the reaction of sodium iodide in acetone with a compound containing covalent bromine or chlorine is a measure of the reactivity of the compound toward bimolecular nucleophilic substitution (S_N2 reaction). The test depends on the fact that sodium iodide is readily soluble in acetone, whereas sodium chloride and bromide are only slightly soluble.

S_N2 reactions are concerted and involve backside nucleophilic displacements which result in inversion of the configuration of the carbon atom at the reaction site.

$$I^- + H-\underset{\underset{R'}{|}}{\overset{\overset{R}{|}}{C}}-X \longrightarrow \left[I^{\delta-}\cdots\underset{\underset{H\quad R'}{}}{\overset{\overset{R}{|}}{C}}\cdots X^{\delta-} \right] \longrightarrow I-\underset{\underset{R}{|}}{\overset{\overset{R}{|}}{C}}-H + X^-$$

As might be expected from the consideration of the transition state, such reactions are relatively insensitive to electronic effects of the R groups, but steric effects are of primary importance. In contrast to S_N1 reactions, the rates of S_N2 reactions follow the order

$$\text{methyl} > \text{primary} > \text{secondary} > \text{tertiary}.$$

It should also be noted that α-halocarbonyl compounds and α-halonitriles are highly reactive under S_N2 conditions. This has been attributed to stabilization of the transition state by a partial distribution of the charge from the entering and leaving groups onto the carbonyl oxygen, etc. Also important may be the fact that the stereochemistry of the carbonyl and nitrile groups provides a favorable environment for nucleophilic attack.

A summary of the results to be expected in the sodium iodide in acetone test follows:

1. Precipitate within 3 min at room temperature.
   (a) primary bromides
   (b) acyl halides
   (c) allyl halides
   (d) α-haloketones, -esters, -amides, and -nitriles.
   (e) carbon tetrabromide
2. Precipitate only when heated up to 6 min at 50°.
   (a) primary and secondary chlorides
   (b) secondary and tertiary bromides
   (c) geminal polybromo compounds (bromoform)
3. Do not react in 6 min at 50°.
   (a) vinyl halides
   (b) aryl halides
   (c) geminal polychloro compounds (chloroform, carbon tetrachloride, trichloroacetic acid)
4. React to give precipitate and also liberate iodine.
   (a) vicinal halides

$$-\underset{\underset{Cl}{|}}{\overset{\overset{|}{}}{C}}-\underset{\underset{Cl}{|}}{\overset{\overset{|}{}}{C}}- \overset{NaI}{\longrightarrow} {>}C{=}C{<} + I_2 + 2NaCl$$

   (b) sulfonyl halides

$$ArSO_2Cl + NaI \longrightarrow ArSO_2I + NaCl \quad \text{(immediate)}$$
$$\overset{NaI}{\Big\downarrow}$$
$$ArSO_2Na + I_2$$

(c) triphenyl methyl halides

$$(C_6H_5)_3CX + NaI \longrightarrow NaX + (C_6H_5)_3CI \longrightarrow$$
$$(C_6H_5)_3C—C(C_6H_5)_3 + I_2$$

12.2.1e Procedure: Sodium iodide in acetone

Reagent: Dissolve 15 g of sodium iodide in 100 ml of pure acetone. The reagent should be stored in a dark bottle. On standing, it slowly discolors. It should be discarded when a definite red-brown color develops.

To 1 to 2 ml of the reagent in a small test tube, add 2 drops (or 0.1 g in a minute volume of acetone) of the compound. Mix well and allow the solution to stand at room temperature for 3 to 4 min. Note whether a precipitate forms and whether the solution acquires a red-brown color. If no change occurs at room temperature, warm the tube in a water bath at 50° for 6 min. Cool to room temperature and record any changes.

12.2.2 Alkyl Halides

12.2.2a Characterization

A number of the commonly encountered alkyl halides may be identified by reference to physical and spectral properties. A good number of alkyl halides are readily available in most organic laboratories. The utility of comparative glc and infrared spectroscopy should not be overlooked. Useful methods for preparing solid derivatives of polyhalogen nonbenzenoid hydrocarbons are not available. Identification in such cases rests solely on spectral and physical data and chromatographic comparisons.

A group of the most readily formed solid derivatives of primary and secondary alkyl halides is the *S-alkylthiuronium picrates*. The initial reaction involves the direct displacement of halide ion by the strongly nucleophilic thiourea; tertiary halides do not react.

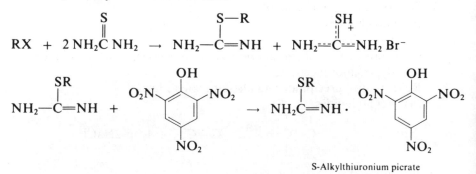

S-Alkylthiuronium picrate

12.2.2b Procedure: S-alkylthiuronium picrates

The alkyl halide (1 mmole) and thiourea (2 mmole) are dissolved in 5 ml of ethylene glycol contained in a small tube or flask fitted with a condenser. The mixture is heated in an oil bath for 30 min. Primary alkyl iodides react at 65°; others require temperatures near 120°. Add 1 ml of a saturated solution of picric acid in ethanol and continue to heat for 15 min more. Cool the reaction mixture and add 5 ml of cold water. Allow the mixture to stand in an ice bath. Collect the precipitate and recrystallize from methanol.

Alkyl 3,5-dinitrobenzoates may be prepared by reaction of alkyl iodides with the silver salt of 3,5-dinitrobenzoic acid.

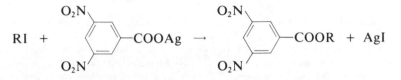

12.2.2c Procedure: 3,5-Dinitrobenzoates

The alkyl iodide and a slight excess of finely powdered silver 3,5-dinitrobenzoate are refluxed in a small volume of alcohol until conversion to silver iodide appears complete. Evaporate the mixture to dryness, and extract the ester with ether. Recrystallize from alcohol.

12.2.2d Derivatives prepared from Grignard reagents

The alkyl halide can be converted to a Grignard reagent, which, in turn, may be treated with an isocyanate to form a substituted amide, with mercuric halide to form an alkyl mercuric halide, or with carbon dioxide to form an acid. The latter should be attempted only in cases where solid acids are obtained.

$$RX \xrightarrow[Et_2O]{Mg} RMgX \begin{cases} \xrightarrow[(2)\ H_2O]{(1)\ ArNCO} RCONHAr \\[2mm] \xrightarrow{HgX_2} RHgX + MgX_2 \\[2mm] \xrightarrow[(2)\ H_3O^+]{(1)\ CO_2} RCOOH \end{cases}$$

Grignard reagents of the allyl or benzyl type are likely to give rearranged and/or dimeric products

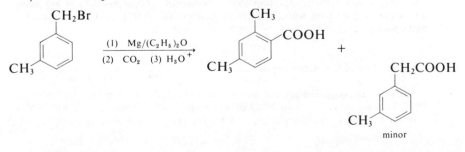

$$CH_3CH{=}CH{-}CH_2Br \xrightarrow[\text{(2) } CO_2 \text{ (3) } H_3O^+]{\text{(1) } Mg/(C_2H_5)_2O} CH_3\underset{\underset{COOH}{|}}{CH}CH{=}CH_2$$

$$CH_2{=}CH{-}CH_2Br \xrightarrow[(C_2H_5)_2O]{Mg} CH_2{=}CH{-}CH_2{-}CH_2{-}CH{=}CH_2$$

12.2.2e Procedure: Preparation of Grignard reagents

Grignard reagent: The choice of apparatus is governed by the steps involved in the reaction of the prepared Grignard reagent. The apparatus should consist of a flask, a tube or a small separatory funnel used as a reaction vessel, a condenser, and a drying tube. Into the carefully dried apparatus, place about 0.5 mmole (120 mg) of finely cut magnesium turnings, 0.55 mmole of the halide, 5 ml of absolute ether, and a small crystal of iodine. If the reaction is slow in starting, warm it with a beaker of warm water. The reaction is normally complete in 5 to 10 min.

12.2.2f Procedure: Alkylmercuric halides

CAUTION: *These materials are highly toxic and should be handled with care.*

The Grignard reaction mixture (equivalent to 0.5 mmole Grignard) is filtered through a microporous disc or a little glass wool into a test tube containing 1 to 2 g of mercuric chloride, bromide, or iodide (corresponding to the halogen of the alkyl halide). Stopper the tube and shake the mixture vigorously. Remove the stopper, warm the tube on the steam bath for a minute or two, and shake again. Evaporate to dryness, add 8 to 10 ml of ethanol, and heat on the steam bath until the alcohol boils. Filter and dilute with $\frac{1}{2}$ volume of water. Cool in ice bath; when crystallization is complete, collect the product and recrystallize from dilute ethanol.

12.2.2g Procedure: N-Aryl amides

Prepare the Grignard solution as above. Immerse the reaction vessel in cool water. Dissolve 0.5 mmole of α-naphthyl, phenyl, or p-tolyl isocyanate in 10 ml of absolute ether. By means of a dropper, add the isocyanate solution in small portions through the condenser into the Grignard solution. Shake the mixture and allow it to stand for 10 min. Pour the reaction mixture into a separatory funnel containing 20 ml of ice water and 1 ml of concentrated hydrochloric acid. Shake well, and discard the lower aqueous layer. Dry the ether solution with anhydrous magnesium sulfate. Evaporate the ether and recrystallize the crude product from alcohol, aqueous alcohol, or petroleum ether.

If the apparatus or reagents are not kept dry, diphenylurea (mp 241°), di-p-tolylurea (mp 268°) or di-1-naphthylurea (mp 297°) is obtained. These isocyanates are liquids and should not be used without purification if crystals (substituted urea formed by hydrolysis) are present in the reagent bottle.

12.2.2h Procedure: Carboxylic acids

(a) Filter the Grignard reagent into a mixture of finely crushed Dry Ice and ether. Pour the reaction mixture into 20 ml of water and 1 ml of concentrated hydrochloric acid contained in a separatory funnel. After the reaction subsides, separate the ether layer.

(b) In areas of high humidity, large amounts of water may condense in the Dry Ice/ether mixture described earlier. In such cases it is advantageous to bubble carbon dioxide into, or over, the Grignard solution with stirring. The carbon dioxide may be obtained from a tank or from Dry Ice.

12.2.3 Alkyl Fluorides and Fluorocarbons

12.2.3a Characterization

Simple alkyl and cycloalkyl fluorides are identified by reference to physical constants and spectral characteristics.

In recent years, the chemistry of highly fluorinated compounds ("fluorocarbons") has attracted considerable attention. Perfluoro compounds (all hydrogen atoms, except those in the functional group, are replaced by fluorine) and compounds containing the perfluoromethyl group are commonplace. The chemistry and physical properties of these materials differ considerably from their nonfluorinated analogues. Some pertinent points are noted below.

The boiling points are usually lower than the hydrogen analogues, e.g., perfluoropentane (29°) and pentane (36°), acetophenone (202°) and perfluoromethyl phenyl ketone (152°).

Highly fluorinated olefins do not add bromine under the usual conditions. They do, however, react with potassium permanganate. Most highly halogenated compounds fail to sustain a flame.

A high degree of fluorination increases the acidity of acids ($pK_a \sim 1$) and alcohols ($pK_a \sim 10$ to 12) and reduces the basicity of amines. Perfluoroaldehydes and ketones exhibit abnormally high carbonyl frequencies in the infrared. They are readily cleaved into acids and highly volatile monohydrogen perfluoroalkanes by the action of hydroxide (haloform reaction). The usual carbonyl derivatives of these compounds may be made by the appropriate carbonyl reagents (semicarbazide hydrochloride, etc.) in alcohol employing long reaction times.

12.2.4 Vinyl Halides

12.2.4a Characterization

Vinyl halides, like aryl halides, are much less reactive than halides at saturated carbon. Identification can usually be made by reference to spectral and physical data. The nature and stereochemistry of substitution at the double bond can be inferred from the C=C stretching and C—H bending bands in the infrared spectrum (Sec. 4.2.1b) and from the coupling constants in the nmr (Sec. 5.7). For vinyl fluorides, J_{FH} may be particularly useful.

Grignard reagents may be made from vinyl chlorides, iodides, and bromides, provided that dry tetrahydrofuran is substituted for ethyl ether (Sec. 12.2.2d).

12.2.5 Aryl Halides

12.2.5a Characterization

As in the case of many other types of aromatic compounds, the best procedures for making solid derivatives of aryl halides involves additional substitution of the aromatic nucleus. Nitration by the procedures outlined for aromatic hydrocarbons (Sec. 12.1.4) most generally is useful.

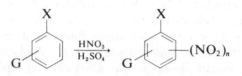

The aryl halides react readily with chlorosulfonic acid to produce sulfonyl chlorides, which in turn may be treated with ammonia to yield sulfonamides.

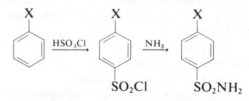

The procedure (Sec. 12.4.1e) outlined under aromatic ethers may be employed. Sometimes the intermediate sulfonyl chloride will serve as a suitable derivative. With polyhalo aromatics, more vigorous conditions may be necessary to produce the sulfonyl chloride; the chloroform solution may be warmed or the reactants warmed neat to 100°. Sometimes side reactions become predominant, products resulting from nuclear chlorination are obtained, or the reaction may provide diaryl sulfones.

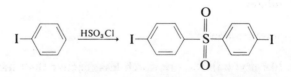

Aryl bromides and chlorides having aliphatic side chains may be oxidized to the corresponding acids. Aryl iodides and bromides may be converted to Grignard reagents and subsequently carbonated to yield acids or reacted with isocyanates to form amides (Sec. 12.2.2g). Aryl chlorides form Grignard reagents only in tetrahydrofuran; in ether, bromochlorbenzenes form the Grignard reagent exclusively at the bromo position.

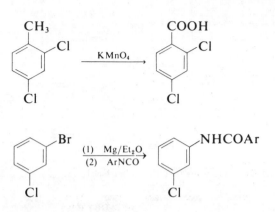

Occasionally aromatic polyvalent iodine compounds are encountered; representative examples are as follows:

$$ArICl_2 \qquad ArI(OAc)_2$$
Iodoarene dichloride   Iodoarene diacetate

$$Ar_2I^+X^- \qquad ArIO \qquad ArIO_2$$
Diaryliodonium salts   Iodosoarene   Iodoxyarene

The iodoarene dichloride acts as a ready source of chlorine. It will add chlorine to olefins. The oxygen derivatives all act as oxidants. With the exception of the diaryliodonium salts, all will liberate iodine from acidified starch iodide paper. The diaryliodonium salts are arylating agents; they may be pyrolyzed to yield ArI and ArX.

12.3 ALCOHOLS AND PHENOLS

In the absence of carbonyl absorption, the appearance of a medium-to-strong band in the infrared spectrum in the 3600 to 3400 cm^{-1} (2.78 to 2.94 μ) region indicates an alcohol, a phenol, or a primary or secondary amine (or possibly a wet sample, see Fig. 12.6 for the infrared spectrum of water). Amines may be distinguished from alcohols and phenols by taking advantage of their basic character. Water-insoluble amines are soluble in dilute hydrochloric acid. The water-soluble amines have a characteristic ammoniacal odor, and their aqueous solutions are basic to litmus. Phenols are considerably more acidic than alcohols and may be differentiated from the latter by their solubility in 5% sodium hydroxide solution and by the colorations usually produced when treated with ferric chloride solution. For additional details, the classification sections on alcohols (12.3.1, 12.3.2) phenols (12.3.3) and amines (12.14.1) should be consulted.

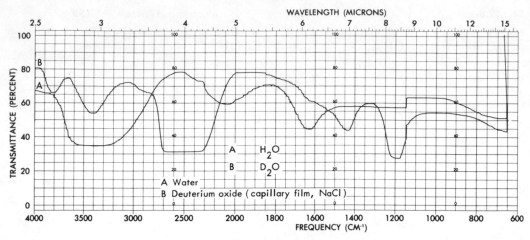

Fig. 12.6. Spectra of water and deuterium oxide.

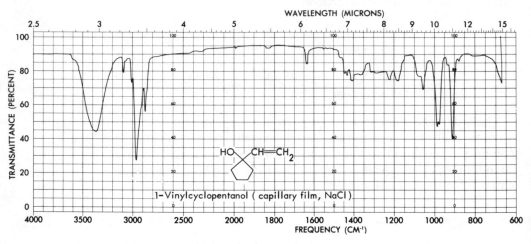

Fig. 12.7. Typical spectrum of an alcohol.

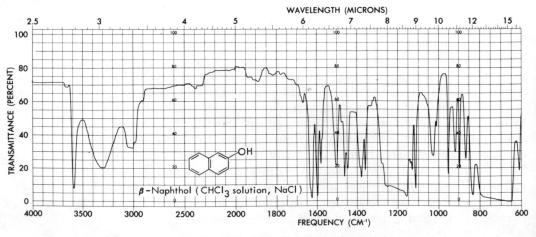

Fig. 12.8. Typical spectrum of a phenol.

12.3.1 Alcohols

12.3.1a Classification

Once it has been established that the compound in question is an alcohol, the next logical step in a structure determination is to distinguish between primary, secondary, and tertiary alcohols. Although there are certain correlations between both the O—H and C—O stretching frequencies and the subclass of alcohols, the exact peak positions and the factors affecting the peak positions are quite variable. One is cautioned against trying to use infrared spectroscopy to determine the subclass. With experience in a given series of alcohols, infrared may be used with some degree of success (Sec. 4.2.7), but it certainly cannot be recommended for general use.

   A solution of zinc chloride in concentrated hydrochloric acid (Lucas Reagent) has been widely used as a classical reagent to differentiate between the lower primary, secondary, and tertiary alcohols. With this reagent, the order to reactivity is typical of S_N1 type reactions. The zinc chloride undoubtedly assists in the heterolytic breaking of the C—O bond as illustrated in the following equation. The test is applicable only to alcohols that dissolve in the reagent.

$$\text{ROH} + \text{ZnCl}_2 \rightleftharpoons \overset{\delta^+}{\text{R}}\!\!-\!\!\overset{\delta^-}{\underset{|}{\text{O}}}\text{-}\text{ZnCl}_2 \xrightarrow{-[\text{HOZnCl}_2]^-} \text{R}^+ \xrightarrow{\text{Cl}^-} \text{RCl}$$
$$\underset{\text{H}}{|}$$

12.3.1b Procedure: Lucas test

   Reagent: Dissolve 16 g of anhydrous zinc chloride in 10 ml of concentrated hydrochloric acid with cooling to avoid a loss of hydrogen chloride.
   Add three to four drops of the alcohol to 2 ml of the reagent in a small test tube. Shake the test tube vigorously, and then allow the mixture to stand at room temperature. Primary alcohols lower than hexyl dissolve; those higher than hexyl do not dissolve appreciably, and the aqueous phase remains clear. Secondary alcohols react to produce a cloudy solution of insoluble alkyl chloride after 2 to 5 min. With tertiary, allyl, and benzyl alcohols, there is almost immediate separation of two phases owing to the formation of the insoluble alkyl chloride. If any question remains about whether the alcohol is secondary or tertiary, the test may be repeated employing concentrated hydrochloric acid. With this reagent, tertiary alcohols react immediately to form the insoluble alkyl chloride, whereas secondary alcohols do not react.

   A second method to differentiate tertiary alcohols from primary and secondary alcohols takes advantage of the inertness of tertiary alcohols toward oxidation with *chromic acid.*

$$3 \ \text{H}\!-\!\!\overset{|}{\underset{|}{\text{C}}}\!-\!\text{OH} + 2\,\text{CrO}_3 + 6\text{H}^+ \longrightarrow 3 \ {\Large{>}}\text{C}\!\!=\!\!\text{O} + 2\,\text{Cr}^{+3} + 6\,\text{H}_2\text{O}$$

12.3.1c Procedure: Chromic acid test

Reagent: Dissolve 1 g of chromic oxide in 1 ml of concentrated sulfuric acid, and carefully dilute with 3 ml of water.

Dissolve 20 mg or one drop of the alcohol in 1 ml of reagent grade acetone, and add one drop of the reagent. Shake the mixture. Primary and secondary alcohols react within 10 sec to give an opaque blue-green suspension. Tertiary alcohols do not react with the reagent. Other easily oxidized substances such as aldehydes, phenols, and enols also react with this reagent.

Nuclear magnetic resonance spectroscopy can be used to classify alcohols with respect to subclasses. In the most common nmr solvents such as

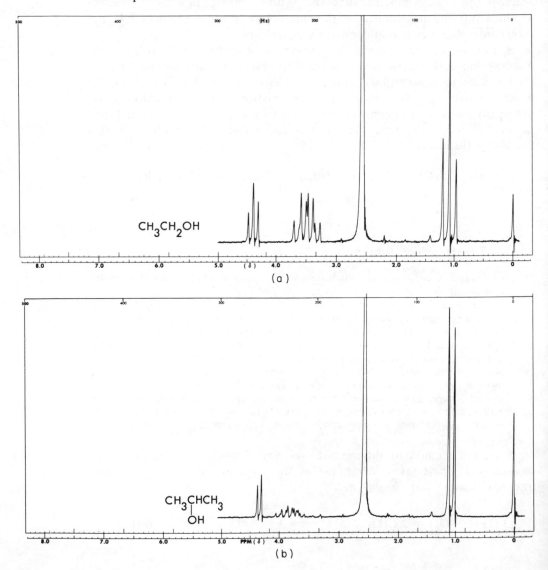

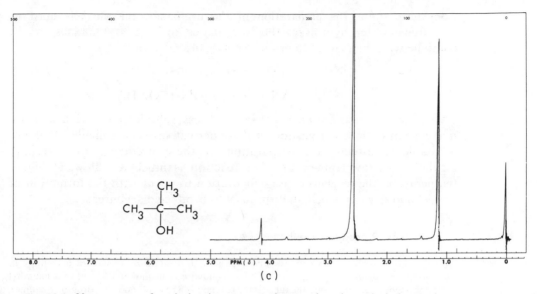

Fig. 12.9. Nmr spectra of typical primary, secondary, and tertiary alcohols run in dimethyl sulfoxide. Absorption at δ2.6 is due to dimethyl sulfoxide.

deuterochloroform or carbon tetrachloride, the hydroxyl hydrogen resonance is not only often obscured, but traces of acid present in the solvents catalyze proton exchange so that spin-spin splitting of the hydroxyl peaks is rarely observed. However, in dimethyl sulfoxide, the strong hydrogen bonding to the solvent shifts the hydroxyl hydrogen resonance downfield to δ4 or lower, and the rate of proton exchange slows sufficiently to permit observation of the carbinol hydrogen-hydroxyl hydrogen coupling (Fig. 12.9).[16] Because of the low-field resonance of the hydroxyl hydrogens, it is possible to use dimethylsulfoxide rather than its deuterated analogue. The hydroxyl hydrogen in methanol gives a quartet; primary, secondary, and tertiary alcohols give clearly resolved triplets, doublets, and singlets, respectively. Polyhydroxy compounds often give separate peaks for each hydroxyl hydrogen. This method is not reliable when acidic or basic groups are present, e.g., amino alcohols, hydroxy acids, or phenols.

12.3.1d Characterization

Phenyl and α-naphthylurethanes.

$$\text{ROH} + \text{ArN}{=}\text{C}{=}\text{O} \longrightarrow \text{ArNH}\overset{\displaystyle\overset{O}{\|}}{\text{C}}{-}\text{OR}$$

Isocyanate   Urethane

The most generally applicable derivatives of primary and secondary alcohols are the urethanes, prepared by reaction of the alcohol with the appropriate

[16]O. L. Chapman and R. W. King, *J. Am. Chem. Soc.*, **86**, 1256 (1964).

isocyanate. **For the preparation of the urethanes, the alcohols must be anhydrous**; water hydrolyzes the isocyanates to give aryl amines, which combine with the reagent to produce disubstituted ureas.

$$ArNCO + H_2O \longrightarrow ArNH_2 + CO_2$$

$$ArNH_2 + ArNCO \longrightarrow ArNHCONHAr$$

These ureas, which are high melting and less soluble than the urethanes, make isolation and purification of these derivatives very difficult. It is not advisable to attempt the preparation of these urethanes from tertiary alcohols. At low temperatures the reaction is much too slow; at higher temperatures the reagents cause dehydration to occur with the formation of an olefin and water which, in turn, react to produce diaryl ureas.

12.3.1e Procedure: Urethanes

Thoroughly dry a small test tube over a flame or in an oven; cork it and allow to cool. By means of a pipette, place 0.2 ml of anhydrous alcohol and 0.2 ml of α-naphthyl or phenyl isocyanate into the tube (if the reactant is a phenol, add one drop of pyridine), and immediately replace the cork. If a spontaneous reaction does not take place, the solution should be warmed in a water bath at 60 to 70° for 5 min. Cool in an ice bath, and scratch the sides of the tube with a glass rod to induce crystallization. The urethane is purified by recrystallization from petroleum ether or carbon tetrachloride. Filter the hot solution to remove the less-soluble urea which may form owing to traces of moisture. Cool the filtrate in an ice bath and collect the crystals. Diphenylurea, di-*p*-tolylurea, and di-1-naphthylurea have melting points of 241, 268, 297°, respectively.

Arenesulfonylurethanes. Benzenesulfonyl and p-toluenesulfonyliso-cyanate, which are considerably more reactive than phenyl and related isocyanates, may be used to advantage in the preparation of urethanes of tertiary and other highly hindered alcohols.[17] Extensive compilations of data are not available.

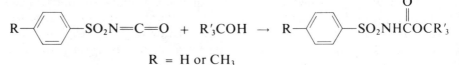

$$R = H \text{ or } CH_3$$

12.3.1f Procedure: Arenesulfonylurethanes

To a solution of 1 mmole of the arenesulfonylisocyanate in 1 ml of toluene, add 1 mmole of alcohol. After 5 min cool to 0°, and collect the precipitate by filtration. If the derivative fails to precipitate, add petroleum ether. Recrystallize from benzene or toluene.

For the preparation of urethanes from phenols, use a slight excess of the phenol, stopper the tube, and heat at 80 to 100° for 1 to 6 hr. Usually 1 hr is sufficient for

[17]J. W. McFarland and J. B. Howard, *J. Org. Chem.*, **30**, 957 (1965); J. W. McFarland, D. E. Lenz, and D. J. Grosse, *Ibid.*, **31**, 3798 (1966); p-toluenesulfonylisocyanate is available from the Upjohn Co., Carwin Organic Chemicals, North Haven, Conn.

normal phenols, but the longer reaction times are necessary for highly hindered phenols.

Esters. Numerous esters have been used to aid in the characterization of alcohols. Among these are a wide variety of phthalic acid esters, xanthates, benzoates, and acetates. The latter two are especially useful for the characterization of glycols and polyhydroxy compounds (Sec. 12.3.2). Among the most generally useful esters are the *3,5-dinitro and p-nitrobenzoates.* These esters may be prepared from the corresponding benzoyl chlorides by either of the two procedures given below. They are to be especially recommended for water-soluble alcohols which are likely to contain small amounts of moisture and hence produce trouble in the formation of the urethane derivatives. The method below employing pyridine as a solvent is one of the most useful methods for making derivatives of tertiary alcohols.

The acyl halides tend to hydrolyze on storage; it is advisable to check the mp of the reagent prior to use [3,5-dinitrobenzoyl chloride, mp 74° (acid, 202°) *p*-nitrobenzoyl chloride, mp 75° (acid 241°)].

12.3.1g Procedure: 3,5-Dinitro- and p-nitrobenzoates

(a) In a small test tube place 200 mg of pure *3,5-dinitrobenzoyl or p-nitrobenzoyl* chloride and 0.1 ml (two drops) of alcohol. Heat the tube for 5 min (10 min if the alcohol boils above 160°), employing a microburner so that the melt at the bottom of the tube is maintained in the liquid state. Do not overheat. Allow the melt to cool and solidify. Pulverize the crystalline mass with a glass rod or spatula. Add 3 or 4 ml of a 2% sodium carbonate solution and thoroughly mix and grind the mixture. Warm the mixture gradually to 50 or 60° and stir thoroughly. Collect the remaining precipitate and wash several times with small portions of water. Recrystallize from ethanol or aqueous ethanol.

(b) In a small test tube or round-bottomed flask, place 100 mg of alcohol or phenol, 100 mg of p-nitrobenzoyl or 3,5-dinitrobenzoyl chloride, 2 ml of pyridine, and a boiling chip. Put a condenser in place and reflux gently for 1 hr or allow to stand overnight. Cool the reaction mixture and add 5 ml of water and two or three drops of sulfuric acid. Shake or stir well and collect the crystals. Suspend the crystals in 5 ml of 2% sodium hydroxide, and shake well to remove the nitrobenzoic acid. Filter, wash several times with cold water, and recrystallize the derivative from alcohol or alcohol-water mixtures.

Oxidation to carbonyl compounds. The oxidation of primary and secondary alcohols to carbonyl compounds often proves to be an exceedingly useful method of identification. From the carbonyl compounds, easily prepared derivatives can be made on a small scale and, equally important, the spectroscopic features of the carbonyl group can provide much valuable structural information.

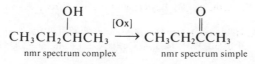

$$\underset{\text{nmr spectrum complex}}{\overset{\overset{\displaystyle \text{OH}}{\underset{\displaystyle |}{}}}{CH_3CH_2CHCH_3}} \xrightarrow{\text{[Ox]}} \underset{\text{nmr spectrum simple}}{\overset{\overset{\displaystyle O}{\underset{\displaystyle \|}{}}}{CH_3CH_2CCH_3}}$$

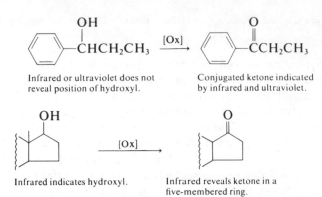

Infrared or ultraviolet does not reveal position of hydroxyl.

Conjugated ketone indicated by infrared and ultraviolet.

Infrared indicates hydroxyl.

Infrared reveals ketone in a five-membered ring.

Numerous methods are available for the oxidation of alcohols to carbonyl compounds. Only a few are included here; they are chosen for their mild conditions and suitability for small-scale operation.

The chromic anhydride-pyridine complex (Sarrett reagent)[18] is particularly useful for the oxidation of substances containing acid-sensitive groups. Alcohols, including allylic and benzylic, may be oxidized to the corresponding aldehydes or ketones.

12.3.1h Procedure: Sarrett reagent

Two ml of pyridine in a small round-bottomed flask equipped with a magnetic stirrer is cooled to 15 to 20°. To the pyridine is added, in portions, 200 mg of chromium trioxide at such a rate that the temperature does not rise above 30°. CAUTION: *If the pyridine is added to the chromium trioxide, the mixture will ignite.* A slurry of the yellow complex in pyridine remains after the last addition.

To the slurry is added 100 mg of the alcohol in 1 ml of pyridine. The flask is stoppered and allowed to stir overnight. The reaction mixture is poured into 20 ml of ether; the precipitated chromium salts are removed by filtration. The filtrate is placed in a separatory funnel and washed at least three times with water to remove the pyridine. The ether layer is dried over sodium sulfate and concentrated to provide the aldehyde or ketone. Low molecular weight, water-soluble ketones may be isolated by pouring the reaction mixture directly into water, neutralizing the pyridine by addition of acid and extraction with an appropriate solvent.

Chromic acid in acetone (Jones reagent)[19] is very convenient for the oxidation of acetone-soluble secondary alcohols to ketones. The reagent does not attack centers of unsaturation. The method is applicable on scales from 1 mmole to 1 mole. The reagent oxidizes primary alcohols and aldehydes to acids.

[18]G. I. Poos, G. E. Arth, R. E. Beyler, and L. H. Sarett, *J. Am. Chem. Soc.*, **75**, 422 (1963).

[19]K. Bodwen, I. M. Heilbron, E. R. H. Jones, and B. C. L. Weedon, *J. Chem. Soc.*, 39, (1946); A. Bowers, T. G. Halsall, E. R. H. Jones, and A. J. Lemin, *Ibid.*, 2555 (1943).

$$\text{RCH}_2\text{OH} \xrightarrow{\text{H}_2\text{CrO}_4} \text{RCHO} \xrightarrow{\text{H}_2\text{CrO}_4} \text{RCOOH}$$

12.3.1i Procedure: Jones reagent

Reagent: Dissolve 6.7 g of chromic anhydride (CrO_3) in 6 ml of concentrated sulfuric acid and carefully dilute with distilled water to 50 ml. One ml of this reagent is sufficient to oxidize 2 mmole of a secondary alcohol to a ketone, 2 mmole of aldehyde to an acid, or 1 mmole of primary alcohol to an acid.

The oxidation is carried out by the addition of the reagent from an addition funnel to a stirred acetone (reagent grade) solution of the alcohol maintained at 15 to 20°. The reaction is nearly instantaneous; the mixture separates into a lower green layer of chromous salts and an upper layer which is an acetone solution of the oxidation product. The reaction mixture may be worked up by the addition of water or by other means, depending on the properties of the oxidation product. Any brown color-ation, caused by an excess of the oxidizing agent remaining in the mixture, in the upper layer may be removed by a small pinch of sodium hydrogen sulfite or a few drops of methanol to destroy the excess reagent.

t-Butyl hypochlorite may be used to oxidize secondary alcohols to ketones; primary alcohols give esters, e.g., *n*-butanol gives *n*-butyl butyrate.

$$\text{H}-\overset{|}{\underset{|}{\text{C}}}-\text{OH} + (\text{CH}_3)_3\text{COCl} \longrightarrow \text{C}=\text{O} + \text{HCl} + (\text{CH}_3)_3\text{COH}$$

Under the conditions given below, chlorine is produced by reaction of the hypochlorite with hydrogen chloride; the hydrogen chloride may be trapped as the pyridine salt by the addition of one equivalent of pyridine to the reaction mixture.

12.3.1j Procedure: *t*-Butyl hypochlorite oxidation

Reagent: At room temperature in subdued light, add 6 ml of acetic acid and 10 ml of *t*-butyl alcohol to 150 ml of Clorox in a 250-ml separatory funnel. Shake well; allow the hypochlorite to separate as an upper layer (approximately 9 g of 99% purity). It may be dried over calcium chloride and distilled at 77 to 78°; however, for most purposes, the drying and distillation will not be necessary. The hypochlorite has long-term stability when stored in the dark in the refrigerator.

Cover the outside of a test tube with aluminum foil to exclude light; fit the test tube with a one holed stopper. Place in the test tube approximately 1 mmole of secondary alcohol in 5 ml of carbon tetrachloride. To this solution add 0.25 ml of *t*-butyl hypo-chlorite. Put the stopper in place, and place the tube in a beaker of water at 50°; allow to stand for 30 min. Pour the reaction mixture into a separatory funnel containing 0.2 g sodium arsenite in 10 ml of water. Shake well, and draw off the lower layer of the ketone in carbon tetrachloride.

[20]J. Meinwald, J. Crandall, and W. E. Hymans, *Org. Syn.*, **45**, 77 (1965).

Ceric ammonium nitrate. Cerium (IV) is a very convenient oxidant for the conversion of benzyl alcohols to benzaldehydes.[21]

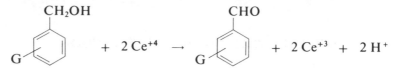

12.3.1k Procedure: Cerium(IV) oxidation of benzyl alcohols

Add a slight excess of 50% aqueous acetic acid solution containing 0.5 M ceric ammonium nitrate to the benzyl alcohol. Then warm the solution on the steam bath for a few minutes if necessary. After the orange cerium (IV) solution has turned to a pale yellow cerium (III) solution, extract the mixture with ether. Wash the ether solution with 1.5 N potassium hydroxide, and dry. Removal of the ether by distillation provides pure aldehyde in greater than 90% yield.

Active manganese dioxide[22,23] is often used to oxidize allylic and benzylic alcohols to aldehydes and ketones. Active manganese dioxide (as opposed to ordinary manganese dioxide) is obtained by the mixing of aqueous solutions of manganous sulfate, sodium hydroxide, and potassium permanganate. The oxidation is accomplished by shaking a suspension of the active manganese dioxide with the alcohol in an inert solvent, usually petroleum ether or chloroform. The oxidation of alcohols with this reagent has been used as an indication that the alcohols were allylic or benzylic. Caution should be used in applying this criterion since other alcohols are occasionally attacked by the reagent.

12.3.2 Polyhydric Alcohols

12.3.2a Classification

The polyhydric alcohols are viscous, high-boiling liquids to high-melting solids: the simpler ones are quite water-soluble; many tend to be more or less insoluble in nonpolar solvents. They may be detected and characterized by the methods already outlined under alcohols. Polyhydric alcohols often give separate peaks in the nmr for each chemically different hydroxyl when dimethyl sulfoxide is employed as solvent.

Compounds containing hydroxyl groups on adjacent carbons have characteristic reactions that are noteworthy. The simplest of these compounds are the 1,2-diols (1,2-glycols). The characteristic reactions of the 1,2-glycols are shared by the 1,2,3-triol, etc., but the situation is often more complicated (see also Sec. 12.9 on carbohydrates).

[21]W. S. Trahanovsky and L. B. Young, *J. Chem. Soc.*, 5777 (1965).

[22]S. Ball, T. W. Goodwin, and R. A. Morton, *Biochem. J.*, **42**, 516 (1948).

[23]J. Attenburrow, A. F. B. Cameron, J. H. Chapman, R. M. Evans, B. A. Hems, A. B. A. Jansen, and T. Walker, *J. Chem. Soc.*, 1094 (1952).

The *borax test* for 1, 2-diols depends on the reversible formation of cyclic borate esters which are much stronger acids than boric acid.

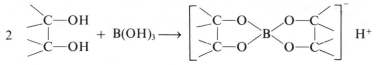

12.3.2b Procedure: Borax test

> *Reagent:* A 1% solution of borax (sodium borate) containing sufficient phenolphthalein to produce a pink coloration.
> Add a drop or two of the polyhydric alcohol to 0.5 ml of the reagent. A 1,2-diol causes the pink color to disappear. It reappears on warming and vanishes again upon cooling.

Periodic acid and 1,2-glycols react to produce carbonyl compounds and iodate (see discussion in Sec. 12.3.2d). The qualitative test depends upon the fact that silver iodate is only sparingly soluble in dilute nitric acid, whereas silver periodate is very soluble. It is important that exact amounts of reagents and nitric acid be employed; if too much nitric acid is present, the silver iodate will not precipitate.

12.3.2c Procedure: Periodate test for glycols

> *Reagent:* Dissolve 0.5 g of paraperiodic acid (H_5IO_6) in 100 ml of distilled water.
> To a test tube containing 2 ml of the reagent, add one drop of concentrated nitric acid and mix thoroughly. Add one drop or a similar amount of the compound to be tested (for water-insoluble compounds, dioxane as a co-solvent may be employed). Shake the mixture for 15 sec and add one to two drops of 5% silver nitrate solution. The instantaneous formation of a white precipitate of silver iodate constitutes a positive test.

12.3.2d Characterization

Periodic acid is a selective oxidant capable of cleavage of the carbon-carbon bond of 1,2 glycols, β-amino-alcohols, α-amino and α-hydroxy aldehydes and ketones, and α-diketones. In certain cases, α-hydroxy acids also react with the reagent.

$$\begin{matrix} R-CH-OH \\ | \\ R-CH-OH \end{matrix} + HIO_4 \longrightarrow 2\,RCHO + H_2O + HIO_3$$

$$\begin{matrix} R'-C=O \\ | \\ R-CH-OH \end{matrix} + HIO_4 \longrightarrow RCHO + R'COOH + HIO_3$$

$$\begin{matrix} R-CH-OH \\ | \\ R'-CH-NH_2 \end{matrix} + HIO_4 \longrightarrow RCHO + R'CHO + NH_3 + HIO_3$$

The reaction proceeds via the rapid and reversible formation of a cyclic periodate ester; the hydroxyl groups must be situated so that formation of the cyclic ester is sterically possible. In the case of carbonyl compounds, the ester is apparently formed from the hydrated carbonyl compounds, i.e., $-C(OH)_2-CHOH-$.

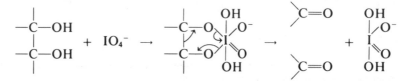

The periodate reaction may be used as a qualitative test (see above) or in quantitative determinations. In the quantitative procedures, excess standard periodate is allowed to react with the compound. The excess periodate is reduced by standard sodium arsenite with iodide catalyst and controlled pH. The arsenite is then back-titrated with standard iodine solution. Detailed procedures are available in numerous standard text and reference books. For example, in a quantitative assay, it can be determined that 1 mole of glyceraldehyde will consume 2 moles of periodate and produce 2 moles of formic acid and 1 mole of formaldehyde. Similar reactions occur with lead tetraacetate.

$$
\begin{array}{c}
\text{CHO} \\
| \\
\text{H}-\text{C}-\text{OH} \\
| \\
\text{CH}_2\text{OH}
\end{array}
+ 2\ IO_4^- \longrightarrow 2\ HCOOH + CH_2O + 2\ IO_3^-
$$

The classic *acid catalyzed rearrangement of a 1,2-glycol* is the pinacol-pinacolone rearrangement.

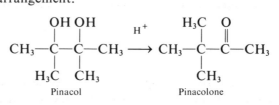

Pinacol   Pinacolone

In such rearrangements, hydrogen "migrates" in preference to an alkyl group.

$$RCHOHCH_2OH \xrightarrow{H^+} RCH_2CHO$$

$$RCHOHCHOHR \xrightarrow{H^+} RCH_2COR$$

The rearrangement, which may be carried out by warming the glycol with any strong acid, has utility in the detection and characterization of glycols. Upon heating with a little potassium hydrogen sulfate ethylene glycol yields acetaldehyde; glycerol gives acrolein (CH_2=CHCHO) which is

responsible for the characteristic odor of burning fat. The standard carbonyl derivatives are useful for the identification of the products of such
rearrangements.

The 1,2-glycols and 1,3-glycols react with aldehydes or ketones to yield
cyclic acetals or ketals which have occasional utility in separation and
identification procedures.

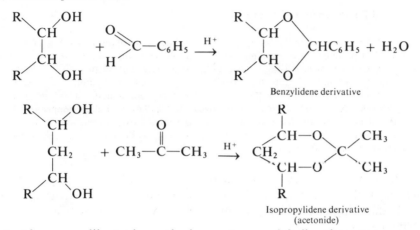

Benzylidene derivative

Isopropylidene derivative
(acetonide)

Acetonides are readily made employing acetone or 2,2-dimethoxypropane.

12.3.2e Procedure: Acetonides

(a) Dissolve 100 mg of the diol in 2 ml of 2,2-dimethoxypropane, and add 5 mg
of p-toluenesulfonic acid. Allow the mixture to stand for 30 min to 2 hr. Strip off the
excess reactant and methanol, and purify the product. In some cases it is advantageous
to drive the equilibrium towards the desired ketal by the addition of benzene to the
reaction mixture and the removal of the methanol by distillation of the benzene-
methanol azeotrope.

(b) Dissolve 0.5 g of the diol in 50 ml of acetone. Add 0.5 ml of 70% perchloric
acid, and allow the mixture to stand for 2 hr to 1 day. Add 1 g of solid sodium carbonate. Filter and remove the excess acetone to obtain the crude acetonide.

Because of their volatility, ketals, acetates, and trimethylsilyl ethers are
very useful derivatives of polyhydroxy compounds for use in glc and/or
mass spectral analysis.

Acetates may be made using acetic acid, acetic anhydride, and sodium
acetate mixture, or with acetic anhydride containing a small amount of
sulfuric acid or pyridine as catalyst. The following procedure gives better
yields under milder conditions.

12.3.2f Procedure: Acetates

The polyhydroxy compound (0.5 g) is added to 5 ml of pyridine. Acetic anhydride
(2 g) is added dropwise with shaking. After the initial reaction has subsided, the

solution is refluxed for 3 to 5 min. The mixture is cooled and poured into 15 ml of ice water. The derivative is collected and washed with cold 2% hydrochloric acid and water. The product is recrystallized from ethanol.

The *benzoates* are often very suitable crystalline derivatives of poly-hydroxy compounds.

12.3.2g Procedure: Benzoates

(a) Gradually add 0.5 ml of benzoyl chloride to a solution of 0.1 g of alcohol in 1 ml of pyridine. Warm the mixture over a low flame for a minute or two; then pour into ice water with stirring. Collect the precipitate and suspend it in 1 ml of 5% sodium carbonate solution. Recollect and recrystallize from ethanol.

(b) In a test tube, place 100 mg of polyhydroxy compound and 0.5 ml of benzoyl chloride. Add 5 ml of 10% sodium hydroxide. Place a rubber stopper on the tube and shake vigorously for 1 min and then occasionally over 5 min. Allow to stand in a cold bath for 30 min. Collect the precipitate and wash thoroughly with water. Dry and recrystallize from ethanol.

The *trimethylsilylation* reaction proceeds smoothly, quickly, and often quantitatively to provide derivatives of mono- and polyhydric alcohols which show high volatility and often are easily separated from closely related compounds by glc.

$$2\,ROH + (CH_3)_3SiNHSi(CH_3)_3 \longrightarrow 2\,ROSi(CH_3)_3 + NH_3$$

12.3.2.h Procedure: Trimethylsilyl derivatives

Reagent: Mix anhydrous pyridine (10 ml), hexamethyldisilazane (2 ml), and tri-methylchlorosilane (1 ml). A reagent of similar composition is marketed by the Pierce Chemical Company under the name TRI-SIL.

Place 10 mg or less of the sample in a small vial. Add 1 ml of the reagent, place the top on the vial, and shake vigorously for 30 sec. Warm if necessary to effect solution. Allow the mixture to stand for 5 min. The mixture may be injected directly into the gas chromatograph.

12.3.3 Phenols

12.3.3a Classification

Phenols are compounds of acidity intermediate between that of carboxylic acids and alcohols. Alcohols do not show acid properties in aqueous systems, whereas acids and phenols react with and are soluble in 5% aqueous sodium hydroxide solution (exceptions are the highly hindered phenols, such as *o*-di-*t*-butylphenol, which are insoluble in alkali). Acids and phenols may be differentiated on the basis of the insolubility of phenols in 5% aqueous sodium hydrogen carbonate solution. Phenols which contain highly electro-negative substituents, such as 2,4-dinitrophenol and 2,4,6-tribromophenol, show increased acidity and are soluble in sodium hydrogen carbonate solution.

Most phenols yield intense red, blue, purple, or green colorations in the *ferric chloride test*. All phenols do not produce color with this reagent; a negative test must not be taken as confirming the absence of the phenol grouping without additional supporting evidence. Other functional groups also produce color changes with ferric chloride: aliphatic acids give a yellow solution; aromatic acids may produce a tan precipitate; enols give a red-tan to red-violet color; oximes, hydroxamic acids, and sulfinic acids give red to red-violet colorations.

12.3.3b Procedure: Ferric chloride test

To 1 ml of a dilute aqueous solution (0.1 to 0.3%) of the compound in question, add several drops of a 2.5% aqueous solution of ferric chloride. Compare the color produced with that of pure water containing an equivalent amount of the ferric chloride solution. The color produced may not be permanent, thus the observation should be made at the time of addition.

Certain phenols do not produce coloration with the foregoing procedure. As an alternative procedure, dissolve or suspend 20 mg of a solid or one drop of a liquid in 1 ml of chloroform, and add two drops of a solution made by dissolving 1 g of ferric chloride in 100 ml of chloroform. To the test solution add one drop of pyridine and observe the color change.

Phenols rapidly react with *bromine water* to produce insoluble substitution products; all available positions ortho and para to the phenol are brominated. Advantage may be taken of this reaction both as a qualitative test for the presence of the phenolic grouping and for the preparation of solid derivatives. The tests should be applied with some discrimination since aniline and its substituted derivatives also react rapidly with bromine water to produce insoluble precipitates.

12.3.3c Procedure: Bromine water test

To a 1% aqueous solution of the suspected phenol, add a saturated solution of bromine water, drop by drop, until bromine color is no longer discharged. A positive test is indicated by the precipitation of the sparingly soluble bromine substitution product and the production of a very strongly acidic reaction mixture. In the case of phenol, the product is 2,4,6-tribromophenol.

In the ultraviolet region, ionization of a phenol by a base increases both the wavelengths and the intensities of the absorption bands.

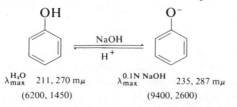

$$\lambda_{max}^{H_2O} \quad 211, 270 \; m\mu \qquad\qquad \lambda_{max}^{0.1N \; NaOH} \quad 235, 287 \; m\mu$$
$$(6200, 1450) \qquad\qquad\qquad (9400, 2600)$$

This shift may be visually observed in the case of *p*-nitrophenol; *p*-nitrophenol is yellow, whereas sodium *p*-nitrophenolate is red.

12.3.3d Characterization

Phenols, like alcohols, react with isocyanates to produce urethanes. The α-naphthylurethanes are the most generally used derivatives for the identification of phenols. The procedure employed is that given for alcohols (Sec. 12.3.1). For highly hindered phenols, benzenesulfonylisocyanate is the reagent of choice.

The melting points of a large number of 3,5-dinitrobenzoates of phenols have also been recorded. These derivatives can be prepared using the pyridine method discussed under alcohols (Sec. 12.3.1).

The preparation of *brominated phenols* is an exceedingly simple procedure, and the bromo-substituted phenols are very useful derivatives. Although saturated bromine water may be used satisfactorily for this bromination, the following procedure usually yields better results on a preparative scale.

12.3.3e Procedure: Bromination

In a test tube or an Erlenmeyer flask, dissolve 1 g of potassium bromide in 6 ml of water. Carefully add 0.6 g of bromine. In a second test tube, place 100 mg of the phenol, 1 ml of methanol, and 1 ml of water. Add 1 ml of the prepared bromine solution and shake. Continue the addition of bromine solution in small portions until the mixture attains a yellow color after shaking. Add 3 to 4 ml of cold water and shake vigorously. Filter the bromophenol, and wash the precipitate well with water. Dissolve the crystals in hot methanol, filter the solution, and add water dropwise to the filtrate until a permanent cloudiness results. Allow the mixture to cool to complete crystallization.

The phenoxide ions, produced by solution of phenols in aqueous alkali, react readily with chloracetic acid to give *aryloxyacetic acids*. These derivatives are very useful; they crystallize well from water, have well-defined melting points, and can be characterized by reference to their neutralization equivalents.

$$\underset{G}{\bigcirc}\!\!-OH \; + \; Cl-CH_2COOH \; \xrightarrow[(2)\ H_3O^+]{(1)\ NaOH} \; \underset{G}{\bigcirc}\!\!-O-CH_2COOH$$

12.3.3f Procedure: Aryloxyacetic acids

Approximately 200 mg of the phenol is dissolved in 1 ml of 6 *N* sodium hydroxide in a small test tube. Additional water should be added if necessary to completely dissolve the sodium phenoxide. To the solution, 0.5 ml of a 50% aqueous solution of chloro-acetic acid is added. The tube is provided with a microcondenser, and the reaction is heated on a water bath at 90 to 100° for 1 hr. The solution is cooled, 2 ml of water added, and the solution acidified with dilute hydrochloric acid. The mixture is extracted several times with small portions of ether. The ether extract is washed with 2 ml of water and then extracted with 5% sodium carbonate solution. The sodium carbonate extract is acidified with dilute HCl to precipitate the aryloxyacetic acid. The derivatives should be recrystallized from hot water. The melting point, and the neutralization equivalent if necessary, are determined.

12.4 ETHERS

[See also epoxides (Sec. 12.5) and acetals and ketals (Sec. 12.6).]

12.4.1a Classification

The identification of the ether grouping in the infrared is complicated by the fact that other functional groups contain the C—O bond and, consequently, have bands in the same region. As a general guide, a relatively strong band in the 1250 to 1100 cm^{-1} (8.0 to 9.9μ) region and the absence of C=O and O—H bands are good indications of an ether (Fig. 12.10). The aliphatic ethers absorb at the lower end of the range; conjugation raises the frequency.

Like hydrocarbons, ethers are quite unreactive, but they may be chemically distinguished from saturated hydrocarbons by the iodine charge-transfer test (Sec. 12.1.1b) and by their solubility in sulfuric acid (except diaryl ethers). Dialkyl ethers are also soluble in concentrated hydrochloric acid, whereas others are not.

In the absence of hydroxyl and carbonyl absorption in the infrared, a methyl (especially with characteristic ringing) in nmr spectra near $\delta 4$ or a methine or methylene at slightly lower field should alert the investigator to the possibility of the presence of an alkyl ether.

CAUTION: *Ethers tend to form highly explosive peroxides on standing, particularly when exposed to air and light.* If peroxides are present, they accumulate in the pot during distillation and may lead to a violent explosion. The presence of peroxides may be detected by means of starch iodide paper that has been moistened with dilute hydrochloric acid. A positive test is indicated by the blue starch-iodine color. Peroxides (as well as water and alcohols) may be removed from ethers by filtering the ether through a short column of highly active alumina (Woelm basic alumina, activity grade 1, is recommended). Hydroperoxides (but not water-insoluble dialkyl or diacyl peroxides) may be removed from organic materials by treatment with ferrous sulfate; wash 10 ml of peroxide containing ether with 3 to 5 ml of 1% ferrous sulfate solution acidified with one drop of sulfuric acid.

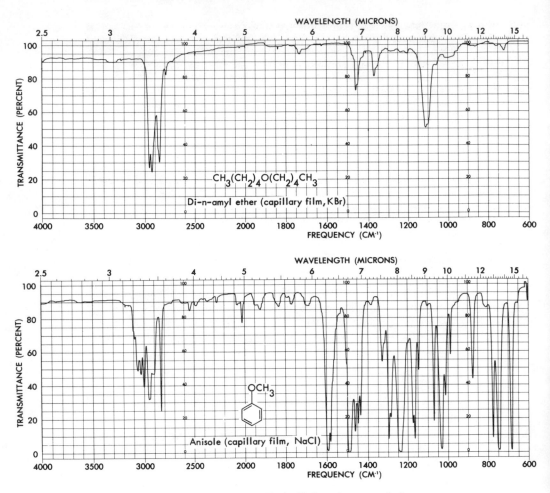

Fig. 12.10. Typical infrared spectra of ethers.

12.4.1b Characterization

Aliphatic ethers. The low reactivity of aliphatic ethers makes the problem of preparation of suitable derivatives a difficult one. As in the case of hydrocarbons, increased reliance must be placed on comparison of physical and spectral properties.

Aliphatic ethers may be cleaved with 3, 5-dinitrobenzoyl chloride and ZnCl₂ as catalyst to yield *3, 5-dinitrobenzoates.*

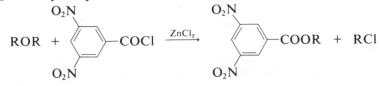

The method, however, is only useful for symmetrical ethers and requires at least 1 ml of sample. With a low boiling aliphatic ether, the reaction may fail owing to loss of the ether before the cleavage is effected. In these cases it may be necessary to use a sealed-tube method to acquire sufficient derivative.

12.4.1c Procedure: 3,5-Dinitrobenzoates

For best results, be sure that equipment is flame dried. Improved results often are obtained with freshly fused zinc chloride. Into a microflask, place 400 to 500 mg of zinc chloride, 0.5 ml of the ether, and 250 mg of 3,5-dinitrobenzoyl chloride. Immediately insert a condenser equipped with a drying tube. Reflux the mixture for 1 to 2 hr. Remove the condenser and allow the volatile material to distill off. Pulverize the residue. Add 5 ml of 5% sodium carbonate, and stir and warm the reaction mixture to 60 to 70° (ca. 1 min on a steam bath). Collect the precipitate by suction filtration. Wash the precipitate with 2 ml of 5% sodium carbonate, and follow by several washings with small volumes of water and allow to dry. The crude ester may be recrystallized from chloroform, carbon tetrachloride, or aqueous ethanol. The yield may be as low as 5 mg or even lower.

Aliphatic ethers undergo oxidative cleavage with aqueous bromine at 25°. The products obtained depend upon the conditions and the ratio of bromine to ether employed.[24] The following examples are illustrative:

$$(CH_3)_2CHOCH(CH_3)_2 \xrightarrow[25°]{Br_2, H_2O} CH_3\overset{\displaystyle O}{\overset{\|}{C}}CH_3$$

$$CH_3CH_2OCH_2CH_3 \xrightarrow[25°]{Br_2, H_2O} CH_3COOH$$

$$C_6H_5CH_2OCH_2C_6H_5 \xrightarrow[25°]{Br_2, H_2O} C_6H_5CHO$$

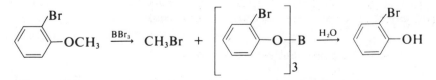

Cleavage of an aliphatic or alkyl aryl ether may also be effected by boron tribromide.[25,26]

[24]N. C. Deno and N. H. Potter, *J. Am. Chem. Soc.*, **89**, 3550 (1967).
[25]F. L. Benton and T. E. Dillon, *Ibid.*, **64**, 1128 (1942).
[26]J. F. W. McOmie and M. L. Watts, *Chem. Ind.* (London) 1658 (1963).

Alkyl and aryl ethers are cleaved by hydriodic acid.

$$R_2O + 2\,HI \longrightarrow 2\,RI + H_2O$$

$$ArOR + HI \longrightarrow ArOH + RI$$

To obtain sufficient product, especially for characterization of the alkyl iodides, it is necessary to use a sample of 4 to 5 g. The alkyl iodides and phenols may be transformed into suitable derivatives.

Vinyl ethers are readily cleaved with dilute mineral acids to alcohols and carbonyl compounds.

Aromatic ethers. The derivatives of aromatic ethers most frequently employed are those obtained by electrophilic aromatic substitution.

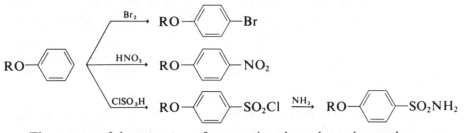

The extent of *bromination* of aromatic ethers depends on the groups already present.

12.4.1d Procedure: Bromination

The ether (5 to 100 mg) is dissolved in glacial acetic acid and placed in an ice bath; a slight excess of bromine is added with cooling (liquid bromine may be used, though for small-scale operations, a solution of bromine in glacial acetic acid is recommended). The reaction mixture is allowed to stand for a short while in the ice bath. It is then removed from the ice bath to stand at room temperature for 10 to 15 min. The bromo compound is separated by the addition of water. The crude product is recrystallized from dilute ethanol or petroleum ether. In some cases it is advantageous to use solvents other than acetic acid for the reaction medium. Carbon tetrachloride, chloroform, and ether have been used in these cases, though a less pure derivative is obtained by evaporation of the solvent.

Aromatic ethers are readily chlorosulfonated with chlorosulfonic acid. The intermediate sulfonyl chlorides may be isolated, but it is usually more convenient to convert them directly to *sulfonamides*.

12.4.1e Procedure: Sulfonamides

A solution of 0.25 g, or 0.25 ml of aromatic ether in 2 ml of chloroform, in a test tube is cooled in an ice bath. About 1 g of chlorosulfonic acid is added, drop by drop, over 5 min. The tube is removed from the ice bath and allowed to stand for 30 min. The reaction mixture is then poured into a small separatory funnel containing 5 ml of ice water. The chloroform layer is separated and washed with water. The chloroform layer is added, with stirring, to 3 ml of concentrated ammonia solution (or ammonia gas is bubbled into the chloroform solution). The solution is stirred for 10 min and then the chloroform is evaporated off. The residue is dissolved in 3 ml of 5% sodium hydroxide, and the solution is filtered to remove insoluble material. The filtrate is acidified with dilute hydrochloric acid and cooled in an ice bath. The sulfonamide is then collected and recrystallized from dilute ethanol.

Nitration can be accomplished by any of the procedures outlined for aromatic hydrocarbons (Sec. 12.1.4). Again the extent of nitration depends on the conditions employed and the groups already present.

Aromatic ethers form *molecular complexes with picric acid.*

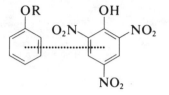

12.4.1f Procedure: Picric acid complexes

Dissolve 1 mmole of the aromatic ether in a small volume of boiling chloroform and 1.05 mmole (0.241 g) of picric acid in 10 ml of boiling chloroform. The hot solutions are mixed and allowed to cool. The crystals are collected, dried rapidly between filter papers, and the melting point immediately determined. Although a number of such picrates decompose on standing, some are stable enough to be recrystallized (from a minimum volume of chloroform).

Aromatic ethers containing an alkyl side chain may be oxidized to substituted benzoic acids with potassium permanganate (see procedure under aromatic hydrocarbons, Sec. 12.1.4). As noted before under aliphatic ethers, alkyl aryl ethers are cleaved by hydriodic acid and boron tribromide. Allyl aryl ethers undergo the Claisen rearrangement, upon heating, to give *o*-allylphenols.

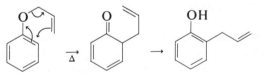

12.5 EPOXIDES

12.5.1a Classification

Epoxides are three-membered cyclic ethers which differ greatly in their chemical reactivity compared with larger-ring cyclic and acyclic ethers. Epoxides, which are often volatile liquids, display broad, moderately intense

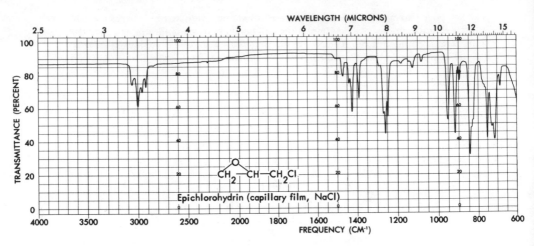

Fig. 12.11. Typical infrared spectrum of an epoxide.

bands in the infrared at 1250 cm^{-1} (8.0 μ) and in the 900 to 800 cm^{-1} (11 to 12.5 μ) region (Fig. 12.11). The nmr spectra of epoxides are quite complex.

The *periodate test* for epoxides is a modification of the test used for the detection of glycols (Sec. 12.3.2c).

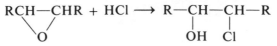

12.5.1b Procedure: Periodate test for epoxides

Add two drops of concentrated nitric acid to 2 ml of 0.5% solution of periodic acid, and then add one or two drops of the epoxide. Water-insoluble compounds may be dissolved in acetic acid. Shake the mixture vigorously, and add two drops of 5% silver nitrate solution. A positive test is indicated by the appearance of a white precipitate of silver iodate. It is advisable to run a blank.

Most epoxides *react with concentrated hydrochloric acid* to produce chlorohydrins which are insoluble in the reagent. This test should be run only after it has been established that the unknown is not an alcohol (see Lucas test, Sec. 12.3.1). Most aliphatic ethers are soluble in, but are not cleaved by, concentrated hydrochloric acid.

$$RCH-CHR + HCl \longrightarrow R-CH-CH-R$$
$$\underset{O}{\diagdown\diagup} \qquad\qquad \underset{OH\ \ Cl}{|\ \ \ |}$$

12.5.1c Procedure: Chlorohydrins

To 0.5 ml of cold concentrated hydrochloric acid in a small test tube, add two drops of the unknown. If the unknown is not immediately soluble in the hydrochloric acid,

add two or three drops of dioxane to produce a homogeneous solution. Allow the reaction mixture to come to room temperature and stand for 10 to 15 min. A positive test
is indicated by the formation of an insoluble oil (chlorohydrin). The chlorohydrins of
less than four carbons are soluble.

12.5.1d Characterization

Epoxides are partially characterized by their boiling points and refractive
indices.

Since it is generally not possible to form suitable derivatives directly from
epoxides, chemical conversions to other functional groupings are usually
necessary. The chemistry of epoxides is characterized by their high reactivity
toward ring-opening reactions which fall into two categories: nucleophilic
and electrophilic. Nucleophilic attack occurs at the least substituted carbon
atom to give a β-substituted alcohol. Electrophilic ring-opening reactions
may occur without rearrangement to give a β-substituted alcohol (note that
a different compound is formed), or with rearrangement to give a carbonyl
derivative via the more stable intermediate carbonium ion.

Nucleophilic ring opening

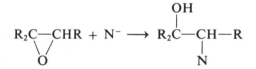

Electrophilic ring opening

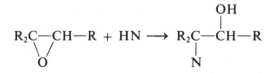

Rearrangements

$$R_2C\overset{\diagup\diagdown}{\underset{O}{}}CHR \xrightarrow{H^+} R_2\overset{+}{C}\underset{OH}{}CHR \longrightarrow R_2CH\underset{O}{\overset{\|}{}}C-R$$

The two most useful reactions for conversion of epoxides to another
functional group for purposes of derivation are the sulfuric or Lewis acid-
catalyzed rearrangements (see rearrangement reactions of glycols under
polyhydric alcohols, Sec. 12.3.2d) and the *lithium aluminum hydride reduction*. The latter is preferred due to fewer side reactions.

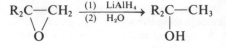

12.5.1e Procedure: Lithium aluminum hydride reduction

To a solution of a 50% excess of lithium aluminum hydride (based on an anticipated molecular weight of the epoxide) in 20 ml of absolute ether is slowly added 300 mg of the unknown epoxide dissolved in 5 ml of ether. The reaction mixture is refluxed for 30 min, and the excess hydride is carefully decomposed by the dropwise addition of saturated aqueous sodium sulfate solution or water (avoid a great excess of water). Sometimes it may be convenient to dissolve the aluminum hydroxide by addition of sodium potassium tartrate, and then to extract the mixture with ether. The reaction mixture is filtered to remove the insoluble inorganic residue. The ether solution is dried over magnesium sulfate, and the ether is removed by distillation. The alcohol may be characterized in the usual manner.

12.6   ACETALS AND KETALS

(See Sec. 12.19.2 on sulfides for dithio- and monothioacetals and ketals.)

12.6.1a Classification

A large portion of acetals and ketals are liquids with reasonably pleasant odors. An acetal or ketal should be suspected if three bands [1190 to 1160, 1195 to 1125, and 1098 to 1063 cm^{-1} (8.40 to 8.62, 8.37 to 8.89 and 9.11 to 9.41 μ, respectively)] appear in the C—O stretching region of the infrared spectrum, especially in the absence of carbonyl and hydroxyl absorptions (Fig. 12.12). In addition, the spectrum of an acetal contains a characteristic strong C—H bending deformation band in the 1116 to 1103 cm^{-1} (8.96 to 9.02 μ) region which, however, may be obscured by C—O peaks.

A compound suspected of being an acetal or ketal should be subjected to hydrolysis by dilute acid, and the hydrolysate should be tested for the presence of an aldehyde or ketone. A simple way to perform such a test is to add several drops of the compound to 2,4-dinitrophenylhydrazine reagent and warm the mixture. The acid contained in the reagent is usually sufficient to cause hydrolysis.

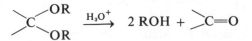

CAUTION: *Acetals, like ethers, form highly explosive peroxides on exposure to air and light.*

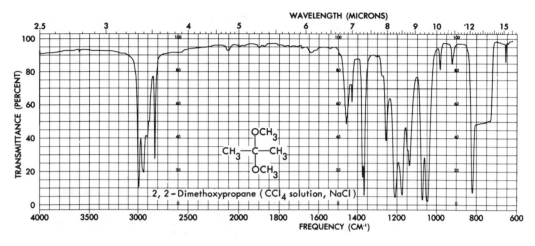

Fig. 12.12. Infrared spectrum of a simple ketal.

12.6.1b Characterization

The preparation of derivatives of acetals and ketals is based upon their hydrolysis to the component alcohol and carbonyl compound. The hydrolysis can usually be accomplished by heating with dilute mineral acid for 10 to 30 min with a co-solvent such as dioxane if necessary to effect solution. The hydrolysate should be divided into two portions. One portion should be used to prepare the 3,5-dinitrobenzoate or other suitable derivative of the alcohol. (In some cases, derivation of the alcohol component may be unnecessary since that portion of the molecule may be unambiguously identified by nmr or mass spectral data obtained on the original acetal or ketal.) The second portion is used to characterize the aldehyde or ketone as the 2,4-dinitrophenylhydrazone or semicarbazone.

12.7 ALDEHYDES AND KETONES

12.7.1a Classification

The presence of an intense band in the 1850 to 1650 cm^{-1} (5.40 to 6.06 μ) region of the infrared indicates a carbonyl compound. The lack of absorption in the 3600 to 3300 cm^{-1} (2.78 to 3.03 μ) region eliminates acids and primary and secondary amides as possibilities. The differentiation between aldehydes and ketones and the other remaining carbonyl compounds (esters, anhydrides, tertiary amides, etc.) can be accomplished by the 2,4-dinitrophenylhydrazine test. From·a cursory inspection of infrared spectra, the

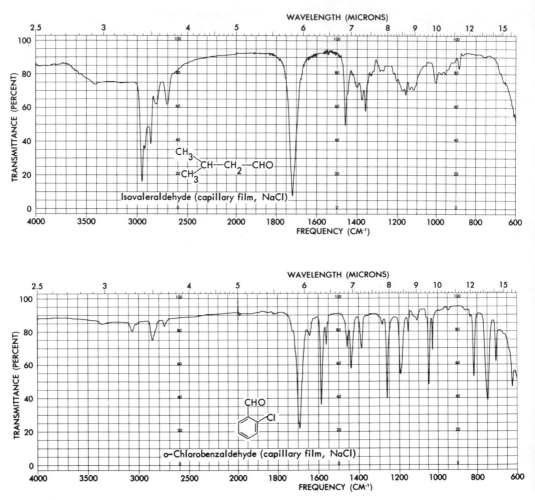

Fig. 12.13. Typical infrared spectra of aldehydes.

most likely compounds to be confused with aldehydes and ketones are the esters. Compare the spectra of aldehydes (Fig. 12.13) and ketones (Fig. 12.14) with those of esters (Fig. 12.21). Note that the two very intense C—O bands in the 1250 to 1050 cm⁻¹ (8.0 to 9.52 μ) region in the ester spectrum, which are absent in the aldehydes and ketones. With the latter, the carbonyl peak is usually the most intense in the spectrum.

12.7.1b 2, 4-Dinitrophenylhydrazine test

The utility of 2,4-dinitrophenylhydrazine lies in the fact that almost all aldehydes and ketones readily yield insoluble, solid 2,4-dinitrophenyl-

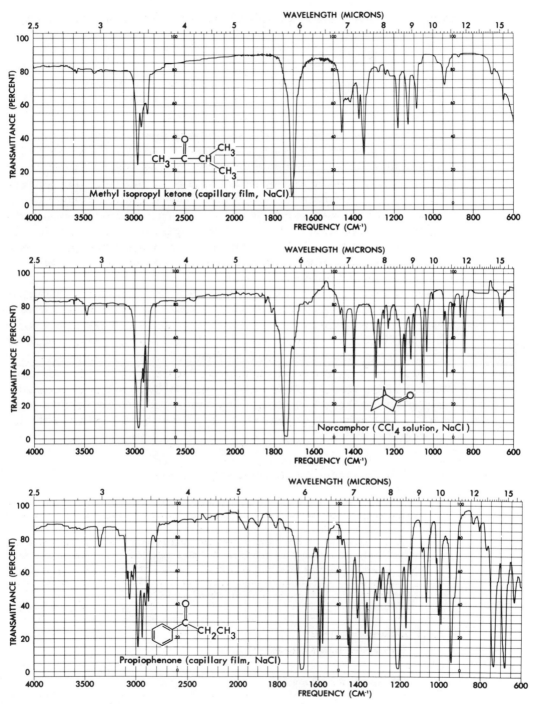

Fig. 12.14. Typical infrared spectra of ketones.

382 Functional Group Classification and Characterization

hydrazones. There are a few exceptions among the long chain aliphatic ketones which yield oils.

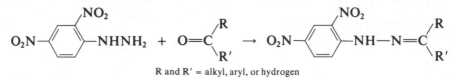

R and R' = alkyl, aryl, or hydrogen

It is wise to apply the test with some discrimination, i.e., only to compounds that give infrared evidence of being an aldehyde or ketone. The reagent may react with anhydrides or highly reactive esters; it is capable of oxidizing certain allylic and benzylic alcohols to aldehydes and ketones, which then give positive tests. The reagent may also give insoluble charge transfer complexes with amines, phenols, etc. These, however, are usually considerably more soluble than the phenylhydrazones and may be distinguished on this basis. Indiscriminate use of the reagent may provide misleading positive tests owing to the hydrolysis of compounds such as ketals or trace impurities of carbonyl compounds.

The color of a 2,4-dinitrophenylhydrazone may provide a qualitative indication about conjugation, or lack thereof, in the starting carbonyl compound. Dinitrophenylhydrazones of saturated aldehydes and ketones are typically yellow. Conjugation with a carbon-carbon double bond or with an aromatic ring changes the color from yellow to orange-red. It should be noted that 2,4-dinitrophenylhydrazone itself is orange-red, and sometimes under the reaction conditions this material precipitates or contaminates the dinitrophenylhydrazone. Any orange-red precipitate that melts near 2,4-dinitrophenylhydrazine (198° d.) should be looked at with due suspicion.

12.7.1c Procedure: 2,4-Dinitrophenylhydrazine test

Reagent: Dissolve 3 g of 2,4-dinitrophenylhydrazine in 15 ml of concentrated sulfuric acid. Add this solution with stirring to 20 ml of water and 70 ml of 95% ethanol. Mix this solution thoroughly and filter.

Place 1 ml of the 2,4-dinitrophenylhydrazine reagent in a small test tube. Add one drop of the liquid carbonyl compound to be tested or an equivalent amount of solid compound dissolved in a minimum amount of ethanol. Shake the tube vigorously; if no precipitate forms immediately, allow the solution to stand for 15 min. The formation of a yellow to orange-red precipitate is considered a positive test.

12.7.2 Differentiation of Aldehydes and Ketones

12.7.2a Spectral methods

Infrared spectra. The infrared spectra of ketones exhibit carbonyl bands in the 1780 to 1665 cm^{-1} (5.62 to 6.01μ) region. Aldehyde carbonyl bands appear in the more restricted 1725 to 1685 cm^{-1} (5.80 to 5.94μ) region. In the aldehyde spectra, additional bands appear near 2820 cm^{-1} (3.55 μ) and 2700 cm^{-1} (3.70 μ) which are not present in the spectra of ketones (Fig. 12.13). Regrettably, these bands are often weak and/or ill-resolved.

Nmr spectra. The very low field resonance of the aldehydic hydrogen (δ9.5 to 10.1) is very diagnostic of an aldehyde. Hydrogens on saturated carbon adjacent to the carbonyl of aldehydes and ketones usually fall in the δ2 to 3 region. In the case of aldehydes, these hydrogens are coupled with the aldehydic hydrogen with a relatively small coupling constant (1–3 Hz). Alkyl methyl ketones exhibit sharp methyl singlets near δ2.2; aryl methyl ketones appear nearer δ2.6.

Mass spectra. If the mass spectrum exhibits a discernible parent peak and a significant parent -1 peak, the spectrum should be examined for other evidence of an aldehyde. In the mass spectrum of a ketone, the molecular ion is usually pronounced. Major fragmentation peaks for acyclic ketones result from alpha bond cleavage to give acylium ions. In the alkyl aryl ketones, the ArCO$^+$ fragment is usually the base peak.

12.7.2b Chemical methods

The liability of aldehydes toward oxidation to acids forms the basis for a number of tests to differentiate between aldehydes and ketones.

Most aldehydes reduce silver ion from *ammoniacal silver nitrate solution* (*Tollen's reagent*) to metallic silver.

$$\text{RCHO} + 2\,\text{Ag(NH}_3)_2\text{OH} \longrightarrow 2\,\text{Ag} \downarrow + \text{RCOO}^-\text{NH}_4^+ + \text{H}_2\text{O} + \text{NH}_3$$

The test is also given by hydroxyketones (reducing sugars as well as certain amino compounds such as hydroxylamines). Normal ketones do not react. The test should be run only after it has been established that the unknown in question is an aldehyde or ketone.

12.7.2c Procedure: Tollen's test

Reagent: Solution A. Dissolve 3 g of AgNO$_3$ in 30 ml of water. *Solution B.* 10% sodium hydroxide. When the reagent is required, mix 1 ml of solution *A* with 1 ml of solution *B* in a *clean* test tube, and add dilute ammonia solution dropwise until the silver oxide is just dissolved.

CAUTION: *Prepare the reagent only immediately prior to use.* Do not heat the reagent during its preparation or allow the prepared reagent to stand, since the *very explosive* silver fulminate may be formed. Wash any residue down the sink with a large quantity of water. Rinse the test tube with dilute HNO$_3$ when the test is completed.

Add a few drops of a dilute solution of the compound to 2 to 3 ml of the prepared reagent. In a positive test, a silver mirror is deposited on the walls of the tube either in the cold or upon warming in a beaker of hot water.

Chromic acid in acetone rapidly oxidizes aldehydes to acids (the reagent also attacks primary and secondary alcohols, Sec. 12.3.1, whereas ketones are attacked slowly or not at all by the reagent. Aliphatic aldehydes are oxidized somewhat faster than aromatic aldehydes; this rate difference has been used to distinguish between aliphatic and aromatic aldehydes.[27]

12.7.2d Procedure: Chromic acid test

Reagent: Dissolve 1 g of chromic oxide in 1 ml of concentrated sulfuric acid, and dilute with 3 ml of water.

Dissolve one drop or 10 mg of the compound in 1 ml of reagent grade acetone (or better, acetone which has been distilled from permanganate). Add several drops of chromic acid reagent. A positive test is indicated by the formation of a green precipitate of chromous salts. With aliphatic aldehydes, the solution turns cloudy within 5 sec, and a precipitate appears within 30 sec. Aromatic aldehydes generally require 30 to 90 sec for the formation of a precipitate.

With aliphatic aldehydes (but not aromatic) and α-hydroxyketones, *Benedict's reagent (a citrate complex of cupric ion)* gives a yellow-to-red precipitate. A yellow suspension in the blue solution of the reagent appears green. In certain cases, the exact composition of the precipitate is unknown; it is usually thought to be cuprous oxide. This is the classical test that has been used to distinguish between aliphatic and aromatic aldehydes.

$$RCHO + 2\,Cu^{++}\,(citrate)_2 + 4\,OH^- \longrightarrow RCOOH + Cu_2O \downarrow + H_2O$$

12.7.2e Procedure: Benedict's test

Reagent: Dissolve 86.5 g of hydrated sodium citrate and 50 g of anhydrous sodium carbonate in about 350 ml of water. Add a solution of 8.65 g of copper sulfate in 50 ml water with stirring. Dilute the resulting solution to 500 ml; filter if necessary. The solution does not deteriorate significantly on storage.

Into a test tube containing 4 ml of Benedict's solution, add two to three drops of liquid unknown or an equivalent amount of solid in a small volume of ethanol or water. Heat the mixture to boiling. A positive test is indicated by the formation of a yellow-to-red precipitate.

Aldehydes react with methone to give dimethone derivatives. Ketones do not react, and the method can serve to differentiate between the two (see next section).

12.7.3 Aldehydes

12.7.3a Characterization

The most frequently employed derivatives for aldehydes include the 2, 4-dinitro-, the *p*-nitro-, and the unsubstituted phenylhydrazones, the semicarbazones, and the oximes. Discussion and procedures for these deriva-

---

[27]J. D. Morrison, *J. Chem. Ed.*, **42**, 554 (1965).

tives, as well as a method for reduction to a primary alcohol, will be presented in Sec. 12.7.5.

The *oxidation of an aldehyde to the corresponding acid* is a particularly useful method of identification, especially in the case of aromatic aldehydes which yield solid acids (characterized by their melting point and neutralization equivalent).

12.7.3b Procedure: Oxidation to acids

(a) The oxidation may be carried out by the chromic acid in acetone method outlined under alcohols (Sec. 12.3.1).

(b) In a test tube place 0.5 g of aldehyde, 5 to 10 ml of water, and several drops of 10% sodium hydroxide. Add saturated aqueous potassium permanganate several drops at a time. Shake the mixture vigorously. Add additional permanganate until a definite purple color remains. Acidify the mixture with dilute sulfuric acid, and add sodium bisulfite solution until the excess permanganate and manganese dioxide have been converted to the colorless soluble manganese sulfate. Remove the acid by filtration or extraction with chloroform or ether. The acid may be recrystallized from water or water-alcohol, or it may be purified by sublimation.

The equations following this paragraph illustrate the reaction between 5,5-dimethylcyclohexane-1,3-dione (methone, dimethyldihydroresorcinol) and an aldehyde. One mole of the aldehyde condenses with 2 moles of the reagent to give the *dimethone*, sometimes called a *dimedone derivative*. As ketones do not yield derivatives the reaction can serve as another method of differentiating between aldehydes and ketones. Aldehydes can be made to react in the presence of ketones. Also, the reaction can serve as a method for the quantitative determination of formaldehyde. The dimethones are generally more suitable derivatives for low molecular weight aldehydes than are 2,4-dinitrophenylhydrazones.

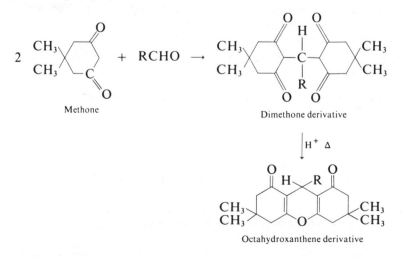

Methone   Dimethone derivative

Octahydroxanthene derivative

12.7.3c Procedure: Dimethone derivatives

To a solution of 100 mg of the aldehyde in 4 ml of 50% ethanol, add 0.3 g of methone and one drop of piperidine. Boil the mixture gently for 5 min. If the solution is clear at this point, add water dropwise to the cloud point. Cool the mixture in an ice bath until crystallization is complete. Crystallization is often slow and one should allow, if necessary, 3 to 4 hr. Collect the crystals by vacuum filtration, and wash with a minimum amount of cold 50% ethanol. If necessary, the derivative may be recrystallized from mixtures of alcohol and water.

If the dimedone derivatives are heated with acetic anhydride or with alcohol containing a small amount of hydrochloric acid, cyclization to the *octahydroxanthene (menthone anhydride)* occurs. The cyclization is rapid and quantitative, and the xanthenes can serve as a second derivative.

12.7.3d Procedure: Octahydroxanthenes

The foregoing dimedone derivative may be converted to the octahydroxanthenes by boiling a solution of 100 mg of the derivative in a small volume of 80% ethanol to which one drop of concentrated hydrochloric acid has been added. The cyclization is complete in 5 min; water is added to the cloud point, and the solution is cooled to induce crystallization. The product is usually pure, and further recrystallization is not necessary.

12.7.4 Ketones

12.7.4a Characterization

The methods described under general derivatives of aldehydes and ketones in the next section are those most often employed to obtain solid derivatives of ketones. In general, low molecular weight ketones are best characterized by 2,4-dinitrophenylhydrazones, *p*-nitrophenylhydrazones, or semicarbazones. Hydrazones, phenylhydrazones, and oximes are usually more suitable for the higher molecular weight ketones.

Other reactions sometimes useful in the structural determination or ketones include:

1. Active hydrogen determination by deuterium exchange or other means (Sec. 11.2.4).

2. The formation of benzylidene derivatives (Sec. 11.2.4).

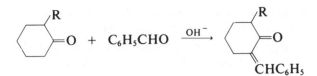

3. The Beckman rearrangement of oximes to amides.[28]

4. The Baeyer-Villiger oxidation of ketones to esters or lactones.[29]

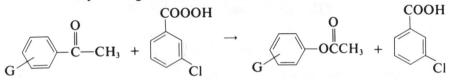

5. The *haloform reaction* is useful in classification and derivation of methyl ketones. The halogenation of aldehydes and ketones is catalyzed by acids and bases. With acid catalysis, the rate-determining step is enol formation, and with base catalysis, it is the formation of the enolate. Hence the reaction is independent of the concentration or kind of halogen. With base catalysis, the initial reaction occurs at the least-substituted carbon.

$$R-\underset{\underset{O}{\|}}{C}CH_3 \xrightarrow[HB]{B^-} R-\underset{\underset{O^-}{|}}{C}=CH_2 \xrightarrow{X_2} R\underset{\underset{O}{\|}}{C}-CH_2X'$$

Since halogen is highly electronegative, successive hydrogens are replaced more readily, and unsymmetrical polysubstitution occurs.

$$R\underset{\underset{O}{\|}}{C}-CH_2X \xrightarrow[X_2]{B^-} R-\underset{\underset{O}{\|}}{C}-CX_3$$

With methyl ketones or acetaldehyde, a trihalogenated product is formed. These trihalocarbonyl compounds are readily cleaved by base to give a haloform and a salt of a carboxylic acid.

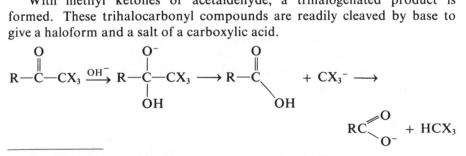

[28]L. G. Donaruma and W. Z. Heldt, *Org. Reactions*, **11**, 1 (1960).
[29]C. H. Hassall, *Ibid.*, **9**, 73 (1957).

The haloform reaction will proceed with aldehydes or ketones that contain the groupings CH_3CO-, CH_2XCO-, CHX_2-CO, or CX_3CO-. The reaction will also proceed with compounds that will react with the reagent to give a derivative containing one of the necessary groupings, e.g.,

$$R-\underset{\underset{OH}{|}}{CH}-CH_3 \xrightarrow[NaOH]{X_2} R-\underset{\underset{O}{\|}}{C}-CH_3 \xrightarrow[NaOH]{X_2} RCOO^- + HCX_3$$

Acetaldehyde is the only aldehyde, and ethyl alcohol is the only primary alcohol, that produces haloform. The principal types of compounds that give the haloform reaction are methyl ketones and alkylmethyl carbinols

$$CH_3-\underset{\underset{O}{\|}}{C}-R \quad and \quad CH_3-\underset{\underset{OH}{|}}{CH}-R$$

R = alkyl, vinyl, or aryl

and compounds which have the structures

$$R-\underset{\underset{O}{\|}}{C}-CH_2-\underset{\underset{O}{\|}}{C}-R, \quad R-\underset{\underset{OH}{|}}{CH}-CH_2-\underset{\underset{OH}{|}}{CH}-R, \quad etc.$$

which are, or which upon oxidation yield, ketoalcohols or diketones that are cleaved by alkali to methyl ketones.

$$R-\underset{\underset{O}{\|}}{C}-CH_2-\underset{\underset{OH}{|}}{CH}-R' \xrightarrow{OH^-} R-\underset{\underset{O}{\|}}{C}-CH_3 + R'CHO$$

Acetic acid and its esters and compounds such as

$$CH_3-\underset{\underset{O}{\|}}{C}-CH_2-COOR, \quad CH_3-\underset{\underset{O}{\|}}{C}-CH_2-CN$$

which substitute on the methylene rather than the methyl, do not produce haloforms.

As a method of preparation of acids from methyl ketones and related substances, sodium hypochlorite (bleaching powder or bleaching solution) is often used. As a qualitative test, the reaction is usually run with alkali and iodine, since iodoform is a water-insoluble crystalline solid readily identified by its melting point (*iodoform test*).

12.7.4b Procedure: Iodoform test

Reagent: Dissolve 20 g of potassium iodide and 10 g of iodine in 100 ml of water.

Dissolve 0.1 g or 4 to 5 drops of compound in 2 ml of water; if necessary, add sufficient dioxane to produce a homogeneous solution. Add 1 ml of 10% sodium hydroxide and the potassium iodide-iodine reagent dropwise, with shaking, until a definite dark color of iodine persists. Allow the mixture to stand for several minutes; if no iodoform separates at room temperature, warm the tube in a 60° water bath. If the color disappears, continue to add the iodine reagent dropwise until its color does not disappear after 2 min of heating. Remove the excess iodine by the addition of a few drops of sodium hydroxide solution. Dilute the reaction mixture with water and allow to stand for 15 min. A positive test is indicated by the formation of a yellow precipitate. Collect the precipitate; dry on filter paper. Note the characteristic medicinal odor, and determine the melting point; iodoform melts at 119 to 121°.

12.7.5 General Derivatives of Aldehydes and Ketones

12.7.5a 2,4-Dinitrophenylhydrazones and p-nitrophenylhydrazones

$$ArNHNH_2 + RCOR' \longrightarrow ArNHN{=}CRR'$$

Ar = 2, 4-dinitrophenyl, or *p*-nitrophenyl

R and R' = alkyl, aryl, or hydrogen

The 2, 4-dinitrophenylhydrazones are excellent derivatives for small-scale operations. It is often possible to obtain sufficient amounts of the derivative for characterization from the precipitate formed in the classification test described earlier (Sec. 12.7.1b). There are certain limitations to the use of nitrophenylhydrazones as derivatives. From derivative tables one will observe that, in many cases, the melting points of the nitrophenylhydrazones are too high for practical utility; hence, the use of other derivatives such as phenylhydrazones or semicarbazones is recommended. The reagent undergoes secondary reactions with α-hydroxy aldehydes and ketones (see discussion under carbohydrates Sec. 12.9). In addition to geometrical isomers about the C=N linkage, the 2,4-dinitrophenylhydrazones tend to form several crystalline modifications of different melting points. One may find it useful to redetermine melting points of 2,4-dinitrophenylhydrazone derivatives after allowing the melt to resolidify.

12.7.5b Procedure: 2,4-Dinitro- and p-nitrophenylhydrazones

Place 100 mg of 2,4-dinitro- or *p*-nitrophenylhydrazine in a test tube or Erlenmeyer flask containing 10 ml of methanol. Cautiously add five drops of concentrated hydrochloric acid, and warm the solution on the steam bath if necessary to complete solution. Dissolve approximately 1 mmole of the carbonyl compound in 1 ml of methanol, and add this to the reagent. Warm the mixture on a water or steam bath for 1 to 2 min and allow to stand for 15 to 30 min. Most of the derivatives precipitate out on cooling, but for complete precipitation, it is advisable to add water to cloudiness. The derivative may be purified by crystallization from alcohol-water mixtures. In the case of less soluble materials, ethyl acetate is often found to be a more suitable solvent.

This procedure is generally suitable for the preparation of both dinitrophenylhydrazones and nitrophenylhydrazones. Numerous alternative procedures are available;[30] some are more suitable for less reactive compounds, whereas others may prevent undesirable side reactions.

Although not recommended as a general procedure, it is sometimes advantageous to isolate carbonyl compounds as the dinitrophenylhydrazone derivatives. Procedures have been developed for regenerating the starting carbonyl compound by hydrazone exchange reaction involving levulinic, pyruvic, and other keto acids.[31]

12.7.5c Semicarbazones

$$\underset{R'}{\overset{R}{>}}C=O + NH_2NHCONH_2 \longrightarrow \underset{R'}{\overset{R}{>}}C=NNHCONH_2 + H_2O$$

Semicarbazide   Semicarbazone

Semicarbazones prepared from carbonyl compounds and semicarbazide hydrochloride are excellent derivatives for ketones and aldehydes above five carbon atoms because of their ease of formation, highly crystalline properties, sharp melting points, and ease of recrystallization. The lower aldehydes react slowly and/or give soluble derivatives. As in the case of most carbonyl derivatives, semicarbazones from unsymmetrical carbonyl compounds are capable of existing in two isomeric forms. A number of substituted semicarbazones and thiosemicarbazones are occasionally used.

12.7.5d Procedure: Semicarbazones

In an 8-in. test tube, place 100 mg of semicarbazide hydrochloride, 150 mg of sodium acetate, 1 ml of water, and 1 ml of alcohol. Add 100 mg (0.1 ml) of the aldehyde or ketone. If the mixture is turbid, add alcohol until a clear solution is obtained. Shake the mixture for a few minutes and allow to stand. The semicarbazone typically will crystallize from the cold solution on standing, the time varying from a few minutes to several hours. In the case of less reactive carbonyl compounds, the reaction may be accelerated by warming the mixture on a water bath for 10 min and then cooling in ice water. The crystals are filtered off and washed with a little cold water; they may usually be recrystallized from methyl or ethyl alcohol alone or mixed with water.

12.7.5e Oximes

$$\underset{R'}{\overset{R}{>}}C=O + NH_2OH \longrightarrow$$

$$\underset{R'}{\overset{R}{>}}C=N_{\diagdown OH} \quad \text{and/or} \quad \underset{R'}{\overset{R}{>}}C=N^{\diagup OH} + H_2O$$

syn- and *anti-*Oximes

[30]H. J. Shine, *J. Org. Chem.*, **24**, 252 (1959); J. A. Maynard, *Australian J. Chem.*, **15** 867 (1962); J. Parrick and J. W. Rasburn, *Can. J. Chem.*, **43**, 3453 (1965).
[31]E. Hershberg, *J. Org. Chem.*, **13**, 542 (1948); C. H. DePuy and B. W. Ponder, *J. Am. Chem. Soc.*, **81**, 4629 (1959).

Oximes do not have as wide a utility as other carbonyl derivatives for the identification of aldehydes and ketones. They are somewhat more difficult to obtain crystalline and are more likely to exist as mixtures of geometrical isomers. However, they occasionally do have utility and should be considered.

12.7.5f Procedure: Oximes

(a) The method described for preparation of semicarbazones may be employed. One substitutes hydroxylamine hydrochloride for the semicarbazide hydrochloride. The heating period is usually necessary.

(b) Reflux a mixture of 0.1 g of the aldehyde or ketone, 0.1 g of hydroxylamine hydrochloride, 2 ml of ethanol, and 0.5 ml of pyridine on a water bath for 15 to 60 min. Remove the solvent at reduced pressure or by evaporation with a current of air in a hood. Add several milliliters of cold water, and triturate thoroughly. Collect the oxime and recrystallize from alcohol, alcohol-water, or benzene.

12.7.5g Phenylhydrazones

Phenylhydrazones are recommended in particular for aryl carbonyl compounds whenever the dinitrophenylhydrazone is not suitable or adequate for identification. The manipulation of these derivatives should be as rapid as possible since they undergo slow decomposition in air.

12.7.5h Procedure: Phenylhydrazones

In a test tube place 100 mg of aldehyde or ketone, 4 ml of methanol, and four drops of phenylhydrazine. Boil the mixture for 1 minute and add one drop of glacial acetic acid and boil gently for 3 min. Add cold water dropwise until a permanent cloudiness results, cool, collect the crystals, and wash with 1 ml of water containing one drop of acetic acid. Recrystallize the product immediately by dissolving in hot methanol, add water to the solution until cloudiness appears, cool, and scratch the sides of the tube if necessary to crystallize. Collect the crystals, wash with a few drops of dilute methanol and determine the melting point as soon as possible. Dry the crystals by pressing between filter paper.

12.7.5i Reduction to alcohols

$$NaBH_4 + 4 \underset{R'}{\overset{R}{>}}C{=}O \longrightarrow NaB\left[O{-}\underset{R'}{\overset{R}{\underset{|}{\overset{|}{C}}}}{-}H\right]_4 \xrightarrow{OH^-}$$

$$4\,H{-}\underset{R'}{\overset{R}{\underset{|}{\overset{|}{C}}}}{-}OH + NaBO_2$$

Alcohols are likely to be liquids, oils, or low-melting solids. However, there are occasions when the alcohols formed by reduction of aldehydes or ketones are well-defined crystalline solids suitable for identification purposes. Metal hydride reductions of carbonyl compounds to the corresponding alcohols generally proceed smoothly and give a high yield of the desired product. Sodium borohydride is less reactive than other hydrides such as lithium aluminum hydride, but it has the advantage of being stable in the presence of air and moisture. Sodium borohydride can be dissolved in cold water without extensive decomposition, whereas lithium aluminum hydride decomposes explosively on contact with water.

A commonly used method of reduction is to dissolve the organic compound in ethanol and to add an aqueous solution of sodium borohydride. Sodium borohydride, although stable in water, reacts at an appreciable rate with ethanol. It is therefore necessary to use an excess of the borohydride in order to offset losses sustained through solvolysis. Excess borohydride remaining in the reaction is destroyed by boiling the reaction mixture for a few minutes. The borate ester formed in the reduction is hydrolyzed during the boiling step by the addition of aqueous base.

12.7.5¡ Procedure: Sodium borohydride reduction

In a small test tube, dissolve 2 mmole of the carbonyl compounds in 3 to 5 ml of ethanol. In a second test tube, place 50 mg of sodium borohydride and dissolve in five drops of distilled water. Add the ethanol solution of the carbonyl compound to the borohydride solution. Shake the reaction mixture several times during a 15-min period. Add 0.5 ml of 6 N aqueous sodium hydroxide solution and a boiling chip to the reaction mixture. Boil it gently on the steam bath for 5 min. Pour the reaction mixture onto a small amount of ice, and collect the resulting precipitate by vacuum filtration. If the alcohol does not crystallize, it may be recovered by extraction with ether and then purified.

12.7.6 Special Methods
         of Separation and Purification
         of Aldehydes and Ketones

Most aldehydes, alkyl methyl ketones, and cyclic ketones react with saturated sodium hydrogen sulfite (bisulfite) solution to form crystalline *bisulfite addition compounds*. Other aliphatic and aromatic ketones do not yield addition compounds.

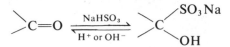

The reaction is reversible, and the aldehyde or ketone can be recovered by reaction of the addition compound with aqueous sodium carbonate or dilute hydrochloric acid. The sequence of formation and decomposition of the addition compounds may be used for the purification and/or separation of aldehydes and certain ketones from other materials.

12.7.6a Procedure: Bisulfite addition compounds

Shake or stir thoroughly the aldehyde- or ketone-containing mixture with a saturated solution of sodium hydrogen sulfite. The temperature will rise as the exothermic addition reaction takes place. Extract the residual organic material into ether and recover. Collect the crystalline precipitate, wash it with a little ethanol followed by ether, and allow it to dry.

To decompose the addition compound, use 10% sodium carbonate solution or dilute hydrochloric acid.

The reagent will undergo Michael addition to α,β-unsaturated carbonyl compounds to yield salts of sulfonic acids.

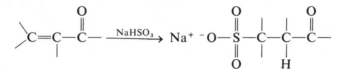

Another chemical method for the separation of aldehydes and ketones from other neutral, water-insoluble compounds involves the use of *Girard's reagent T*. The reagent reacts with carbonyl compounds to yield water-soluble quaternary ammonium hydrazones. The water-insoluble compounds are then separated by ether extraction. The aldehyde or ketone is regenerated by hydrolysis with dilute acid.

$$(CH_3)_3\overset{+}{N}CH_2CONHNH_2 \quad Cl^- + \;\;>C{=}O \;\longrightarrow$$

Girard's reagent T

$$>C{=}NNHCOCH_2\overset{+}{N}(CH_3)_3 \quad Cl^- \xrightarrow[\Delta]{HCl} \;>C{=}O$$

water soluble

12.7.6b Procedure: Girard's reagent T

A solution of 0.5 g of impure aldehyde or ketone, 0.5 g (or slight excess) of Girard-T reagent, and 0.5 ml of acetic acid in 5 ml of 95% ethanol is refluxed for 30 min. Cool and pour the reaction mixture into a separatory funnel. Add water, ether, and sufficient saturated sodium chloride solution to avoid emulsification. Shake well. Separate the layers and recover the ether-soluble material. Treat the aqueous layer with 1 ml of concentrated hydrochloric acid and heat on the steam bath to effect hydrolysis of the derivative back to the aldehyde or ketone. Isolate the aldehyde or ketone by extraction.

Girard-P reagent can be employed in a like manner.

$$\overset{+}{N}CH_2CONHNH_2 \quad Cl^-$$

Girard's reagent P

12.8.1a Classification

Quinones are colored (most are yellow) crystalline compounds with a pene-trating odor. Most *o*- and *p*-quinones have a carbonyl band in the infrared near 1670 cm^{-1} (6.0 μ) (Fig. 12.15). Quinones which have carbonyls in

WAVELENGTH (MICRONS)

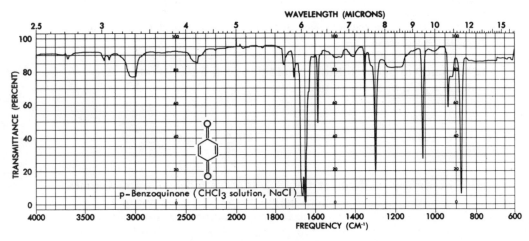

Fig. 12.15. Infrared spectrum of *p*-benzoquinone.

different rings, e.g., 2,6-naphthoquinone, appear near 1645 cm^{-1} (6.08 μ). Some *p*-benzoquinones exhibit doublet carbonyl absorptions. The ultra-violet spectrum is often useful in establishing the presence and type of quinone as well as its substitution pattern (Sec. 3.5.2a).

Most quinones liberate iodine from acidified potassium iodide solutions.

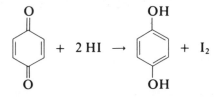

12.8.1b Characterization

Quinones form oximes and semicarbazones, usually from an aqueous alcohol medium. The carbonyl derivatives may not be of the usual struc-ture, e.g.,

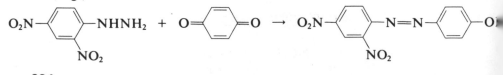

394

Quinones may be readily *reduced to hydroquinones* with a variety of reducing reagents, e.g., zinc and dilute hydrochloric acid or sodium dithionate.

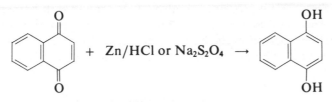

12.8.1c Procedure: Reduction to hydroquinones

(a) Suspend the quinone in dilute hydrochloric acid and add a little zinc dust. When the solution has turned colorless, neutralize with sodium hydrogen carbonate and extract the hydroquinone into ether. Evaporate the solvent and identify the hydroquinone (phenol-type derivatives may be made if necessary).

(b) Dissolve or suspend the quinone (0.5 g) in 5 ml of benzene or ether in a small separatory funnel. Add a solution of 1 g of sodium dithionate ($Na_2S_2O_4$) in 10 ml of 1 *N* sodium hydroxide. Shake the mixture until the characteristic color of the quinone has disappeared. Separate the aqueous alkaline solution, cool in an ice bath, and carefully acidify with concentrated hydrochloric acid. Collect the product (extract with ether if necessary) and recrystallize from dilute ethanol.

Like other aromatic diketones, *o*-quinones react with *o*-phenylene-diamine to afford *quinoxalines*.

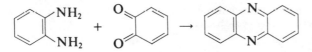

12.8.1d Procedure: Quinoxalines

Dissolve the *o*-quinone in alcohol or glacial acetic acid. Add an equivalent amount of *o*-phenylenediamine in alcohol. Warm the mixture on the steam bath for 15 to 20 min, cool, and dilute with water to crystallize. Recrystallize from aqueous ethanol.

12.9 CARBOHYDRATES

12.9.1a Classification

Mono- and disaccharides are colorless solids or syrupy liquids. They are readily soluble in water but are almost completely insoluble in most organic solvents. The solids typically melt at high temperatures (over 200°) with decomposition (browning and producing a characteristic caramel-like odor), and they char when treated with concentrated sulfuric acid. The polysaccharides possess similar properties but are generally insoluble or only slightly soluble in water.

The infrared spectra of carbohydrates obtained as mulls or potassium bromide pellets exhibit prominent hydroxyl absorption with correspondingly strong absorption in the C—O stretching region. Almost all sugars of four or more carbons exist in cyclic hemiacetal, acetal, hemiketal, or ketal forms and hence exhibit no carbonyl band in the infrared (Fig. 12.16).

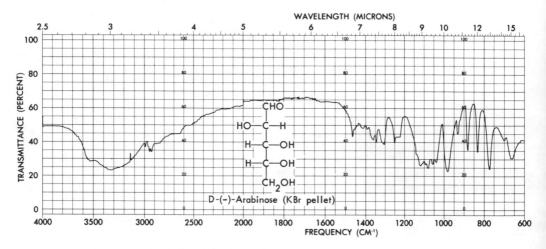

Fig. 12.16. Typical carbohydrate infrared spectrum. Note absence of carbonyl band. Such compounds exist as cyclic hemiacetals.

Sugars containing an aldehyde or α-hydroxy ketone grouping present in the free or hemiacetal (but not acetal) form are oxidized by Benedict's and Tollen's reagents (Sec. 12.7.2c). Such sugars are referred to as "reducing sugars."

Pentoses and hexoses are dehydrated by concentrated sulfuric acid to form furfural or hydroxymethylfurfural, respectively. In the *Molisch test*, these furfurals condense with 1-naphthol to give colored products.

12.9.1b Procedure: Molisch test

Place 5 mg of the substance in a test tube containing 0.5 ml of water. Add two drops of a 10% solution of 1-naphthol in ethanol. By means of a dropper, allow 1 ml of concentrated sulfuric acid to flow down the side of the inclined tube so that the heavier acid forms a bottom layer. If a carbohydrate is present, a red ring appears at the interface of the two liquids. A violet solution is formed on mixing. Allow the mixed solution to stand for 2 min then dilute with 5 ml of water. A dull-violet precipitate will appear.

12.9.1c Characterization

Since sugars decompose on heating, they do not have well-defined characteristic melting points. The same is true of some of their derivatives, e.g., ozazones. Fortunately, the number of sugars normally encountered in identification work is relatively small. Authentic samples of most are readily available for comparative purposes.

Thin-layer and paper chromatographic comparison[32] of an unknown with authentic samples provides an excellent means for tentative identification. For simple free sugars, paper chromatography is generally superior to thin-layer chromatography. Recommended solvent systems for paper chromatography of sugars include:

1. *n*-butanol, ethanol, water (40, 11, 19 parts by volume).
2. *n*-butanol, pyridine, water (9, 5, 4).
3. isopropanol, pyridine, water (7, 7, 5).
4. *n*-butanol, acetic acid, water (2, 1, 1).

For the *detection of reducing sugars* on paper chromatograms, the authors suggest the following spray.

12.9.1d Procedure: Spray for reducing sugars

The dried chromatogram is sprayed with a 3% solution of *p*-anisidine hydrochloride in *n*-butanol and heated to 100° for 3 to 10 min. Aldo- and keto-hexoses, as well as other sugars, give different colored spots with this reagent.

The structure and stereochemistry of a number of sugar and sugar derivatives has been examined by nmr and mass spectrometry. The primary literature should be consulted for details. The relationship between dihedral angle and coupling constants for hydrogens has been of significant utility in determining stereochemical relationship of the various ring hydrogens in cyclic sugar derivatives.

The specific rotation of sugars and derivatives is a useful means of identification. Rotations must be measured under specified conditions of concentration, solvent, and temperature, employing pure samples.

12.9.1e Reaction of sugars with phenylhydrazine

Sugars containing an aldehyde or keto group (as the hemiacetal or hemiketal) react with an equivalent of phenylhydrazine in the cold to produce

---

[32]E. Heftmann, *Chromatography*, 2nd ed., Reinhold Publishing Corporation, New York, 1967; R. J. Block, E. L. Durram, and G. Zweig, *Paper Chromatography and Paper Electrophoresis*, Academic Press, New York, 1958; E. Lederer and M. Lederer, *Chromatography*, 2nd ed., Elsevier Publishing Company, Amsterdam, 1957.

the corresponding phenylhydrazones. These derivatives are water soluble and do not precipitate. Heating of these sugars with excess (3 to 4 equivalents) phenylhydrazine produces *osazones* and polyzones.[33]

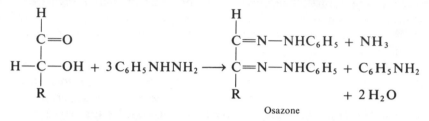

Osazone

It should be noted that, in the formation of osazones, one carbinol grouping is oxidized, and hence a number of isomeric sugars give the same osazone, for example, *D*-glucose, *D*-mannose, and *D*-fructose all produce the same osazone. The osazones exist in chelated structures in equilibrium with a nonchelated isomer.[33]

12.9.1f Procedure: Osazones

Place a 0.1-g sample of the unknown sugar and 0.2-g sample of a known sugar (suspected to be the unknown) in separate test tubes. To each sample, add 0.2 g of phenylhydrazine hydrochloride, 0.3 g of sodium acetate, and 2 ml of distilled water. Stopper the test tubes with vented corks, and place them together in a beaker of boiling water. Note the time of immersion and the time of precipitation of each osazone. Shake the tubes occasionally. The time required for the formation of the osazone may be used as evidence for the identification of the unknown sugar.

Under these conditions, fructose osazone precipitates in 2 min, glucose in 4 to 5 min, xylose in 7 min, arabinose in 10 min, and galactose in 15 to 19 min. Lactose and maltose osazones are soluble in hot water.

After 30 min remove the tubes from the hot water and allow them to cool. Carefully collect the crystals, and compare the unknown with the known crystals under a low-power microscope. The melting points (decomposition) of osazones depend on the rate of heating and lie too close together to be of value.

Osazones may be converted to osotriazoles which have sharp melting points.[34]

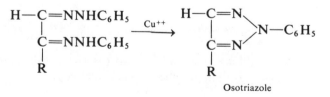

Osotriazole

---

[33]O. L. Chapman, R. W. King, W. J. Welstead, Jr., and T. J. Murphy, *J. Am. Chem. Soc.*, **86**, 4968 (1964); O. L. Chapman, *Tetrahdron Letters*, 2599 (1966).

[34]W. T. Haskins, R. M. Hahn, and C. S. Hudson, *J. Am. Chem. Soc.*, **69**, 1461 (1947) and previous papers.

Other derivatives of value in identification of sugars include acetates, and acetates of thioacetals, benzoates, acetonides, and benzylidene derivatives. The trimethylsilyl ethers are useful for mass spectral and glc analysis. Several procedures are given under polyhydric alcohols (Sec. 12.3.2).

12.10 CARBOXYLIC ACIDS

12.10.1a Classification

Acidic compounds containing only carbon, hydrogen, and oxygen are either carboxylic acids, phenols, or possibly enols. An indication of whether a water-insoluble compound is an acid or a phenol can be obtained from simple solubility tests. Both classes of compounds are soluble in sodium hydroxide, but only carboxylic acids are soluble in 5% sodium hydrogen carbonate with the liberation of carbon dioxide (for exceptions, see Sec. 10.1.4). Water-soluble acids also liberate carbon dioxide from sodium hydrogen carbonate solution. Other classes of compounds which liberate carbon dioxide from sodium hydrogen carbonate solution include salts of primary and secondary amines (these may be differentiated readily on the basis of the liberation of free amines, their melting point behavior, and their elemental analysis); sulfinic and sulfonic acids (differentiated on the basis of elemental analysis, infrared spectra, and their more acidic character); and a variety of substances such as acid halides and acid anhydrides which can be readily hydrolyzed to provide acidic materials. Phenols and enols give positive color tests with ferric chloride solution. Fortunately, all these categories of acidic compounds can be readily differentiated from carboxylic acids on the basis of the extremely characteristic infrared spectrum of a carboxylic acid. Carboxylic acids give rise to a very broad and characteristic O—H band and a carbonyl band near 1700 cm^{-1} (5.88μ)(Fig. 12.17). The chemist with even minimum spectroscopic experience will soon learn to distinguish other compounds from carboxylic acids by infrared spectroscopy.

12.10.1b Characterization

One of the simplest and most informative ways to characterize a carboxylic acid is to determine its *neutralization equivalent* (see Sec. 10.3). The neutralization equivalent of an acid or an acidic compound is its equivalent weight; the molecular weight may be determined from the neutralization equivalent by multiplying that value by the number of acidic groups in the molecule.

Since the pK_a of both the organic acid and the indicator are sensitive to solvent changes, one should employ only enough ethanol to dissolve the

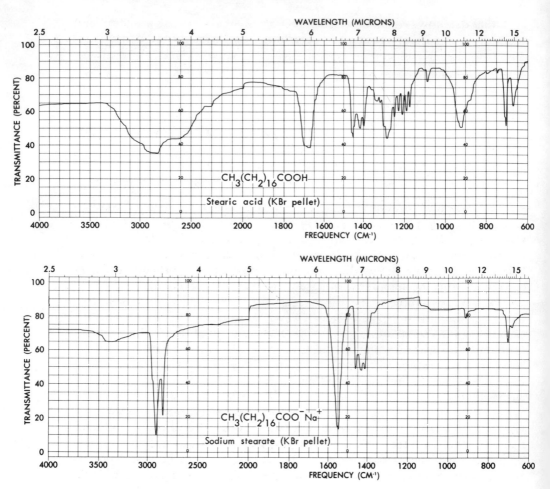

Fig. 12.17. Infrared spectra of a typical acid and its sodium salt. Note the shift in the carbonyl frequency.

organic acid. With high concentrations of ethanol, sharp endpoints are not obtained with phenolphthalein. If it is necessary to employ absolute or 95% ethanol, bromthymol blue should be used as the indicator.

For accurate results, a blank should always be run on the solvent, and one should take care that the neutralization equivalent is determined from a substance that is pure and anhydrous. Neutralization equivalents can be obtained with an accuracy of 1% or less.

With good technique, one may obtain an accurate neutralization equivalent with samples as small as 5 mg, employing more dilute standard alkali in burettes designed for greater accuracy with small volumes.

12.10.1c Procedure: Neutralization equivalents

Approximately 200 mg of the acid is accurately weighed and dissolved in 50 to 100 ml of water or aqueous ethanol. This solution is titrated with standard sodium hydroxide solution (approximately 0.1 N) employing phenolphthalein as the indicator or employing a pH meter.

$$\text{Neutralization equivalent} = \frac{\text{weight of sample in mg}}{\text{ml of base} \times \text{normality}}$$

If an acid has been well characterized, it is often sufficient for identification purposes to obtain the neutralization equivalent and the melting point of a carefully chosen derivative. More than 70 types of derivatives have been suggested at various times for the identification of carboxylic acids. The majority of these derivatives fall in the categories of amides, esters, and salts of organic bases. Representative and recommended examples from each of these categories will be discussed in the following paragraphs.

12.10.1d Amides and substituted amides

Acids or salts of acids may be converted directly to acid chlorides by the action of thionyl chloride. The acid chlorides are converted to the *amides* and *substituted amides* by reaction with excess ammonia or an amine, as illustrated below.

$$RCOOH + SOCl_2 \longrightarrow RCOCl + SO_2 + HCl$$

$$RCOCl + H_2NR' \longrightarrow RCONHR' + HCl$$

12.10.1e Procedure: Amides from acids

(a) *Acid chloride.* In a 25-ml, round-bottomed flask fitted with a condenser and a calcium chloride tube or a cotton plug in the top of the condenser to exclude moisture, place 0.5 to 1.0 g of the anhydrous acid or anhydrous sodium salt and add 2.5 to 5 ml of thionyl chloride. Reflux the mixture gently for 30 min. Arrange for distillation, and distill off the excess thionyl chloride (bp 78°). For acids below four carbon atoms, the bp of the acid chloride may be too near that of thionyl chloride to afford practical separation by distillation. In this event, the excess of the reagent can be destroyed by the addition of formic acid.

$$HCOOH + SOCl_2 \longrightarrow CO + SO_2 + 2\,HCl$$

(b) *Primary amides.* For the preparation of primary amides, it is unnecessary to distill off the excess thionyl chloride. The entire reaction mixture may be cautiously poured into 15 ml of ice-cold concentrated ammonia. The precipitated amide is collected by vacuum filtration and purified by recrystallization from water or aqueous ethanol.

As an alternate method, the acid chloride is dissolved in 5 to 10 ml of dry benzene, and the excess ammonia gas is passed through the solution. If the amide does not precipitate, it may be recovered by evaporation of the solvent.

(c) *Anilides and substituted anilides.* Dissolve the acid chloride in 5 ml of benzene, and add a solution of 2 g of pure aniline, *p*-bromoaniline, or *p*-toluidine in 15 ml of benzene. It may be convenient to run this reaction in a small separatory funnel. Shake the reaction mixture with 5 ml of dilute hydrochloric acid to remove the excess aniline, wash the benzene layer with 5 ml of water, evaporate the solvent, and recrystallize the anilide from water or aqueous alcohol.

Anilides and p-toluides may also be prepared directly from the acids or from alkali metal salts of the acids by heating them directly with the aniline or the hydrochloride salt of the aniline.

$$RCOOH + C_6H_5NH_2 \longrightarrow$$

$$RCOO^{-+}NH_3C_6H_5 \xrightarrow{\Delta} RCONHC_6H_5 + H_2O$$

$$RCOONa + C_6H_5NH_3{}^+Cl^- \xrightarrow{\Delta} RCONHC_6H_5 + NaCl + H_2O$$

12.10.1f Procedure: Anilides and p-toluides

Place 0.5 g of the acid and 1 g of aniline or *p*-toluidine in a small, dry, round-bottomed flask. Attach a short air condenser and heat the mixture in an oil bath at 140 to 160° for 2 hr. Caution should be used to avoid heating the mixture too vigorously, thus causing loss of the acid by distillation or sublimation. If the material is available as the sodium salt, use 0.5 g of the salt and 1 g of the amine hydrochloride. If there is evidence that the substance is a diacid, use double the quantity of the amine and increase the reaction temperature to 180 to 200°. At the end of the reaction time, cool the reaction mixture and triturate it with 20 to 30 ml of 10% hydrochloric acid, or dissolve the residue in an appropriate solvent and wash it with hydrochloric acid, dilute sodium hydroxide, and water. Then evaporate the solvent. The amides usually may be recrystallized from aqueous ethanol.

12.10.1g Esters

Solid esters form a second series of compounds useful in the identification and characterization of acids. The most commonly used are the *p*-nitrobenzyl and *p*-bromophenacyl esters, although the phenacyl, *p*-chlorophenacyl, and *p*-phenylphenacyl are also occasionally used. The halides corresponding to these esters, e.g., the *p*-nitrobenzyl halides and the phenacyl halides, all undergo facile S_N2 displacement reactions. These esters are prepared in aqueous alcoholic solution by displacement of the corresponding halide with the weakly nucleophilic carboxylate anions, as illustrated in the following equation.

$$RCOO^-Na^+ + BrCH_2\overset{\overset{\displaystyle O}{\|}}{C}\!-\!\!\!\left\langle\underset{}{\bigcirc}\right\rangle\!\!\!-Br \longrightarrow$$

$$RCOOCH_2CO\!-\!\!\!\left\langle\underset{}{\bigcirc}\right\rangle\!\!\!-Br + NaBr$$

The method is particularly advantageous because it does not require the acid to be anhydrous, and it can be run with equal facility with the alkali salts of the acids.

Although the phenacyl and benzyl esters are extremely useful for making derivatives of acids which cannot be easily obtained in anhydrous conditions (e.g., from saponification of esters of lower carboxylic acids) or directly from alkali metal salts, there are certain disadvantages which make them undesirable for routine use in identification. *The benzyl and phenacyl halides all have severe lachrymatory and blistering properties.* The formation of the esters is generally slow, and any unreacted halide is often difficult to separate from the ester. For this reason, less than one equivalent of the halide should be used, and the reaction should be continued for 1.5–2 hr to ensure completeness. The traces of the halides remaining with the esters impart irritating properties to the esters and severely depress the melting point.

12.10.1h Procedure: Phenacyl and *p*-nitrobenzyl esters

In a small test tube or a small, round-bottomed flask equipped with a reflux condenser, place 1 mmole of the acid and 1 ml of water. Add one drop of phenolphthalein, and carefully neutralize by the dropwise addition of 10% sodium hydroxide solution until the color of the solution is just pink. Add one or two drops of dilute hydrochloric acid to discharge the pink color of the indicator. Add an alcoholic solution of 0.9 mmole of the halide dissolved in 5 to 8 ml of alcohol. Reflux the solution from 1.5 to 2.5 hr, cool, add 1 ml of water, and scratch the sides of the tube. When precipitation is complete, collect the ester by filtration; wash with a small amount of 5% sodium carbonate solution and then several times with small quantities of cold water. The esters usually may be recrystallized from aqueous ethanol. A frequently used procedure is to dissolve the crystals in hot alcohol, filter, and add water to the hot filtrate until a cloudiness appears. Rewarm the solution until the cloudiness disappears, and then cool. Scratch the sides of the test tube, if necessary, to induce crystallization. *Avoid handling or contact of the crystals with the skin.*

If the original acid is available as an alkali salt, dissolve one equivalent of it in a minimum amount of water, add a drop of phenolphthalein, and adjust the acidity as above.

Methyl esters are also occasionally used as derivatives. In certain cases these are solids, but in the majority of the cases they appear as oils. When the methyl esters are relatively high melting solids, they are often very respectable derivatives for the purposes of melting point determinations. Frequently, one may wish to make the methyl ester for purposes of vapor phase chromatographic comparison of retention times of unknown esters with known methyl esters. The methyl esters have the advantage that they are often more volatile, more easily separated, and they elute as sharper peaks from the vapor phase chromatograph than do the corresponding acids. Methyl esters may be made quantitatively on a very small scale, employing diazomethane. They may also be made conveniently on a small scale employing 2,2-dimethoxypropane with a small amount of *p*-toluenesulfonic or concentrated hydrochloric acid.

$$RCOOH + CH_2N_2 \longrightarrow RCOOCH_3 + N_2$$

$$RCOOH + CH_3\overset{\overset{\displaystyle OCH_3}{|}}{\underset{\underset{\displaystyle OCH_3}{|}}{C}}CH_3 \xrightarrow{H^+} RCOOCH_3 + CH_3OH + CH_3COCH_3$$

12.10.1i Salts

Alkylthiuronium halides prepared by heating alkyl halides with thiourea react with the sodium or potassium salts of carboxylic acids in aqueous ethanol to yield the corresponding alkyl thiuronium carboxylates. These salts are easily made and are formed in good yield and high purity. Regrettably, they possess melting points which are very close together. The most commonly used is the S-benzylthiuronium salt; sometimes the S-p-bromo- and S-p-chlorobenzylthiuronium, and S-1-naphthylmethylthiuronium salts are made.

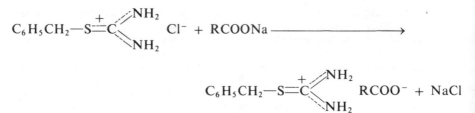

12.10.1j Procedure: S-Benzylthiuronium salts

Reagent: S-Benzylthiuronium chloride, although commercially available, may be easily prepared by refluxing a mixture of 2 g of benzyl chloride, 1.2 g of thiourea, and 3 ml of methanol for 30 minutes. Cool in an ice bath. Collect the product and wash several times with small portions of ethyl acetate.

Dissolve or suspend 0.25 g of the acid or its sodium or potassium salt in 5 ml of water. Adjust the pH to phenolphthalein endpoint with N sodium hydroxide. Add several drops of 0.1 N hydrochloric acid until the solution shows a very pale pink coloration (almost neutral). Add a solution of 1 g of S-benzylthiuronium chloride in 5 ml of water or 10 ml of methanol. Cool in an ice bath. Recrystallize the salt from dilute ethanol.

A second series of salts which has had some utility in the identification of organic acids are the phenylhydrazonium salts. Salts of phenylhydrazine are obtained from the stronger aliphatic acids such as α-chlorocarboxylic acids, from sulfonic acids, and from aliphatic dibasic acids when they are warmed with a benzene solution of phenylhydrazine. Phenylhydrazine may also be used to convert acids directly into phenylhydrazides. At the boiling point of phenylhydrazine (243°), simple aliphatic acids form phenylhydrazides.

$$RCOOH + C_6H_5NHNH_2 \longrightarrow C_6H_5NH\overset{+}{N}H_3RCOO^-$$
Phenylhydrazonium salt

$$RCOOH + C_6H_5NHNH_2 \xrightarrow{\Delta} RCONHNHC_6H_5$$
Phenylhydrazide

12.10.2 Salts of Carboxylic Acids

12.10.2a Classification

Metallic salts of carboxylic acids—the most commonly encountered are the sodium and potassium salts—will generally be suspected from their solubility, melting point, and ignition-test behavior. They are quite water soluble, providing slightly alkaline solutions from which free acids sometimes precipitate upon neutralization with mineral acid. They are insoluble in most organic solvents. The salts melt or decompose only at very high temperatures. They leave a light colored alkaline residue of oxide or carbonate upon ignition. If necessary, the identity of the metal may be made by the standard tests outlined in inorganic qualitative analysis texts or by flame photometry.

Occasionally one encounters the carboxylic acid salts of ammonia or amines. The ammonia or amine is liberated from the salt on dissolution in aqueous alkali. Place 20 to 30 mg of the sample in a small test tube containing 0.5 ml of dilute sodium hydroxide. Ammonia or a low molecular weight amine may be detected by placing a piece of dampened litmus paper in the vapor produced by warming the tube with a small flame. For a more sensitive test, dampen filter paper with a 10% solution of copper sulfate and hold the paper in the vapors. A positive test is indicated by the blue color of the ammonia or amine complex of copper sulfate. The amine may be isolated by extraction of a basic solution of the salt and identified as described in Sec. 12.14.

The infrared spectra of acid salts, usually taken as Nujol mulls, show strong carbonyl bands at 1610 to 1550 cm^{-1} (6.21 to 6.45 μ) and near 1400 cm^{-1} (7.14 μ) (Fig. 12.17).

12.10.2b Characterization

For characterization of carboxylic acid salts, it is necessary to rely on the chemistry and physical constants of the acid. Addition of sulfuric or other strong mineral acids to an aqueous solution of a salt of a carboxylic acid liberates the free acid which may be isolated by extraction, steam distillation, or filtration. Suitable derivatives are then made from the free acid. The sodium or potassium salts may be converted directly to thiuronium salts or phenacyl esters (Sec. 12.10.1).

12.11.1a Classification

The combination of high reactivity and unique spectral properties greatly facilitates the classification and identification of acid halides and acid anhydrides.

Acid anhydrides characteristically exhibit two bands in the carbonyl region of the infrared spectrum (Fig. 12.18); in linear aliphatic anhydrides, these bands appear near 1820 and 1760 cm^{-1} (5.49 and 5.68 μ). They show

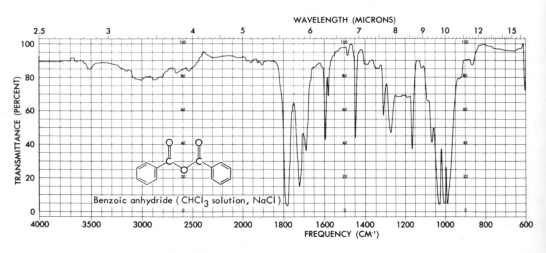

Fig. 12.18. Infrared spectrum of benzoic anhydride. Note the benzoic acid impurity which shows up as a broad weak OH band and a carbonyl band near 1690 cm^{-1}.

the usual variation with unsaturation and ring size. The relative intensity of the two bands is variable; the higher frequency band is stronger in acyclic anhydrides, and the lower frequency band is usually stronger in cyclic anhydrides. Diacyl peroxides also exhibit double carbonyl bands in the infrared (Sec. 12.12).

Acid chlorides, which are by far the most common of the acid halides, have a strong infrared band near 1800 cm^{-1} (5.56 μ) (Fig. 12.19). The band is only slightly shifted to lower frequency on conjugation; in aroyl halides, a prominent shoulder usually appears on the lower-frequency side of the carbonyl band. As would be predicted, acid fluorides are shifted to higher frequencies, whereas the bromides and iodides are at lower frequencies. The presence of this strong carbonyl band and the absence of bands in the O—H, N—H, and C—O regions of the spectrum, the presence of reactive halogen (silver nitrate test, Sec. 12.2.1b), and the high reactivity of the

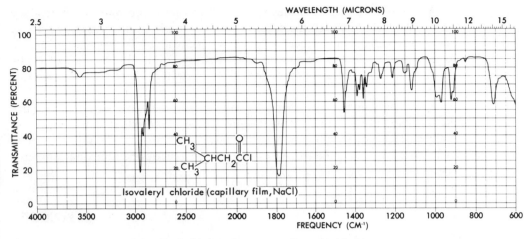

Fig. 12.19. Typical infrared spectrum of an acid chloride.

substance with water, alcohols, and amines provide sufficient diagnosis for the acid halide grouping.

12.11.2b Characterization

The most commonly encountered anhydrides are the simple symmetrical acetic and benzoic anhydrides and the cyclic anhydrides of succinic, maleic, and phthalic acids. Mixed anhydrides formed from two carboxylic acids or from a carboxylic acid and a sulfonic acid are known, but are seldom encountered.

Acid halides and anhydrides are most often converted directly to amides for identification purposes. Recall that, in the identification of a carboxylic acid, the acids are usually converted to the acid chlorides in order to prepare amines. The anilides are highly recommended (Sec. 12.10.1d).

$$RCOCl + 2\,NH_2C_6H_5 \longrightarrow RCONHC_6H_5 + C_6H_5\overset{+}{N}H_3Cl^-$$

$$(RCO)_2O + 2\,NH_2C_6H_5 \longrightarrow RCONHC_6H_5 + C_6H_5\overset{+}{N}H_3\,CH_3COO^-$$

Cyclic anhydrides, on reaction with amines, may give the monamide or the imide, depending upon the conditions of the reaction.

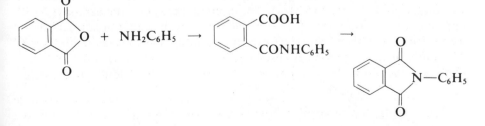

Acid halides and anhydrides react with alcohols and phenols to produce esters. With acid halides, the reaction is often rapid and exothermic. These esters may serve for identification purposes if the alcohol or phenol is chosen so that the product is an appropriate melting solid. In the case of cyclic anhydrides, acid esters are produced; phthalic anhydride reacts with alcohols to give alkyl hydrogen phthalates.

The acid halides and anhydrides are hydrolyzed with water to produce the corresponding acids which, if solid, serve as excellent derivatives. In many cases, some warming may be necessary to complete the reaction. The hydrolysis may be effected by a quantitative reaction employing standard alkali.

12.12 PEROXIDES

12.12.1a Classification

All organic peroxides should be regarded as being potentially explosive. Safety glasses, gloves, and shields should be used at all times when working with organic peroxides.

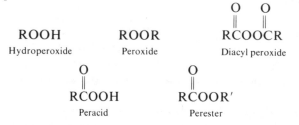

Most organic peroxides[35] can be detected by acidified starch-iodide paper. The dialkyl peroxides can be detected only after they have been hydrolyzed to hydroperoxides by strong acid.

In the infrared region, the —O—O— stretching frequency appears near 877 cm^{-1} (11.4 μ). Because of the symmetry of the grouping, the absorption is generally weak. It is often obscured by, or confused with, other skeletal vibrations which occur in the same region. The O—H absorption in hydroperoxides is very similar to that of alcohols. The peracids exhibit an intramolecularly hydrogen-bonded O—H band at 3280 cm^{-1} (3.05 μ) and a carbonyl near 1750 cm^{-1} (5.71 μ). In aliphatic peresters, the carbonyl band appears near 1770 cm^{-1} (5.65 μ), somewhat higher than that of normal esters. The symmetrical diacyl peroxides exhibit a doublet carbonyl in the infrared. The aliphatic diacyl peroxides absorb near 1815 cm^{-1} (5.51 μ) and 1790 cm^{-1} (5.59 μ); the aromatic compounds absorb at slightly lower frequencies

---

[35]A. G. Davies, *Organic Peroxides*, Butterworths, London, 1961.

WAVELENGTH (MICRONS)

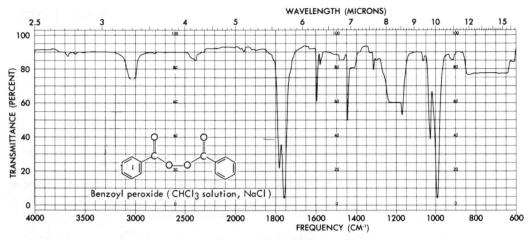

Fig. 12.20. Infrared spectrum of benzoyl peroxide. Compare with the spectrum of benzoic anhydride (Fig. 12.18).

(Fig. 12.20). In linear diacylperoxides, the lower-frequency band is stronger and sharper (see anhydrides, Sec. 12.11).

12.12.1b Characterization

Identification of peroxides is usually accomplished by quantitative reduction procedures. The reduced products, acids, alcohols, etc., may be characterized by the standard methods. Iodometric methods are usually employed for the *quantitative estimation of peroxides*.

$$ROOH + 2I^- + H^+ \longrightarrow ROH + H_2O + I_2$$

12.12.1c Procedure: Titration of peroxides

The peracid or hydroperoxide (0.1 to 0.2 g) is accurately weighed in an iodine flask. The sample is dissolved in 10 ml of chloroform and the flask flushed with nitrogen. Saturated sodium iodide solution (2 ml) is added, followed by 15 ml of glacial acetic acid. The flask is stoppered and permitted to stand in the dark for 5 to 10 min. Water (50 ml) is added, and the solution is titrated to starch endpoint with 0.1 N thiosulfate. A blank determination should be run on the reagents.

This procedure may be adapted for use with peresters and diacyl peroxides, provided the acetic acid employed contains 0.002% ferric chloride hexahydrate as a catalyst for the reduction.[36]

Dialkyl peroxides can be estimated if they are treated with hydrogen iodide in acetic acid at 60° for 45 to 60 min.

$$\text{Percent "active oxygen"} = \frac{\text{ml} \times N \times 0.8}{\text{weight of sample in grams}}$$

---

[36]L. S. Silbert and D. Swern, *Anal. Chem.*, **30**, 385 (1958); *J. Am. Chem. Soc.*, **81**, 2364 (1959).

12.13.1a Classification

The majority of esters are liquids or low-melting solids, many with characteristic flowery or fruity odors. The presence of the ester function may usually be established by infrared spectroscopy. Esters have strong carbonyl bands in the infrared region in the 1780 to 1720 cm^{-1} (5.62 to 5.81 μ) region accompanied by two strong C—O absorptions in the 1300 to 1050 cm^{-1} (7.69 to 9.52 μ) region (Fig. 12.21). The higher-frequency C—O band is usually stronger and broader than the carbonyl band. As a general rule, in

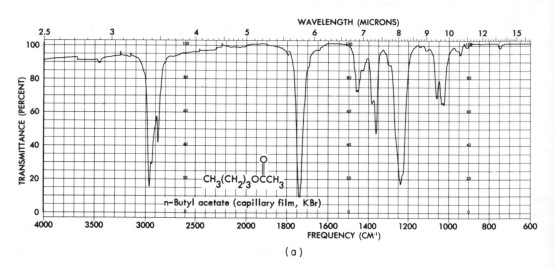

(a)

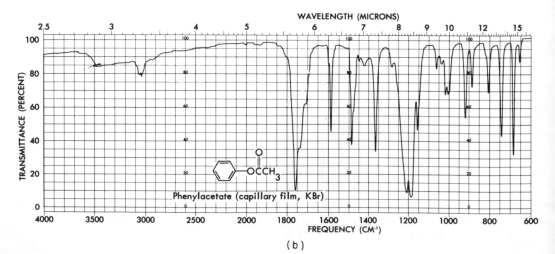

(b)

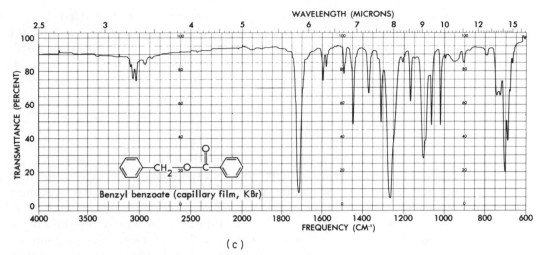

Fig. 12.21. Typical infrared spectra of esters.

the spectra of aldehydes, ketones, and amides, the carbonyl band is the strongest in the spectrum (*cf.* Figs. 12.13, 12.14, 12.21, and 12.26).

Carbonyl compounds whose infrared spectra have the foregoing characteristics, which lack NH and OH absorption, and which do not give a positive 2,4-dinitrophenylhydrazine test, are most likely esters. An ester spectrum is most likely to be mistaken for that of a ketone. If the nmr spectrum of the suspected ester is available, evidence for an ester may be provided by the chemical shifts and coupling patterns of aliphatic hydrogens attached to the ethereal oxygen (near $\delta 4$) and hydrogens alpha to the carbonyl group (near $\delta 2$). Esters, like other carbonyl compounds, exhibit very characteristic fragmentation patterns in mass spectrometry.

12.13.1b *Characterization*

The principal procedure for the characterization of an ester involves the identification of the parent acid and alcohol. Because it is often difficult to separate and purify the hydrolysis products of esters on a small scale, it is usually advantageous to prepare derivatives of the acid and alcohol portion by reactions on the original ester. The sample should be apportioned accordingly. Fortunately, a large number of the esters encountered are derivatives of simple acids (e.g., acetates) or alcohols (e.g., methyl or ethyl esters); it is often possible to unequivocally establish the presence of an acetyl or methyl or ethyl grouping by spectroscopic means.

Aromatic esters may sometimes be identified through solid derivatives prepared directly by aromatic substitution (nitration, halogenation, etc.), thus eliminating the necessity of preparing separate derivatives of both the alcohol and acid portion.

The *saponification equivalent* (the molecular weight divided by the number of ester moieties in the molecule) is determined on a known weight of an ester by hydrolysis with an excess of standard alkali; the excess alkali is then back-titrated with standard acid. The method is exceedingly useful in limiting the number of structural possibilities for an unknown; regrettably, it is tedious and time consuming.

$$\text{Saponification equivalent} = \frac{\text{weight of ester in milligrams}}{(\text{ml} \times \text{normality})_{\text{alkali}} - (\text{ml} \times \text{normality})_{\text{acid}}}$$

Of the two methods given here for determination of the saponification equivalent, the diethylene glycol method is often more advantageous because of faster reaction times, a more stable standard solution, and less likelihood of loss of esters because of transesterification to the more volatile ethyl esters. However, the viscous glycol solution is somewhat more difficult to transfer with accuracy. With either procedure, ground glass equipment should be employed. If such equipment is not available, rubber stoppers which have been boiled with alkali and then thoroughly washed with distilled water may be employed.

12.13.1c Diethylene glycol method

Reagent: Dissolve 6.0 g of potassium hydroxide pellets in 30 ml of diethylene glycol with the aid of heat. Use a thermometer as a stirring rod, and keep the temperature below 130° to prevent discoloration of the solution. Transfer the solution to a glass-stoppered bottle, and add 70 ml of diethylene glycol. Mix the solution thoroughly, and allow it to cool. Standardize with 0.5 N hydrochloric acid. Pipette 10 ml of the reagent (ca. 1 N) into a 125-ml Erlenmeyer flask, add 15 ml of water, and titrate to phenolphthalein endpoint. Because of the viscosity of the solution, a large-bore pipette is desirable. Alternatively, the standardization and saponification may be conducted with accurately weighed quantities of the glycol solution.

Procedure: For the determination of the saponification equivalent of the ester, transfer approximately 10 ml of the standardized reagent, accurately weighed or measured, into a 50-ml, glass-stoppered Erlenmeyer flask. Quantitatively transfer about 0.5 g of the ester, accurately weighed, into the flask, and put the stopper in place. Mix the ester and the reagent by swirling the flask. While holding the stopper in place with swirling, heat the mixture in an oil bath that has been preheated to 120 to 130°. After 2 or 3 min, remove the flask from the bath, shake well, allow to drain, and carefully remove the stopper to allow the internal pressure to escape. Replace the stopper and continue heating in the oil bath until the temperature of the reaction mixture is about 120°. If the ester is high boiling, the stopper may be removed and a thermometer inserted. After several minutes at this temperature, cool the flask to 80 to 90°; remove the stopper and wash it well with distilled water, allowing the rinsing to drain into the flask. Add 10 to 15 ml of distilled water to the flask and several drops of phenolphthalein indicator, and titrate with standard hydrochloric or sulfuric acid.

12.13.1d Ethanolic potassium hydroxide method

Reagent: Place 8 g of potassium hydroxide in 250 ml of ethyl alcohol (95%), and shake to effect dissolution. Allow the mixture to stand for 24 hr; then decant or filter the clear solution from the residue of potassium carbonate. It is necessary to standardize the ethanolic potassium hydroxide solution (approximately 0.5 N) immediately before use. Either standard hydrochloric or sulfuric acid, using phenolphthalein as the indicator, is suitable.

Procedure: Place an accurately weighed sample of the ester (about 0.5 g) in a 100-ml, round-bottomed flask. Add 25 ml of standard 0.5 N alcoholic potassium hydroxide by means of a pipette (CAUTION: *Be sure to use a bulb pipette control*). Attach an efficient reflux condenser and reflux gently on a steam bath for 1.5 to 2 hr, or until the hydrolysis is complete. Allow the mixture to cool. Introduce about 25 ml of water through the condenser. Remove the condenser, add two to three drops of phenolphthalein, and titrate with standard acid.

12.13.2 Derivatives of the Acyl Moiety

12.13.2a Hydrolysis to acids

If the acid obtained by hydrolysis of the ester is a solid, it will serve as an excellent derivative (mp and neutralization equivalent). The acid may be obtained by quantitative or qualitative application of the procedures for the determination of saponification equivalents, or by *hydrolysis with aqueous sodium hydroxide*. The potassium hydroxide/diethylene glycol procedure is especially effective for water insoluble esters boiling above 200°.

12.13.2b Procedure: Hydrolysis of esters

Into a small, round-bottomed flask place 0.2 to 1 g of the ester; add 2 to 10 ml of 25% aqueous sodium hydroxide and a boiling chip. Attach a condenser and reflux for 30 min for esters boiling below 110°, 1 to 2 hr for esters boiling between 110 and 200°, or until the oily layer and/or the characteristic ester odor disappears. Cool the flask and acidify with dilute acid. Phosphoric is recommended because of the high solubility of the sodium phosphate. Recover the acid by filtration or extraction, and purify by suitable means.

12.13.2c Conversion to amides

Esters may be heated with aqueous or alcoholic ammonia to produce amides. Some simple esters react on standing or at reflux; however, most must be heated under pressure.

$$RCOOCH_3 + NH_3 \longrightarrow RCONH_2 + CH_3OH$$

Many esters may be converted to crystalline *N-benzylamides* by refluxing with benzylamine in the presence of an acid catalyst. This is often the most preferred method of making a derivative of the acyl moiety.

$$RCOOR' + CH_3OH \xrightarrow[\text{or HCl}]{CH_3O^-} RCOOCH_3 + R'OH$$

$$RCOOCH_3 + C_6H_5CH_2NH_2 \xrightarrow{NH_4Cl} RCONHCH_2C_6H_5 + CH_3OH$$

12.13.2d Procedure: *N*-Benzylamides from esters

Into a test tube or a 10-ml round-bottomed flask equipped with a reflux condenser, place 100 mg of ammonium chloride, 1 ml or 1 g of the ester, and 3 ml of benzylamine and a boiling stone. Reflux gently for 1 hr. Cool and wash the reaction mixture with water to remove the excess benzylamine and soluble salts. If crystallization does not occur, add a drop or two of dilute hydrochloric acid. If crystallization still does not occur, it may be due to excess ester. It is often possible to remove the ester by boiling the reaction mixture for a few minutes in an evaporating dish. Collect the solid *N*-benzylamide by filtration, wash it with a little ligroin, and recrystallize from aqueous alcohol or acetone.

The aminolysis of esters higher than ethyl is usually quite slow. In these cases, it is best to convert the ester to the methyl ester before attempting to make the benzylamide. This may be done by refluxing 1 g of ester with 5 ml of absolute methanol in which about 0.1 g of sodium has been dissolved, or by refluxing in 3% methanolic hydrogen chloride. Remove the methanol by distillation, and treat the residue as described above.

Occasionally, it is expedient to convert the ester to the corresponding *p*-toluide by the reaction sequence shown below.[37] The lithium salt of the amine, formed by reaction of *p*-toluidine with *n*-butyl lithium, may also be used.

$$CH_3CH_2MgBr + CH_3-\bigcirc-NH_2 \rightarrow CH_3-\bigcirc-NHMgBr + CH_3CH_3$$

$$RCOOR' + 2CH_3-\bigcirc-NHMgBr \rightarrow R-\underset{OMgBr}{\underset{|}{C}}-(NH-\bigcirc-CH_3)_2 + R'OMgBr$$

$$\xrightarrow{2HCl} RCONH-\bigcirc-CH_3 + CH_3-\bigcirc-\overset{+}{N}H_3Cl^- + MgBrCl$$

Hydrazine reacts with methyl and ethyl esters to produce hydrazides. The ester is refluxed with hydrazine hydrate, a small amount of ethanol is added, and the refluxing continued for several hours.

$$RCOOCH_2CH_3 + NH_2NH_2 \longrightarrow RCONHNH_2 + CH_3CH_2OH$$

12.13.3 Derivatives of the Alcohol Moiety

Occasionally, the alcohol derived from the ester by hydrolysis is a solid, in which case it and subsequent derivatives may serve to identify that portion of the ester. In general, alcohols above four carbons may be recovered from aqueous hydrolysates by extraction procedures. In many cases, it is much more convenient to identify alcohol moiety by a transesterification reaction with 3,5-dinitrobenzoic acid to produce the *alkyl 3,5-dinitrobenzoate*.

[37]C. F. Koelsch and D. Tannenbaum, *J. Am. Chem. Soc.*, **55**, 3049 (1933); D. V. N. Hardy, *J. Chem. Soc.*, 398 (1936).

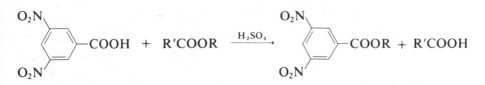

12.13.3a Procedure: 3,5-Dinitrobenzoates from esters

Mix about 0.5 g of the ester with 0.4 g of 3,5-dinitrobenzoic acid, and add one drop of concentrated sulfuric acid. If the ester boils below 150°, heat to reflux with the condenser in place. Otherwise, run the reaction in an open test tube placed in an oil bath at 150°. Stir the mixture occasionally. If the acid dissolves within 15 min, heat the mixture for 30 min; otherwise, continue the heating for 1 hr. Cool the reaction mixture, dissolve it in 20 ml of ether, and extract thoroughly with 10 ml of 5% sodium carbonate solution. Wash with water, and remove the ether. Dissolve the crystalline or oily residue in 1 to 2 ml of hot ethanol and add water slowly until the 3,5-dinitrobenzoate begins to crystallize. Cool, collect, and recrystallize from aqueous ethanol if necessary.

12.14   AMINES

12.14.1a General classification

Most simple amines are readily recognized by their solubility in dilute mineral acids. However, many substituted aromatic amines (e.g., diphenylamine) and aromatic nitrogen heterocycles, which might formally be classified as amines, fail to dissolve in dilute acids. (The parent ring system of the latter may often be determined from their characteristic ultraviolet spectra.) Water-soluble amines may be detected by their basic reaction to litmus or other indicators, or by the *copper ion test*.

12.14.1b Procedure: Copper ion test

Add 10 mg or a small drop of the unknown to 0.5 ml of a 10% solution of copper sulfate. A blue to blue-green coloration or precipitate is indicative of an amine. The test may be run as a spot test on filter paper that has been treated with the copper sulfate solution. Ammonia also gives a positive test.

Infrared spectroscopy can be very useful in the recognition and classification of amines (Figs. 12.22 and 12.23). Primary amines, both aliphatic and aromatic, exhibit a weak but recognizable doublet in the 3500 to 3300 cm^{-1} (2.86 to 3.03 μ) region (symmetric and asymmetric NH stretch) and strong absorption due to NH bending in the 1640 to 1560 cm^{-1} (6.10 to 6.41 μ) region. The latter band is somewhat broad in aliphatic amines, but reasonably sharp in aromatic amines. Secondary amines exhibit a single band in

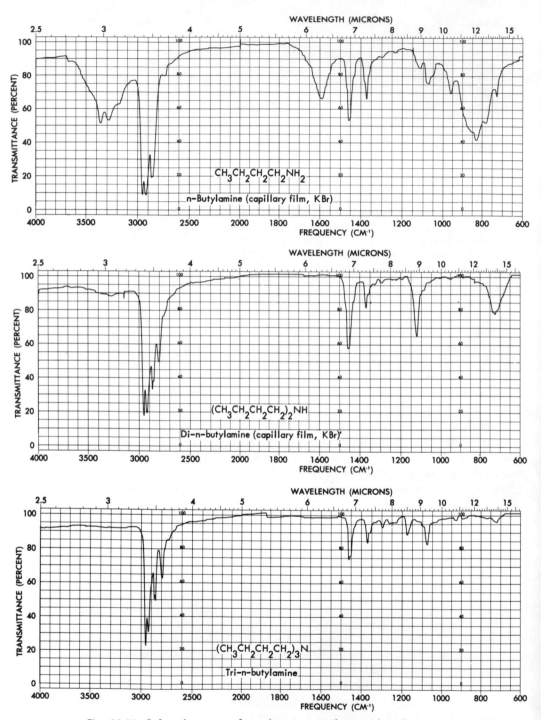

Fig. 12.22. Infrared spectra of a primary, secondary, and tertiary amine. In the spectrum of the primary amine note the strong NH stretching, bending, and deformation bands. The NH stretching and bending bands of secondary amines are frequently very weak and may not be observed at the usual cell thickness.

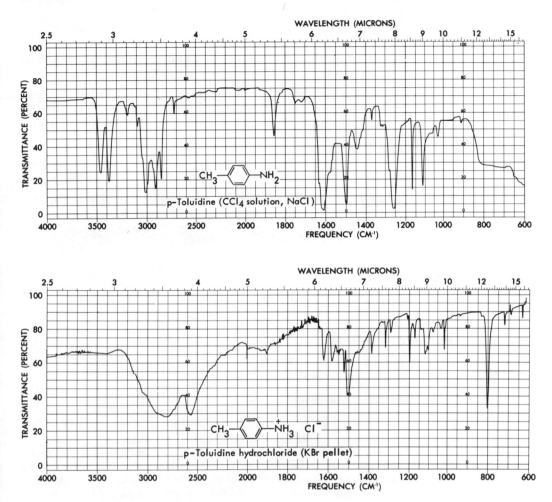

Fig. 12.23. Typical infrared spectra of an aromatic amine and its hydrochloride. In the spectrum of p-toluidine note the strong 1,4-disubstitution bands in the 2000 to 1660 cm^{-1} (5 to 6 μ) region.

the 3450 to 3310 cm^{-1} (2.90 to 3.02 μ) region; aromatic amines absorb nearer the higher end of the range, and the aliphatic absorb at the lower end. The NH bending band in secondary amines in the 1580 to 1490 cm^{-1} (6.33 to 6.71 μ) region is weak and generally of no diagnostic value. Tertiary amines have no generally useful characteristic absorptions. (However, the absence of characteristic NH absorptions is a helpful observation in the classification of a compound known to be an amine.) The presence of a tertiary amine grouping may be established by infrared examination of the hydrochloride or other proton-acid salt of the amine; $\equiv\overset{+}{N}H$ absorption occurs in the 2700 to 2250 cm^{-1} (3.70 to 4.45 μ) region.

12.14.1c Hinsberg test

This test is based on the fact that primary and secondary amines react with arenesulfonyl halides to give *N*-substituted sulfonamides. The tertiary amines do not give derivatives. The sulfonamides from primary amines react with alkali to form soluble salts (the sodium salts of certain primary alicyclic amines and some long chain alkyl amines are relatively insoluble and give confusing results).

Primary amines

$$\text{ArSO}_2\text{Cl} + \text{RNH}_2 + 2\,\text{NaOH} \longrightarrow [\text{ArSO}_2\text{NR}]^-\text{Na}^+ + \text{NaCl} + 2\,\text{H}_2\text{O}$$
Soluble

$$[\text{ArSO}_2\text{NR}]^-\,\text{Na}^+ + \text{HCl} \longrightarrow \text{ArSO}_2\text{NHR} + \text{NaCl}$$
Insoluble

Secondary amines

$$\text{ArSO}_2\text{Cl} + \text{R}_2\text{NH} + \text{NaOH} \longrightarrow \text{ArSO}_2\text{NR}_2 + \text{H}_2\text{O} + \text{NaCl}$$
Insoluble

Tertiary amines

$$\text{ArSO}_2\text{Cl} + \text{R}_3\text{N} + \text{NaOH} \longrightarrow \text{tertiary amine recovered}$$

12.14.1d Procedure: Hinsberg test

> In a test tube, add 0.1 ml or 0.1 g of the amine, 0.2 g of *p*-toluenesulfonyl chloride (0.2 ml of benzenesulfonyl chloride may be used, but the products tend to be oils), and 5 ml of 10% sodium hydroxide solution. Stopper the tube, and shake intermittently for 3 to 5 min. Remove the stopper, and warm the tube, with shaking, for 1 min. If no reaction has occurred, the substance is probably a tertiary amine. If a precipitate is present in the alkaline solution, dilute the reaction mixture with 5 ml of water and shake; if the precipitate does not dissolve, an *N,N*-disubstituted sulfonamide is present, indicating that the starting amine was secondary. If the solution is clear, acidify it cautiously with dilute hydrochloric acid. A precipitate of an *N*-substituted sulfonamide is indicative of a primary amine.

This procedure may be scaled up and used to separate mixtures of primary, secondary, and tertiary amines. Unreacted tertiary amines may be recovered by solvent extraction or steam distillation. The sulfonamides of primary and secondary amines may be separated by taking advantage of the solubility of the primary derivative in dilute sodium hydroxide. The original primary and secondary amines may be recovered by hydrolysis of the sulfonamides with 10 parts of 25% hydrochloric acid, followed by neutralization and extraction. The sulfonamides of primary amines require 24 to 36 hr of reflux; those of secondary amines require 10 to 12 hr for hydrolysis.

12.14.1e Nitrous acid

The following equations illustrate the reactions of nitrous acid with amines. The reactions are the basis for a classical method of distinguishing

Primary amines

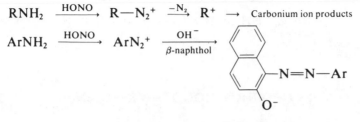

$$RNH_2 \xrightarrow{\text{HONO}} R-N_2^+ \xrightarrow{-N_2} R^+ \longrightarrow \text{Carbonium ion products}$$

$$ArNH_2 \xrightarrow{\text{HONO}} ArN_2^+ \xrightarrow[\beta\text{-naphthol}]{OH^-}$$

$$-N=N-Ar$$

$$O^-$$

Red azo dye

Secondary amines

$$R_2NH \xrightarrow{\text{HONO}} R_2N-NO \quad \text{(yellow oils or solids)}$$

Tertiary amines

$$R_3N \xrightarrow{\text{HONO}} R_3\overset{+}{N}-NO \xrightarrow{H_2O} R_2NNO + \text{aldehydes, ketones}$$

$$-NR_2 \xrightarrow{\text{HONO}} ON-\!\!\!\!-\!\!\!\!-NR_2$$

various types of amines. The details of test procedures may be found in most organic texts and organic qualitative analysis texts.

The greatest utility is the demonstration of the presence of primary aliphatic amino group (by the loss of N_2) and primary aromatic amine group (through coupling with reactive aromatic nuclei such as β-naphthol to form azo dyes).

12.14.1f Procedure: Nitrous acid test

(a) Dissolve 0.1 g of the primary aliphatic amine in 3 ml of 2 N hydrochloric acid; cool in an ice bath and add 1 ml of ice-cold 10% aqueous sodium nitrite solution. Warm gently; nitrogen will be rapidly evolved.

(b) Place three small test tubes in an ice-salt bath. In one tube, place 50 mg of the primary aryl amine, 1 ml of water, and four drops of concentrated sulfuric acid. In the second tube, place 1 ml of 10% aqueous sodium nitrite. In the last tube, make a solution of 100 mg of β-naphthol in 2 ml of 10% sodium hydroxide. When the solutions are thoroughly cooled, add the sodium nitrite dropwise and with shaking to the amine solution. Then add the sodium β-naphthoxide solution dropwise to the mixture. A red color or precipitate of an azo dye is indicative of a primary aryl amine.

It should be established that a simple amino functional group is present before the nitrous-acid test is employed. Nitrous acid will react with many other functional groups, which may lead to erroneous interpretation of the results. A number of such reactions are given below.

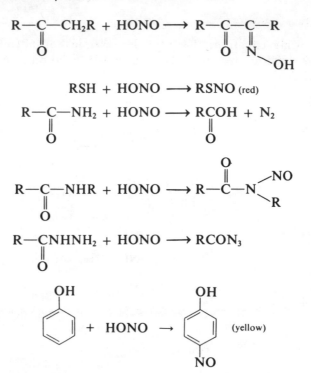

12.14.1g Characterization

Note: A number of the procedures for the preparation of derivatives of amines are quick and simple methods that can be run on a very small scale. For this reason, it is often expedient to use certain of these reactions as a combination classification-derivation procedure.

In many cases, with simple amines the combination of boiling or melting point, solubility in acid, and infrared spectrum provides sufficient information to allow the investigator to choose an appropriate derivative. In other cases, it is wise to examine selected classification tests and to obtain additional physical (e.g., pK_b) and spectral (nmr or ultraviolet) data before proceeding. See Sec. 5.4.1b for NH chemical shifts. In some cases, the NH proton signals are broadened by quadruple interactions with the nitrogen to the extent that the signal can barely be differentiated from the base line. In other cases, the signal is much sharper. If there is any doubt in the assignment of the peak due to protons on nitrogen, the investigator may find it helpful to exchange the nitrogen protons for deuterium (deuterium oxide with an acid or base catalyst) or to examine the spectrum of a salt of the amine. A simple technique for the latter is to use trifluoroacetic acid as solvent; with this solvent and tertiary amines, the $\overset{+}{\equiv}\text{NH}$ shows up between $\delta 2.3$ and 2.9.

In ultraviolet spectra of amines, a dramatic change is observed in going from the free base to the protonated salt if the nitrogen is part of the chromophore, e.g.,

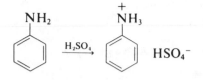

λ_{max} 230, 280 mμ λ_{max} 203, 254 mμ

12.14.2 Derivatives of Primary and Secondary Amines

12.14.2a Amides and related derivatives

The most commonly employed derivatives are the *benzene- and p-toluene-sulfonamides*. These may be made by the procedure outlined earlier for the Hinsberg test. The product should be recrystallized from 95% ethanol.

Benzyl-(α-toluene-), p-bromobenzene-, m-nitrobenzene-, α-naphthalene-, and methanesulfonamides are also occasionally used.

12.14.2b Procedure: Sulfonamides

Reflux for 5 to 10 min a mixture of 1 mmole of the sulfonyl chloride and 2.1 mmole of the amine in 4 ml of dry benzene. Allow the mixture to cool. Filter off the amine hydrochloride that precipitates. Evaporate the benzene filtrate to obtain the crude sulfonamide. Recrystallize from 95% ethanol.

Primary and secondary amines react rapidly at room temperature with 2,4-dinitrobenzenesulfenyl chloride to yield *2,4-dinitrobenzenesulfenamides*.

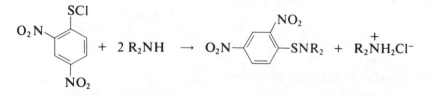

12.14.2c Procedure: 2,4-Dinitrobenzenesulfenamides

To a solution of 2 mmole of the amine in 4 ml of ether, add dropwise a solution of 2,4-dinitrobenzenesulfenyl chloride in ether (ca. 1 mmole (235 mg) in 4 ml of ether) until precipitation of the amine hydrochloride is complete. Filter off the hydrochloride, and wash with ether. The combined ether solutions are evaporated and the crude sulfenamide recrystallized several times from ethanol.

Primary and secondary amines react with isothiocyanates to yield *substituted thioureas*. The most useful derivatives are obtained from phenyl- and 2-naphthylisothiocyanate; these reagents have the advantage of being insensitive to water and alcohols.

$$RNH_2 + ArN{=}C{=}S \longrightarrow ArNH{-}\overset{\displaystyle S}{\overset{\displaystyle \|}{C}}{-}NHR$$

12.14.2d Procedure: Thioureas

(a) Reflux a solution of 1 mmole of the isothiocyanate and 1.2 mmoles of the amine in 2 ml of alcohol for 5 to 10 min. Add water dropwise to the hot mixture until a permanent cloudiness occurs. Cool, and scratch with a glass rod if necessary, to induce crystallization. Collect the solid. Wash with 1 to 2 ml of hexane. Recrystallize from hot alcohol.

(b) Mix equal amounts of the amine and isothiocyanate in a test tube and shake for 2 min. If no reaction occurs, heat the mixture gently over a small flame for 1 to 2 min. Cool in an ice bath, and purify by recrystallization.

Primary and secondary amines react with aryl isocyanates to produce substituted ureas. In contrast to the isothiocyanates, the isocyanates react rapidly with water and alcohols. Traces of moisture in the amines result in the formation of diaryl ureas which are often difficult to separate from the desired product.

$$R_2NH + ArN{=}C{=}O \longrightarrow ArNHCONR_2$$

$$2\,ArN{=}C{=}O + H_2O \longrightarrow ArNHCONHAr + CO_2$$

Both aliphatic and aromatic primary and secondary amines readily react with acid anhydrides or acid chlorides to form substituted amides. The reaction of benzoyl chloride with amines under Schotten-Baumann conditions (aqueous sodium hydroxide) produces *benzamides* which are generally excellent derivatives. The *p-nitro and 3,5-dinitrobenzamides* are also useful derivatives.

12.14.2e Procedure: Benzamides

CAUTION: *Benzoyl chloride is a lacrymator and should be used and stored in a hood.*

(a) About 1 mmole of the amine is suspended in 1 ml of 10% sodium hydroxide, and 0.3 to 0.4 ml of benzoyl chloride is added dropwise, with vigorous shaking and cooling. After about 5 to 10 min the reaction mixture is carefully neutralized to about pH 8 (pH paper). The *N*-substituted benzamide is collected, washed with water, and recrystallized from ethanol-water.

(b) Benzoyl, *p*-nitrobenzoyl, or 3,5-dinitrobenzoyl chloride (2 to 3 mmoles) is dissolved in 2 to 3 ml of benzene; 1 mmole of the amine is added, followed by 1 ml of

10% sodium hydroxide. The mixture is shaken for 10 to 15 min. The crude amide is obtained by evaporation of the benzene layer. Alternately, the amine and aroyl chloride are mixed with 2 ml of pyridine, and the mixture is refluxed for 30 min. The reaction mixture is poured into ice water and the derivative collected and recrystallized from ethanol-water.

Most water-insoluble primary and secondary amines form crystalline *acetamides* upon reaction with acetic anhydride.

$$RNH_2 + (CH_3CO)_2O \xrightarrow{\text{NaOH}} CH_3CONHR$$

12.14.2f Procedure: Acetamides

Dissolve about 0.2 g of water-insoluble amine in 10 ml of 5% hydrochloric acid. Add sodium hydroxide (5%) with a dropping pipette until the mixture becomes cloudy, and remove the turbidity by the addition of a few drops of 5% hydrochloric acid. Add a few chips of ice and 1 ml of acetic anhydride. Swirl the mixture and add 1 g of sodium acetate (trihydrate) dissolved in 2 ml of water. Cool in an ice bath, and collect the solid; recrystallize from ethanol-water.

Primary and secondary amines react with 3-nitrophthalic anhydride to provide phthalamic acids; those from primary amines dehydrate when heated to 150° to form *N*-alkyl 3-nitrophthalimides.

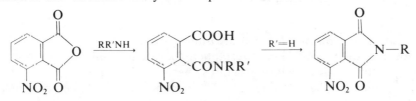

12.14.2g Salts

The amine salts of a number of proton acids are useful for purposes of identification or purification. Some salts have reproducible melting or decomposition points; others may be useful only as solid derivatives which may be carefully purified for purposes of microanalysis, spectroscopic studies, X-ray crystallography, etc.

Hydrochlorides may be prepared by passing dry hydrogen chloride (from a tank or from a sodium chloride/sulfuric acid generator) into an ether, benzene, or isopropyl alcohol solution of primary, secondary, or tertiary amines. The hydrobromides and hydroiodides are usually more hygroscopic than the hydrochlorides.

The *fluoroborates* (BF_4^-) and *perchlorates* (ClO_4^-) are often well-defined crystalline substances. *All perchlorates should be handled as potentially explosive substances.* Melting points of perchlorates should be taken with very small samples and probably should not be determined for perchlorates melting over 230°.

12.14.2h Procedure: Fluoroborate and perchlorate salts

CAUTION: *Do not allow ether or alcohol-ether solutions of perchloric acid to stand for any length of time.* The highly explosive ethyl perchlorate may form. Wash all such solutions down the sink as soon as the reaction has been completed.

To a small volume of ether in a test tube, add 70% fluoroboric or perchloric acid. Mix and allow the layers to separate. Using a dropping pipette, draw off a small volume of the upper ethereal layer, and add it dropwise to an ether solution of the amine until precipitation is complete. Alternatively, the fluoroboric or perchloric acid may be added dropwise to an ether-alcohol solution of the amine until precipitation is complete. Recrystallize the salts from alcohol or alcohol-ether mixtures.

Picrates are frequently employed as derivatives of the tertiary amines. Picric acid forms addition compounds with primary and secondary amines as well, but extensive data are available only for the tertiary amines. (Picric acid also forms crystalline complexes with amine oxides, certain aromatic hydrocarbons, and aryl ethers).

12.14.2i Procedure: Picrates

Dissolve 1 mmole of the amine in 5 ml of 95% ethanol. Add 5 ml of a saturated solution of picric acid in 95% ethanol. Heat the solution to boiling and allow to cool slowly. The yellow crystals of the picrate may be recrystallized from ethanol or methanol if necessary.

Ammonium tetraphenylborates are quite water insoluble. The addition of an aqueous solution of sodium tetraphenylborate to a solution of an amine salt (usually the hydrochloride) causes the immediate precipitation of the amine as the tetraphenylborate salt. The method can be adapted as a means of detection of amines and amine salts, including quaternary ammonium salts. The melting points of a number of ammonium tetraphenylborates have been recorded.[38]

$$RNH_2 \xrightarrow{\quad HCl \quad} RNH_3^+Cl^- \xrightarrow{\quad NaB(C_6H_5)_4 \quad} RNH_3^+ B(C_6H_5)_4^-$$

$$R_4N^+ X^- \xrightarrow{\quad NaB(C_6H_5)_4 \quad} R_4N^+ B(C_6H_5)_4^-$$

12.14.2j Procedure: Tetraphenylborates

Dissolve or suspend the amine (0.1 g) in 5 to 10 ml of water. Add sufficient hydrochloric acid to dissolve the amine and to adjust the solution to pH 2 or 3. Slowly add, with stirring, an aqueous solution of sodium tetraphenylborate until precipitation of the salt is complete. Collect the white precipitate, wash well with distilled water, and dry below 60°.

---

[38] F. E. Crane, Jr., *Anal. Chem.*, **28**, 1794 (1956).

Other salts which are sometimes found to be useful include the Rein-eckates[39] [salts of $HCr(NH_3)_2(SCN)_4$] and the *chloroplatinates*.

12.14.2k Procedure: Chloroplatinates

To a solution of 50 mg of the amine in 1 ml of 10% hydrochloric acid, add dropwise 1 ml of a 25% aqueous solution of chloroplatinic acid ($H_2PtCl_6 \cdot 6H_2O$). Collect and wash the chloroplatinate with dilute hydrochloric acid. If necessary, the salt may be recrystallized from ethanol containing a drop of concentrated hydrochloric acid. Chloroplatinates are easily reduced to metallic platinum; avoid contact with metal spatulas.

12.14.3 Derivatives of Tertiary Amines

For solid derivatives of tertiary amines, salts are usually prepared. Quaternary ammonium salts, formed by alkylation of the tertiary amine with methyl iodide, benzyl chloride, or methyl *p*-toluenesulfonate, have been used. The most extensive data are available on the *methiodides*.

$$R_3N + CH_3I \longrightarrow R_3\overset{+}{N}CH_3I^-$$

12.14.3a Procedure: Methiodides

(a) A mixture of 0.5 g of the amine and 0.5 g of methyl iodide is warmed in a test tube over a low flame or in a water bath for a few minutes and then cooled in an ice bath. The tube is scratched with a glass rod, if necessary, to induce crystallization. Recrystallize from alcohol or ethyl acetate.

(b) A tertiary amine-methyl iodide mixture dissolved in benzene is refluxed until precipitation of the methiodide is complete.

The salts formed between tertiary amine and protonic acids form an important category of derivatives. By far the most important are the *picrates*. Extensive data are also available on the *hydrochlorides* and *chloroplatinates*. The procedures are those given under derivatives of primary and secondary amines.

For structure determinations of complex tertiary amines, it is often necessary to employ degradation methods. The following equations illustrate the most commonly used reactions.

von Braun reaction[40]

$$R_2NR' + BrCN \longrightarrow R_2NCN + R'Br$$

[39] B. F. Aycock, E. J. Eisenbraun, and R. W. Schrader, *J. Am. Chem. Soc.*, **73**, 1351 (1951).
[40] H. A. Hageman, *Org. Reactions*, **7**, 198(1953).

Amine oxide pyrolysis[41]

$$R_2NCH_2CH_2R \xrightarrow{H_2O_2} R_2\overset{+}{\underset{|}{N}}CH_2CH_2R \xrightarrow{\Delta} R_2NOH + CH_2{=}CHR$$
$$\hspace{4.5cm} \overset{O^-}{}$$

Hoffman elimination[41]

$$R_2NCH_2CH_2R \xrightarrow[\text{2) Ag}_2\text{O}]{\text{1) CH}_3\text{J}} R_2\overset{+}{\underset{^-OH}{N}}{-}CH_2CH_2R \xrightarrow{\Delta} R_2NCH_3 + CH_2{=}CHR$$
$$\hspace{4.5cm} \overset{CH_3}{\underset{|}{}}$$

A special category of tertiary amines are the vinyl amines or *enamines.*[42] Primary and secondary vinyl amines are unstable and rearrange to imines. The enamines are important synthetic intermediates whose reactivity closely resembles that of enolates.

Enamines form stable iminium salts by *C*-protonation; the perchlorates are usually nicely crystalline salts.[43]

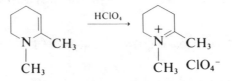

Most enamines may be hydrolyzed to secondary amines and carbonyl compounds which may be separated and identified by the usual methods.[44]

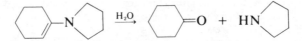

Enamines are reduced to the saturated tertiary amines by catalytic hydrogenation or formic acid.[45]

12.14.4 Amine Salts

Salts of amines fall into two categories—the salts of primary, secondary, and tertiary amines with proton acids, and the quaternary ammonium salts.

[41]A. C. Cope and E. R. Trumbull, *Ibid.*, **11,** 317 (1960).

[42]J. Szmuszkovicz, *Advan. Org. Chem.*, **4,** 1 (1963).

[43]N. J. Leonard and K. Jann, *J. Am. Chem. Soc.*, **84,** 4806 (1962).

[44]J. L. Johnson, M. E. Herr, J. C. Babcock, A. E. Fonken, J. E. Stafford, and F. W. Heyl, *Ibid.*, **78,** 430 (1956); J. Joska, J. Fajkos, and F. Sorm, *Collection Czech. Chem. Comm.*, **26,** 1646 (1961).

[45]P. L. DeBenneville and J. H. Macartney, *J. Am. Chem. Soc.*, **72,** 3073 (1950).

Both types are generally water-soluble, and insoluble in nonpolar solvents such as ethers and hydrocarbons. Many are soluble in alcohols, methylene chloride, and chloroform. See Sec. 4.2.3 and Fig. 12.23 for characteristic infrared absorptions of $-\overset{+}{N}H_3$, $-\overset{+}{N}H_2$, and $-\overset{+}{N}H$. There are no characteristics bands for quaternary ammonium salts (Fig. 12.24). Inorganic anions may be identified by the standard methods of inorganic qualitative analysis.

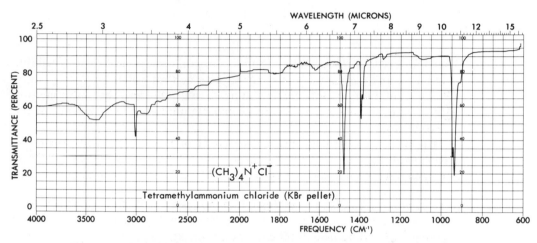

Fig. 12.24. Typical infrared spectrum of a quaternary ammonium salt.

The *salts of proton acids* usually give slightly acidic solutions; exceptions are the salts of weak acids such as acetic, propionic, etc. Upon treatment with dilute alkali, the amine will separate if it is insoluble or sparingly soluble in water. Even if separation does not occur, the presence of the free amine will be recognized by its characteristic odor. It is often convenient to recover the free amine from the alkaline solution by extraction with ether. In other cases, it may be more convenient to adjust the pH of the solution and make a sulfonamide, benzamide, or acetamide according to the directions given under amines.

Quaternary ammonium salts form neutral-to-basic solutions. The hydroxides and alkoxides are very strong bases. These salts may be converted to other salts which may be suitable for identification purposes by anion exchange reactions, as illustrated in the following equations.

$$R_4N^+I^- + AgBF_4 \longrightarrow R_4N^+BF_4^- + AgI$$

$$R_4N^+\,Cl^- + Na\,B(C_6H_5)_4 \longrightarrow R_4N^+\,B\,(C_6H_5)_4^- + NaCl$$

$$R_4N^+\,SO_4H^- \xrightarrow[\text{column}]{\text{Ion exchange}} R_4N^+\,OH^- \xrightarrow{\text{HClO}_4} R_4N^+\,ClO_4^-$$

They may also be degraded to tertiary amine by the Hoffman elimination reaction illustrated for tertiary amines.

12.14.5 α-Amino Acids

12.14.5a Classification

Amino acids are commonly encountered in identification work as the free amino acids, the hydrochlorides, or peptides. The free amino acids exist as the internal salts or zwitter ions, i.e.,

$$\underset{\overset{+}{N}H_3}{R\!-\!\overset{\displaystyle |}{C}H\!-\!COO^-}$$

They are very slightly soluble in nonpolar organic solvents, sparingly soluble in ethanol, and very soluble in water to give neutral solutions. They show increased solubility in both acidic and basic solutions, and they have melting points or decomposition points between 120 and 300° which depend on the rate of heating.

The principal infrared absorption bands of free amino acids are those arising from the carboxylate [asymmetric stretching near 1600 cm^{-1} (6.25 μ) and symmetric stretching near 1400 cm^{-1} (7.05 μ)] and the ammonium group [$\overset{+}{N}H_3$ stretching bands overlapping CH bands near 3000 cm^{-1} (3.33 μ); overtones sometimes present in the 2600 to 1900 cm^{-1} (3.85 to 5.26 μ) region; asymmetric bending near 1600 cm^{-1} (6.25 μ); and symmetric bending near 1500 cm^{-1} (6.67 μ)]. (See Fig. 12.25.) In the hydrochloride salts, the very broad —OH band of the carboxylic acid obscures the —$\overset{+}{N}H_3$ and CH stretching region. The other —$\overset{+}{N}H_3$ bands appear as above, along with carboxylic carbonyl near 1720 cm^{-1} (5.81 μ) (Fig. 12.25). Because of low

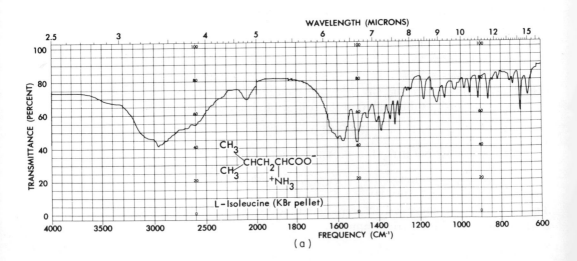

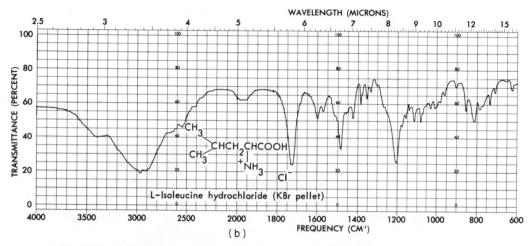

Fig. 12.25. Infrared spectra of a typical amino acid and its hydrochloride.

solubility, the free acids and salts are usually run as potassium bromide discs or as mulls. Under these conditions, the spectra of the racemic and optically active forms may show considerable differences. The solution spectra, however, are identical.

Upon warming an aqueous solution of an α-amino acid with a few drops of 0.25% *ninhydrin* (indane-1,2,3-trione hydrate), a blue-to-violet color appears. Proline and hydroxyproline give a yellow color. (Ammonium salts also give a positive test.)

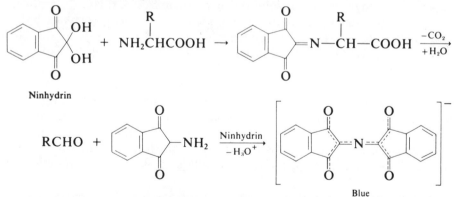

Addition of an aqueous solution of an α-amino acid to a *copper sulfate* solution produces a deep blue coloration. (Ammonia and amines also give a blue color with copper sulfate.)

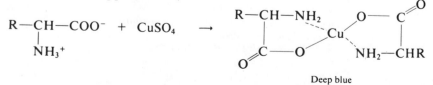

Amino acids yield hydroxy acids and nitrogen when treated with *nitrous acid*. The amount of nitrogen present in amino acids or peptides as primary amino groups can be quantitatively determined by measurement of the volume of nitrogen evolved.

12.14.5b Characterization

The twenty common amino acids found as constituents of proteins are optically active (except glycine) and have the L configuration. A number of less common amino acids are found elsewhere in nature, some of which have the D configuration. The specific rotations are valuable in identification.

Paper[46] and thin-layer chromatography[47] are excellent techniques for the tentative identification of amino acids. *Tables for Identification of Organic Compounds* list the R_F values on paper using a variety of solvent systems. Whenever possible, direct comparison should be made using a known, the unknown, and a mixture of the two. Two-dimensional paper chromatography has proven especially useful in the separation of amino acid mixtures.[46]

With paper or thin-layer chromatography, the spots are usually detected by developing the chromatogram with ninhydrin. The chromatogram is sprayed with a 0.25% solution of ninhydrin in acetone or alcohol, and developed in an oven at 80 to 100° for 5 minutes.

An interesting technique for determining the absolute configuration of microgram quantities of amino acids has been developed. By spraying a paper chromatogram with a solution of D-amino acid oxidase (a commercially available enzyme isolated from pig kidneys) in a pyrophosphate buffer and incubating, only D-amino acids are destroyed;[48] the α-keto acids, thus formed, may be detected with an acidified 2,4-dinitrophenylhydrazine spray.[49]

The qualitative and quantitative composition of amino acid mixtures (as obtained from the hydrolysis of peptides) can be achieved by commerically available automated amino acid analyzers. These instruments employ buffer solutions to elute the amino acids from ion-exchange columns. The effluent from the column is mixed with ninhydrin, and the intensity of the blue color is measured photoelectrically and plotted as a function of time.

Crystalline derivatives of the amino acids are usually obtained by reactions at the amino group by treatment with conventional amine reagents

---

[46]R. J. Block, E. L. Durrum, and G. Zweig, *Paper Chromatography and Paper Electrophoresis*, 2nd ed., Academic Press, New York, 1958.

[47]E. Stahl, *Thin-Layer Chromatography*, Springer-Verlag, Berlin, 1965.

[48]T. S. G. Jones, *Biochem. J.*, **42,** LIX (1948).

[49]J. L. Auclair and R. L. Patton, *Rev. Can. Biol.*, **9,** 3 (1950).

in alkaline solution. Extensive data are available on the *N-acetyl, N-benzoyl, and N-p-toluenesulfonyl* derivatives.

$$R-\underset{\underset{NH_3^+}{|}}{CH}-COO^- + NaOH \longrightarrow R-\underset{\underset{NH_2}{|}}{CH}-COO^-Na^+ + H_2O$$

$$R-\underset{\underset{NH_2}{|}}{CH}-COO-Na^+ \xrightarrow[\text{(2) HCl}]{\text{(1) R'COCl}} R-\underset{\underset{NHCOR'}{|}}{CH}-COOH$$

12.14.5b Procedure: *N*-Acetyl, *N*-benzoyl, and
**   *N*-p-toluenesulfonyl derivatives**

(a) Aqueous alkaline solutions of amino acids react with acetic anhydride or benzoyl chloride to yield the acetyl or benzoyl derivatives (Sec. 12.14.2e and f). In order to precipitate the derivative, the solution should be neutralized with dilute mineral acid. In the benzoyl case, any benzoic acid which precipitates can be removed from the desired derivative by washing with a little cold ether.

(b) The *p*-toluenesulfonyl derivatives may be conveniently made by stirring an alkaline solution of 1 equivalent of amino acid with an ether solution of a slight excess of *p*-toluenesulfonyl chloride for 3 to 4 hr. The aqueous layer is acidified to Congo red with dilute hydrochloric acid, and cooled if necessary. The crude sulfonamide may be recrystallized from dilute ethanol.

The *2,4-dinitrophenyl derivatives* are yellow crystalline materials of relatively sharp melting points. Mixture of these "DNP" derivatives of amino acids can be separated by thin-layer chromatography on silica gel. This method has been used for *N*-terminal amino acid analysis in proteins and peptides; the free amino groups on the end of proteins or peptides are reacted with the DNP reagent prior to hydrolysis.[50] The DNP derivatives of simple primary and secondary amines are sometimes useful in identification work.

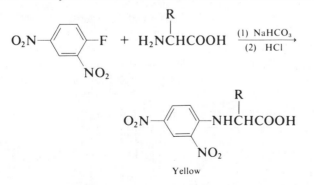

Yellow

---

[50]F. Sanger, *Biochem. J.*, **45**, 563 (1949) and previous papers.

12.14.5c Procedure: N-2,4-Dinitrophenyl derivatives

Dissolve or suspend 0.25 g of the amino acid in 5 ml of water containing 0.5 g of sodium hydrogen carbonate. Add a solution of 0.4 g of 2,4-dinitrofluorobenzene in 3 ml of ethanol. Stir the mixture for 1 hr. Add 3 ml of saturated salt solution, and extract several times with small volumes of ether to remove any unchanged reagent. Pour the aqueous layer into 12 ml of cold 5% hydrochloric acid, with stirring. The mixture should be distinctly acid to Congo red. The derivative may be recrystallized from 50% ethanol. These derivatives are light-sensitive; the reaction should be run in the dark (cover flask with aluminum foil), and the products should be kept in a dark place or in dark containers. As a substitute for the 2,4-dinitrofluorobenzene, the chloro compound may be used; in this case, however, the reaction mixture should be refluxed for 4 hr.

12.15 AMIDES AND RELATED COMPOUNDS

12.15.1a Classification

Amides, imides, ureas, and urethanes are nitrogen-containing compounds which exhibit a carbonyl band in the infrared and do not give a positive 2,4-dinitrophenylhydrazine test or show evidence for other nitrogen functions such as nitrile, amino, and nitro. Almost all are colorless crystalline solids; notable exceptions are certain N-alkylformamides and tetraalkyl-ureas, which are high boiling liquids.

Infrared spectra can be of great diagnostic value in the classification of amides (Fig. 12.26). Attention should be given to the presence of NH stretch, the exact position of the carbonyl band (amide I band), and the presence and position of the NH deformation (amide II band). Refer to Chapter 4 on infrared spectroscopy for details.

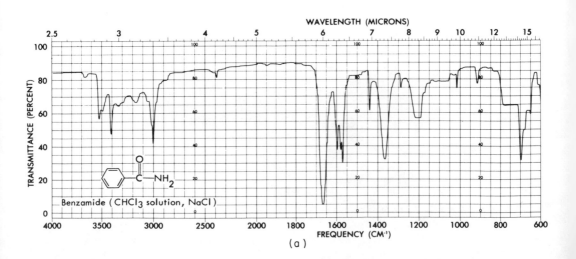

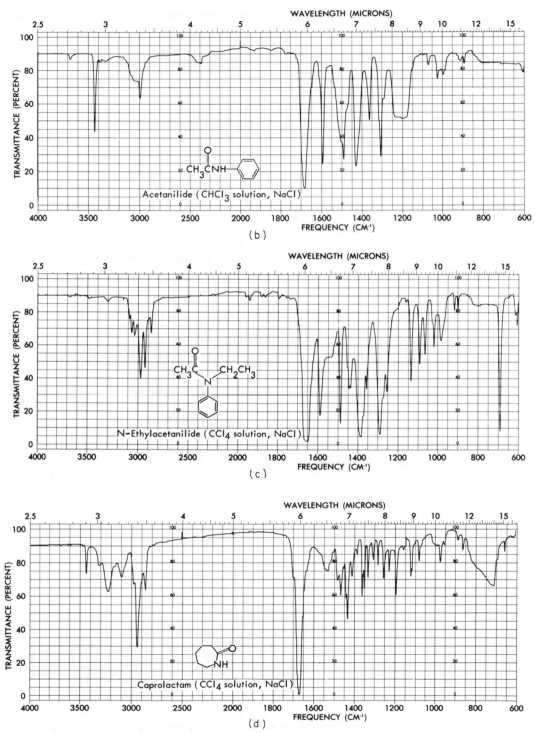

Fig. 12.26. Infrared spectra of typical amides. Note the absence of the amide II band in the spectrum of the lactam.

Occasionally, the investigator may have difficulty in deciding from infrared analysis whether a compound in question is an amide. In such cases it may be useful to test for liberation of an amine by treatment of a small sample of the unknown with refluxing 20% sodium hydroxide or fusion with powdered sodium hydroxide. A number of color tests have been developed for various classes of amides, ureas, etc.[51,52]

In the nmr, amide hydrogens exhibit broad absorptions in the region of $\delta 1.5$ to 5.0. The spectra of tertiary amides indicate magnetic nonequivalence of the N-alkyl groups due to the partial double-bond character of the C—N bond.

12.15.1b Characterization

The only completely general method for the preparation of derivatives of amides and related compounds is *hydrolysis* and identification of the products.

$$R'CONR_2 \xrightarrow{H_2O} R'COOH + R_2NH$$

Amides

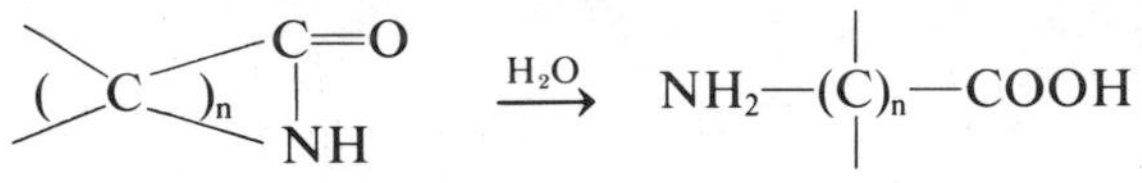

Lactams

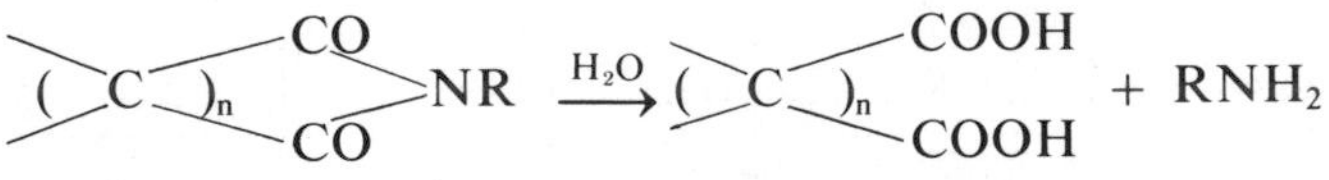

Imides

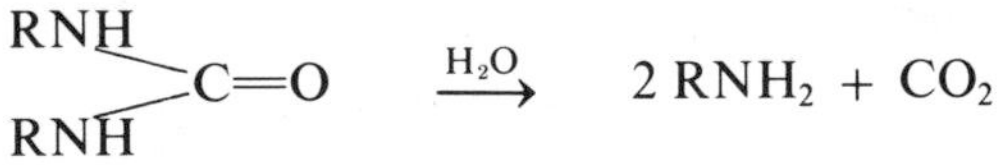

Ureas

Urethanes

Alkaline hydrolysis produces the free amine and the salt of the carboxylic acid; acidic hydrolysis yields the free acid and a salt of the amine. A special case is presented by the lactams, which yield amino acids upon hydrolysis; these amino acids can be converted to derivatives utilizing the general principles discussed in Sec. 12.14.5 for the α-amino acids.

---

[51] N. D. Cheronis, J. B. Entrikin, and E. M. Hodnett, *Semimicro Qualitative Organic Analysis*, 3rd ed., Interscience Publishers, Inc., New York, 1965.

[52] F. Feigl, *Spot Tests in Organic Analysis*, Elsevier, Amsterdam, 1960.

The hydrolysis may be effected by refluxing with 6 N hydrochloric acid or 10 to 20% sodium hydroxide. The latter is usually faster. Resistant amides may be hydrolyzed by boiling with 100% phosphoric acid [(prepared by dissolving phosphorous pentoxide (400 mg) in 85% phosphoric acid (1 g)], or by heating at 200° in a 20% solution of potassium hydroxide in glycerol. In any case, appropriate precautions should be taken to trap any volatile acids or amines.

Primary amides may be converted to acids by reaction with nitrous acid.

$$RCONH_2 + HONO \longrightarrow RCOOH + N_2 + H_2O$$

Xanthydrol reacts with most unsubstituted amides, imides, ureas, and barbiturates to form *N-xanthylamides (9-acylamidoxanthenes).* Oxamides, trichloroacetamides, and salicylamide do not react.

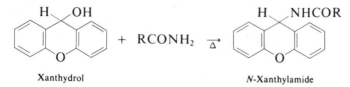

Xanthydrol   N-Xanthylamide

12.15.1c Procedure: N-Xanthylamides

Dissolve 400 mg of xanthydrol in 5 ml of glacial acetic acid (filter off any undissolved material) and add 1 to 1.5 mmole of the amide. Warm the mixture for 10 to 45 min. The *N*-xanthylamide which crystallizes upon cooling may be recrystallized from aqueous dioxane or aqueous alcohol.

Mercuric oxide reacts with some amides to form mercuric salts.

$$2\,RCONH_2 + HgO \longrightarrow (RCONH)_2Hg + H_2O$$

The Hoffman degradation is of occasional value in structural work. Under appropriate conditions urethanes or symmetrical ureas may be obtained from the intermediate isocyanates.

$$RCONH_2 \xrightarrow{Br_2,\ NaOH} RNCO \xrightarrow{H_2O} RNH_2 + CO_2$$

12.16 NITRILES

12.16.1a Classification

Aliphatic nitriles are usually liquids; the simpler aromatic nitriles are liquids or low melting solids. Both types have a characteristic odor reminiscent of cyanide.

In the infrared, the nitrile absorption appears as a weak band in the 2260 to 2210 cm^{-1} (4.42 to 4.52 μ) region. This region somewhat overlaps that of an isocyanate, but these may be readily differentiated on the basis of their reactivity with water, alcohols, etc., and the fact that the isocyanate absorption is much more intense than that of a nitrile (Fig. 12.28). Further preliminary tests for the nitrile grouping are usually not necessary. Other texts have recommended the conversion of the nitrile to a hydroxamic acid by reaction with hydroxylamine whose presence may be indicated by color test with ferric chloride. Alternatively, a small amount of the supposed nitrile may be heated with potassium hydroxide in glycerol and the evolution of ammonia detected by odor or by the litmus or copper sulfate test (see Sec. 12.14 on amines). Both of these methods, however, depend on the prior establishment of the absence of other functional groups such as esters and amides.

12.16.1b Characterization

Nitriles may be *hydrolyzed to acids* under acidic or alkaline conditions. If the resulting acid is solid, it may be employed as such a derivative; if it is a liquid or water-soluble, it is usually converted directly to a solid derivative.

$$RCN + 2\,H_2O \longrightarrow RCOOH + NH_3$$

12.16.1c Procedure: Hydrolysis to acids

(a) Acid hydrolysis. In a small flask equipped with a reflux condenser, place 4 ml of 85% phosphoric acid, 1 ml of 75% sulfuric acid, and 0.2 to 0.5 g of nitrile. Add a boiling chip and gently reflux the mixture for 1 hr. Cool the mixture and pour onto a small amount of crushed ice. If the acid is a solid, it may precipitate at this point. It should be collected on a filter. If amide impurities are present, the precipitate may be taken up in base, filtered, and reacidified to recover the acid. If the acid does not precipitate, the reaction mixture should be extracted several times with ether, the ether evaporated to recover solid acid. If the acid is a liquid, it may be converted to conventional acid derivatives such as the *p*-toluidide, *p*-bromobenzyl ester, or *S*-benzylthiuronium salt.

(b) Alkaline hydrolysis. In a small flask equipped with a reflux condenser, place 2 g of potassium hydroxide, 4 g of glycerol or ethylene glycol, and 250 to 500 mg of nitrile. Reflux for 1 hr. Dilute with 5 ml of water, cool, and add several milliliters of ether. Shake and allow the layers to separate. Decant the ether layer and discard. Acidify the aqueous solution by slow addition of 6 *N* hydrochloric or sulfuric acid. Extract the acid with several portions of ether, and handle the extract as indicated above under acid hydrolysis.

For those nitriles which yield solid, water-insoluble amides, partial *hydrolysis to amides* is often a very satisfactory method of derivation.

$$RCN \xrightarrow[H_2O]{H_2SO_4} RCONH_2$$

12.16.1d Procedure: Hydrolysis to amides

A solution of the nitrile (0.1 g) in 1 ml of concentrated sulfuric acid is warmed on the steam bath for several minutes. The mixture is cooled and poured into cold water. The precipitated crude amide is collected and suspended in a small volume of bicarbonate solution to remove any acid formed. The insoluble amide is recollected and recrystallized (aqueous alcohol). The amide may be further converted to the acid by action of nitrous acid.

Nitriles, especially aromatic nitriles, may be smoothly *converted to amides by alkaline hydrogen peroxide.*

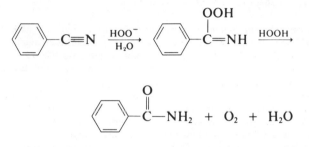

12.16.1e Procedure: Conversion to amides by alkaline hydrogen peroxide

In a small test tube, place 0.1 g of nitrile, 1 ml of ethanol, and 1 ml of 1 N sodium hydroxide. To this mixture add dropwise (cooling if the reaction foams too vigorously) 1 ml of 12% hydrogen peroxide. After addition is complete, maintain the solution at 50 to 60° in a water bath for an additional 30 min to 1 hr. Dilute the reaction mixture with cold water, and collect the solid amide. Recrystallize from aqueous alcohol.

Nitriles react with mercaptoacetic acid in the presence of dry hydrogen chloride to yield crystalline *α-iminoalkylmercaptoacetic acid hydrochlorides.*

$$RCN + HSCH_2COOH + HCl \longrightarrow RC \underset{S-CH_2COOH}{\overset{\overset{+}{N}H_2}{\big\langle}} \quad Cl^-$$

12.16.1f Procedure: α-Iminoalkylmercaptoacetic acid hydrochlorides

In a test tube place 0.5 g of the nitrile, 1.0 g of mercaptoacetic acid, and 7 ml of absolute ether. Cool the mixture in an ice-salt bath, and saturate with dry hydrogen chloride. Stopper the tube and allow it to stand in the ice bath until crystals separate (ca. 1 hr for aliphatic nitriles and up to 1 day for aromatic nitriles). Collect the product, wash with ether, and dry in a vacuum desiccator. Determine the decomposition point and neutralization equivalent (thymol blue as an indicator or potentiometrically).

Nitriles may be *reduced to primary amines* by lithium aluminum hydride or by sodium dissolving in alcohol. The conversion can also be made catalytically, but secondary amine by-products may result; addition of ammonium carbamate suppresses the formation of secondary amine. The

amines obtained by reduction are usually liquids and require further transformation to a solid derivative.

$$\text{RCN} \xrightarrow[\text{or Na/ROH}]{\text{LiAlH}_4} \text{RCH}_2\text{NH}_2$$

$$\text{RCN} \xrightarrow{\text{H}_2 \text{ cat.}} \text{RCH}_2\text{NH}_2 \xrightarrow{\text{RCN}} \text{RCH}_2\text{NH}\overset{\overset{\displaystyle \text{NH}}{\|}}{\text{C}}\text{—R} \xrightarrow{\text{H}_2 \text{ cat.}}$$

$$(\text{RCH}_2)_2\text{NH} + \text{NH}_3$$

12.16.1g Procedure: Reduction to amines

(a) *Reduction with lithium aluminum hydride.* In a small dry flask equipped with a
dropping funnel and magnetic stirrer, place 0.1 g of lithium aluminum hydride and
10 ml of anhydrous ether. To the hydride solution add dropwise 0.2 g of nitrile in 10 ml
of anhydrous ether.   When the reaction has subsided, add water dropwise and
cautiously with cooling, until hydrogen evolution ceases. Several drops of water should
be sufficient.   Then add to the reaction mixture 10 ml of a 20% solution of sodium
potassium tartrate. Transfer the mixture to a separatory funnel. Separate the ether
layer and extract the aqueous layer one or more times with small volumes of ether.
The ether should be removed and amine derivatives made in the usual manner. The
phenylthiourea is a particular suitable derivative for small scale reactions.

(b) *Reduction with sodium in alcohol.* Place 0.5 g of the nitrile and 10 ml of absolute
ethanol in a dry 100-ml round-bottomed flask equipped with a reflux condenser. Add
through the top of the condenser approximately 0.75 g of clean, finely cut sodium at
such a rate that the reaction remains under control. When the sodium has dissolved,
cool the reaction mixture in an ice bath and cautiously add 10 ml of concentrated
hydrochloric dropwise through the condenser. The solution should be acid to litmus.
Transfer the reaction mixture to a beaker, and place it on the steam bath to remove the
ethanol. Cool the reaction mixture in an ice bath, and cautiously make the mixture basic
with concentrated sodium hydroxide. Extract the amine with ether, and use it to make
an amide or thiourea derivatives. If the amine is particularly volatile, distill the amine
directly from the solution made basic in the distillation flask into water (the tip of the
delivery tube should be under the water surface—however be sure to remove the
delivery tube before reducing the heat to the distillation pot!) Add 20 drops of phenyl
isothiocyanate and shake the reaction mixture vigorously for several minutes.   If
necessary, cool and scratch the flask to induce crystallization.

The transformation of nitriles to *aldehydes by partial reduction and
hydrolysis*, or to ketones by reaction with Grignard reagents, is a particularly
suitable identification method for lower molecular weight nitriles which
would yield highly volatile amines on reduction, liquid acids on hydrolysis,
or water-soluble amides on partial hydrolysis. Solid derivatives of the aldehydes or ketones are readily made. Sometimes it may be necessary to isolate
the aldehydes or ketones by use of the Grignard *T* reagent (Sec. 12.7.5).

In the Grignard reaction, the best yields are obtained by the use of phenyl
magnesium bromide in large excess.[53] The reaction can easily be adapted to
small-scale use.

---

[53]R. L. Shriner and T. A. Turner, *J. Am. Chem. Soc.*, **52**, 1267 (1930).

$$RC{\equiv}N + C_6H_5MgBr \longrightarrow R-\overset{\overset{\displaystyle NMgBr}{\|}}{C}-C_6H_5 \xrightarrow{H_3O^+} R-\overset{\overset{\displaystyle O}{\|}}{C}-C_6H_5$$

Aldehydes may readily be prepared from nitriles by use of complex hydride reducing agents. Reagents of particular utility for this reduction are diisobutylaluminum hydride,[54] and triethoxy lithium aluminum hydride.[55]

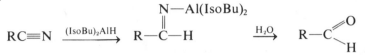

$$RC{\equiv}N \xrightarrow{(IsoBu)_2AlH} R-\overset{\overset{\displaystyle N-Al(IsoBu)_2}{\|}}{C}-H \xrightarrow{H_2O} R-C\overset{\displaystyle\nearrow O}{\underset{\displaystyle\searrow H}{}}$$

12.17   NITRO COMPOUNDS

12.17.1a Classification

Most nitroalkanes are liquids, colorless when pure; but they develop a yellow tinge on storage. The liquid nitro compounds have characteristic odors; they are insoluble in water and have densities greater than unity. Many of the aromatic compounds are crystalline solids. The aromatic nitro compounds are generally yellow; the color increasing markedly on poly-nitration.

When the nitro group is the primary functionality in the molecule, its presence is readily revealed on casual inspection of the infrared spectrum; strong bands appear near 1560 and 1350 cm^{-1} (6.41 and 7.41 μ) (Fig. 12.27). More frequently, the group appears as a substituent in other classes of compounds, and its presence is revealed by the physical constants of the compound and its derivation and/or by more careful inspection of the infrared spectrum. Like halogen substituents, when the nitro group is present along with a more reactive functionality, e.g., carbonyl, hydroxyl etc., derivation is made at the more reactive site.

The high electronegativity of the nitro group results in extensive de-shielding of adjacent aliphatic hydrogens in nmr spectra; nitromethane appears at $\delta 5.72$, —CH$_2$—NO$_2$ at $\delta 5.6$, and $>$CH—NO$_2$ at $\delta 5.3$. The characteristic chemical shift along with coupling patterns allow distinction to be made between primary and secondary aliphatic nitroalkanes.

For the classification of a material as a nitroalkane or nitroarene, it is usually not necessary to resort to chemical methods; however, the following simple tests may have occasional utility.

---

[54]A. I. Zakhaikin and I. M. Kharlina, *Polk. A Rad. Nauk SSSR*, **116**, 422(1957), *Chem. Abstracts*, **52**, 8040 (1958).

[55]H. C. Brown and C. P. Garg, *J. Am. Chem. Soc.*, **86**, 1085 (1964).

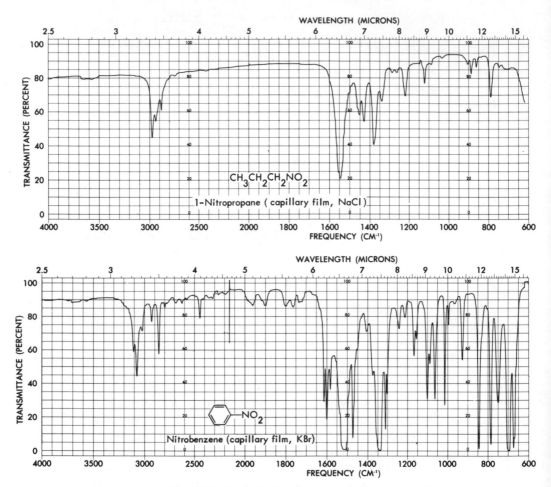

Fig. 12.27. Typical infrared spectra of nitro compounds.

Most nitro compounds (exceptions are nitroethane and 1-nitropropane) will oxidize *ferrous hydroxide* to ferric hydroxide, as will nitroso compounds, quinones, hydroxylamines, nitrates, and nitrites.

$$RNO_2 + 4H_2O + 6\ Fe(OH)_2 \longrightarrow RNH_2 + 6\ Fe(OH)_3$$

12.17.1b Procedure: Ferrous hydroxide test

In a small test tube, mix about 10 to 20 mg of the compound with 1.5 ml of freshly prepared 5% solution of ferrous ammonium sulfate. Add 1 drop of 3 *N* sulfuric acid and 1 ml of 2 *N* potassium hydroxide in methanol. Stopper the tube quickly and shake well. In a positive test, a red-brown precipitate of ferric hydroxide appears within 1 min.

Nitrous acid can be used to differentiate between primary, secondary, and tertiary aliphatic nitro compounds.

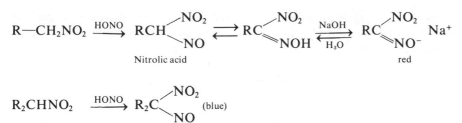

Nitrolic acid   red

$$R_2CHNO_2 \xrightarrow{HONO} R_2C\begin{smallmatrix}NO_2\\ \\NO\end{smallmatrix} \text{(blue)}$$

Pseudonitrole

$$R_3CNO_2 \xrightarrow{HONO} \text{No reaction}$$

12.17.1c Procedure: Nitrous acid test for nitro compounds

Add five drops of the nitro alkane to 2 ml of 10% sodium hydroxide. Allow the mixture to stand for 3 min. Add 1 ml of 10% sodium nitrite solution; then add 10% sulfuric acid dropwise. With primary nitroalkanes, an intense red color of the salt of the nitrolic acid appears and fades upon acidification. With secondary nitro compounds, a blue or blue-green color of the pseudonitrole appears. With tertiary nitro compounds there is no reaction.

12.17.1d Characterization

Many of the simple nitroalkanes may be identified by reference to physical and spectral properties. Relatively few vinyl nitro compounds are known; nitroethylene and other lower nitroolefins readily polymerize.

Both aromatic and aliphatic nitro compounds may be converted to *primary amines by reduction with tin and hydrochloric acid.* The reduction may also be effected by catalytic hydrogenation. The amine thus obtained is identified by standard procedures. This provides the only general method for the conversion of nitroalkanes to solid derivative.

$$RNO_2 \xrightarrow{Sn/HCl} RNH_2$$

12.17.1e Procedure: Reduction with tin and hydrochloric acid

In a 50-ml round-bottomed flask fitted with a reflux condenser, place 1 g of the nitro (nitroso, azo, azoxy, or hydrazo) compound, and 2 to 3 g of granulated tin. If the compound is highly insoluble in water, add 5 ml of ethanol. Add 20 ml of 10% hydrochloric acid in small portions, with shaking after each addition. After the addition is complete, warm the mixture on the steam bath for 10 min. Decant the hot solution into 10 ml of water and add sufficient 40% sodium hydroxide to dissolve the tin hydroxide. Extract the mixture several times with ether. Dry the ether over magnesium sulfate; remove the ether and identify the amine by conversion to one or more crystalline derivatives.

For highly volatile amines, the procedure must be modified. As the solution is made alkaline, the amine may be distilled into dilute hydrochloric acid.

With base catalysis, primary and secondary nitroalkanes give *adducts with carbonyl and α,β-unsaturated carbonyl and related compounds.* Such adducts have only occasional use in identification.

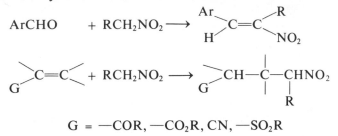

$$ G = -COR, -CO_2R, CN, -SO_2R $$

Primary and secondary nitroalkanes may be *oxidized to the corresponding carbonyl compounds* by potassium permanganate. Magnesium sulfate solution is added to act as a buffering agent.[56]

$$ R-CH_2NO_2 + KMnO_4 \xrightarrow{MgSO_4} RCHO $$

Aromatic mononitro compounds may be characterized by conversion to *di- or trinitro derivatives* (follow procedures for aromatic hydrocarbons, Sec. 12.1.4). Many polynitro aromatics form stable *charge-transfer complexes* with naphthalene and other electron-rich aromatic compounds.[57]

Nitro derivatives of alkylbenzenes may be oxidized to the corresponding nitrobenzoic acids with dichromate (preferred) or permanganate (follow procedures for alkyl benzenes, Sec. 12.1.4). Bromination of the ring often affords a suitable derivative.

12.18 MISCELLANEOUS NITROGEN COMPOUNDS

(See Tables 4.5 and 4.6 for characteristic infrared absorptions of miscellaneous nitrogen compounds.)

12.18.1 Isocyanates and Carbodiimides

The isocyanate group can readily be detected by infrared spectroscopy—a very strong band in the 2275 to 2250 cm^{-1} (4.40 to 4.04 μ) region—much stronger than a nitrile which appears in the same region (Fig. 12.28). Isocyanates may also be distinguished from nitriles in their ready reactivity with water, alcohols, and amines to yield amides, urethanes, and ureas, respectively, all of which may serve to identify the isocyanate.

$$ 2\ RN=C=O \xrightarrow{Cat.} R-N=C=N-R + CO_2 $$

[56]H. Shechter and F. T. Williams, Jr., *J. Org. Chem.* **27**, 3699 (1962).
[57]O. C. Dermer and R. B. Smith, *J. Am. Chem. Soc.*, **61**, 748 (1939).

Carbodiimides, readily available from isocyanates, are used in peptide synthesis and other condensation reactions that involve the removal of water. They are highly reactive (with amines, alcohols, acids, etc.) and may be readily identified by their reaction with water to produce symmetrical ureas. The most commonly encountered carbodiimide is dicyclohexycarbodiimide (Fig. 12.28).

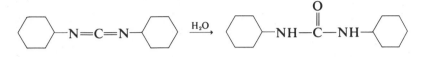

12.18.2 Azo, Azoxy, and Hydrazo Compounds

Azo and azoxy compounds are highly colored substances. Ultraviolet and visible spectroscopy, as well as comparative thin-layer chromatography, should be extensively used as aids in their identification.

The following equation illustrates types of reactions which may be useful in preparing derivatives.

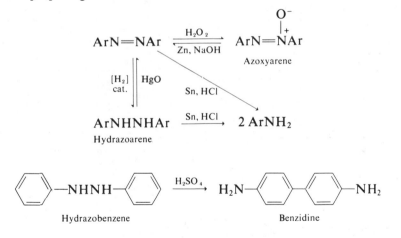

12.18.3 Hydrazines

$$RNHNH_2 \qquad R_2NNH_2 \qquad RNHNHR$$

Hydrazines are weak-to-intermediate strength bases; they dissolve in mineral acids to form salts. The mono and unsymmetrically disubstituted hydrazines may be detected and identified by condensation with carbonyl compounds to yield hydrazones.

The N—N bond of hydrazines may be easily cleaved by tin and hydrochloric acid (method for nitro compounds, Sec. 12.17.1e) or in some cases with aluminum amalgam.

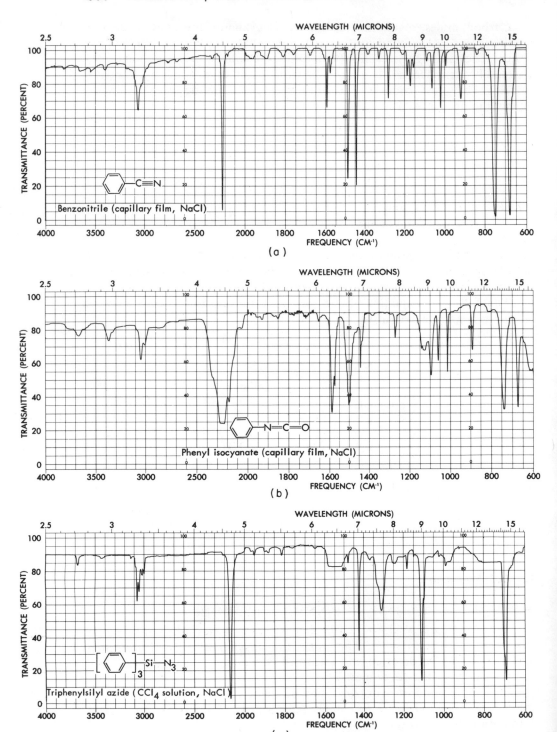

(a)

(b)

(c)

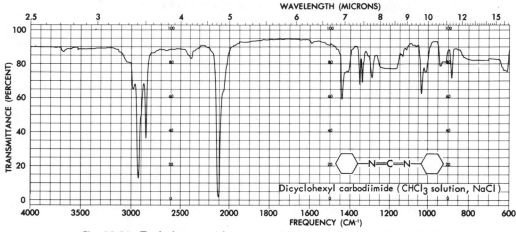

Fig. 12.28. Typical spectra of representative nitrogen compounds having bands in the 2100 to 2200 cm^{-1} region.

12.18.4 Oximes, Hydrazones, and Semicarbazones

These compounds may be hydrolyzed by the action of concentrated hydrochloric acid to yield the hydrochloride of hydroxylamine, the hydrazine, or the semicarbazine and the carbonyl compound. The original carbonyl compound may in some cases be regenerated by exchange reactions,[58] e.g.,

$$R_2C{=}NNHCONH_2$$

$$+ \quad \xrightarrow[\substack{AcOH \\ NaOAc \\ H_2O}]{} \quad R_2C{=}O + CH_3{-}\underset{\underset{COOH}{|}}{C}{=}NNHCONH_2$$

$$\underset{\substack{\| \\ O}}{CH_3{-}C{-}COOH}$$

Semicarbazones also react with nitrous acid to regenerate aldehydes or ketones.

Oximes may be identified and their stereochemistry determined by the Beckman rearrangement to amides.[59]

$$R{-}\underset{\underset{}{\overset{\overset{N{-}OH}{\|}}{}}}{C}{-}R' \xrightarrow{H^+} RNH{-}\underset{\underset{}{\overset{\overset{O}{\|}}{}}}{C}{-}R'$$

12.18.5 Azides

Azides (Fig. 12.28) may be reduced to *primary amines* by sodium borohydride, lithium aluminum hydride, or by catalytic hydrogenation.

$$R{-}N_3 \xrightarrow{[H]} R{-}NH_2$$

[58] E. B. Hershber, *J. Org. Chem.*, **13**, 542 (1948).
[59] L. G. Donaruma and W. Z. Heldt, *Org. Reactions*, **11**, 1 (1960).

12.18.5a Procedure: Borohydride reduction of azides

To 0.3 g of the azide in 5 ml of isopropanol, add 0.2 to 0.3 g of sodium borohydride and reflux on a steam bath overnight. Evaporate the isopropanol, add water, and reflux until effervescence has subsided. Isolate the amine by extraction with ether.

Alkyl and aryl azides will undergo 1,3-dipolar additions with reactive olefins[60] and acetylenes.[61]

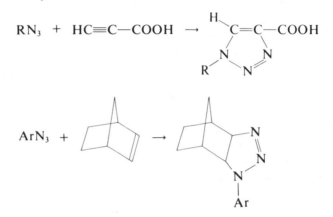

12.18.6 Isocyanides

Isocyanides are very uncommon. They have a very characteristic vile odor. They may be identified by acid hydrolysis to the corresponding primary amine and formic acid, or by reduction to secondary amines.

$$RN{=}C\colon \xrightarrow{H_3O^+} RNH_3^+ + HCOOH$$

$$RN{=}C\colon \xrightarrow{H_2} RNHCH_3$$

12.18.7 Nitramines

Primary nitramines are acidic and form salts with aqueous alkali. These salts may be alkylated with reagents such as methyl iodide to give secondary nitramines which are neutral.

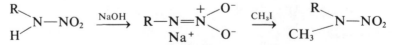

Most alkylnitramines eliminate the nitro group upon treatment with acid. Arylnitramines, like the nitrosoamines, undergo isomerization in the presence of mineral acids; e.g., phenyl nitramine gives a mixture of *o*- and

---

[60] R. Huttle, *Chem. Ber.*, **74B**, 1680 (1941).
[61] P. Scheiner, J. H. Schumaker, S. Deming, W. J. Libbey, and G. P. Nowack, *J. Am. Chem. Soc.*, **87**, 306 (1965).

p-nitroanilines. Nitramines may be reduced to the corresponding primary amines.

12.18.8 C-Nitroso Compounds

The simple nitrosoalkanes and arenes generally exist as dimers which are white crystalline solids, but which assume the blue color of the monomer when fused or vaporized.

Nitroso compounds are reduced to primary amines by zinc, iron, or tin in acid solution, or by catalytic hydrogenation. Nitrosobenzene and its derivatives readily condense with aniline to form azo compounds.[62]

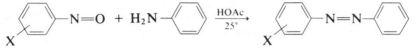

12.18.9 N-Nitroso Compounds

Only the N-nitroso derivatives of secondary amines are stable; they are neutral to weakly basic compounds. Arylnitrosoamines rearrange to the *p*-nitrosoarylamines in the presence of mineral acid. Nitrosoamines may be reduced to hydrazines by zinc and acetic acid, or to secondary amines by tin and hydrochloric acid. The N-nitrosoamides yield diazo compounds upon treatment with sodium hydroxide.

12.18.10 Nitrites and Nitrates

$$R—O—N=O \qquad R—O—\overset{+}{N}\underset{O^-}{\overset{O}{\diagup}}$$

Nitrite   Nitrate

These materials are esters of inorganic acids. *They are toxic and should be handled with care.* Both nitrites and nitrates give a positive ferrous hydroxide test. They may be hydrolyzed in alkaline solution to yield alcohols and salts of nitric or nitrous acid. The common nitrites are ethyl, *n*-butyl, and isoamyl.

12.18.11 Amine Oxides

$$R_3N \xrightarrow{H_2O_2} R_3N^+—O^-$$

Amine oxides, formed by the oxidation of tertiary amines by peroxides, are usually highly hygroscopic, viscous syrups. Some form crystalline hydrates.

---

[62]H. D. Anspon, *Org. Syn.*, **Coll. Vol. 3,** 711 (1955); C. M. Atkinson, *J. Chem. Soc.*, 2023 (1954).

Amine oxides show an intense N—O stretching band at 970 to 950 cm^{-1} (10.31 to 10.53 μ). In heteroaromatic amine oxides, e.g., pyridine N-oxide, this band appears in the 1300 to 1200 cm^{-1} (7.69 to 8.33 μ) region. These bands are shifted to lower frequencies upon hydrogen bonding. Amine oxides form crystalline picrates when treated with alcoholic or aqueous solutions of picric acid.

Reduction to the tertiary amine may be accomplished with a variety of reagents including triphenyl phosphine, triethyl phosphite, and metal-acid mixtures. The former is especially useful for aromatic amine oxides. The reactants are mixed neat and heated until the tertiary amine distills out.[63]

$$R_3N^{+}{-}O^- + (C_6H_5)_3P \longrightarrow R_3N + (C_6H_5)_3P{=}O$$

Amine oxides with β-hydrogens which are sterically accessible to the oxide are easily pyrolyzed to hydroxylamines and olefins.[64]

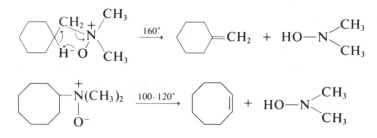

12.18.12 Nitrones

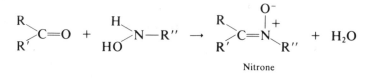

Nitrone

Nitrones,[65] formed by the condensation of aldehydes or ketones with N-substituted hydroxylamines, may be solids or liquids. Unless aryl groups are present, nitrones tend to be quite water-soluble. The nitrones exhibit a strong band in the infrared in the 1615 to 1560 cm^{-1} (6.20 to 6.40 μ) region. Aliphatic nitrones have a characteristic ultraviolet absorption near 229 to 235 mμ (ϵ = 9.000) which moves to longer wavelengths on conjugation.

Nitrones may be reduced to hydroxylamines with sodium borohydride or lithium aluminum hydride, and to secondary amines with zinc and mineral acid or catalytic hydrogenation.

[63]L. Horner and H. Hoffman, *Angew. Chem.*, **68**, 480 (1956). E. Howard, Jr. and W. F. Olszewski, *J. Am. Chem. Soc.*, **81**, 1483 (1959).

[64]A. C. Cope and E. R. Trumbull, *Org. Reactions*, **11**, 317 (1960).

[65]G. R. Delpierre and M. Lamchen, *Quart. Revs.*, **19**, 329 (1965).

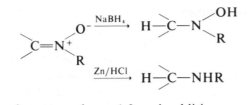

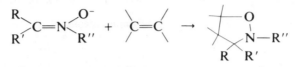

With olefins, nitrones undergo 1,3-cycloaddition reactions to provide isoxazolidines. The ease of such reactions is considerably enhanced by phenyl or carbonyl substituents on the olefin.

12.19 SULFUR COMPOUNDS

A comprehensive discussion of the methods of detection and identification of the many types of organic sulfur compounds is far beyond the scope of this text. In general, organic sulfur chemistry follows along the lines of organic oxygen chemistry; many of the same general types of functional groups occur, but the variety is greatly increased by the variable oxidation states of sulfur as indicated by the following examples.

Oxygen series

ROH   Alcohol   ROR   Ether   ROOR Peroxide

Sulfur series

RSH   Thiol   RSR   Thioether (sulfide) RSSR   Disulfides

$$\overset{O}{\overset{\|}{RSR}}$$

RSO$_2$H   Sulfinic Acid and Derivatives   RSR   Sulfoxide

RSO$_3$H   Sulfonic Acid and Derivatives $$\overset{O}{\underset{O}{\overset{\|}{\underset{\|}{RSR}}}}$$ Sulfones

In this chapter, the most commonly occurring organic sulfur functions will be discussed in some detail; many others will be mentioned only briefly, perhaps if only to make the reader aware of their existence. The most comprehensive listing of the properties of organic bivalent sulfur compounds is found in the series by Reid.[66]

---

[66] E. E. Reid, *Organic Chemistry of Bivalent Sulfur*, Vol. I–V, Chemical Publishing Co., Inc., New York.

12.19.1 Thiols (Mercaptans)

The classification of a compound as a thiol seldom awaits the interpretation of the infrared spectrum. For the uninitiated, most thiols have an objectionable odor reminiscent of hydrogen sulfide. Thiols are generally insoluble in water, but they react with aqueous alkali to form soluble salts.

$$RSH + OH^- \longrightarrow RS^- + H_2O$$

In the infrared, the S—H band appears in the 2600 to 2550 cm^{-1} (3.85 to 3.92 μ) region (Fig. 12.29). This is a relatively weak band and, in some

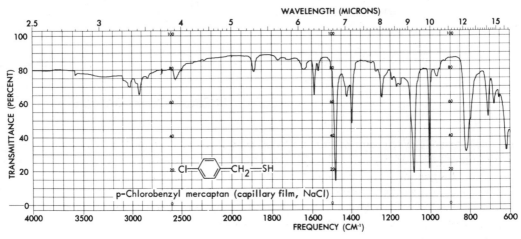

Fig. 12.29. Typical infrared spectrum of a thiol. Note the very weak SH band at 2560 cm^{-1}.

cases, may be overlooked especially if the compound is run as a dilute solution. This may result in misassignment of the compound as a sulfide. Sulfides and thiols may be differentiated on the basis of solubility of the latter in dilute alkali and by the *lead mercaptide test*; the former is insoluble in base and does not precipitate in the lead mercaptide test.

$$2\ RSH + Pb^{++} + 2H_2O \longrightarrow Pb(SR)_2 \downarrow + 2\ H_3O^+$$

12.19.1a Procedure: Lead mercaptide test

Add two drops of the thiol to a saturated solution of lead acetate in ethanol; yellow lead mercaptide precipitates.

Perhaps the most useful derivatives of the alkyl and aryl thiols are the *2,4-dinitrophenyl sulfides* and the corresponding sulfones.

RSH $\xrightarrow{\text{NaOH}}$ RSNa $\xrightarrow{\text{ArCl}}$ RS—⟨ ⟩—NO$_2$ + NaCl

12.19.1b Procedure: 2,4-Dinitrophenyl sulfides

In a test tube dissolve 250 mg of the thiol in 5 ml of methanol. Add 1 ml of 10% sodium hydroxide solution. Add this solution to 3 ml of methanol containing 0.5 g of 2,4-dinitrochlorobenzene in a small flask arranged for reflux. Reflux the mixture for 10 min, cool, and recrystallize from methanol or ethanol.

Occasionally it may be expedient to identify thiols by conversion to simple sulfides or sulfones. The methyl and benzyl sulfides, produced by alkylation of the sodium thiolates with methyl iodide or benzyl bromide in alcohol, are the most useful. These sulfides, as well as the 2,4-dinitrophenyl sulfides may be oxidized to sulfones by procedures outlined under sulfides in the next section. The 2,4-dinitrophenylsulfones are particularly valuable because they exhibit a wide range of melting points.

The *3,5-dinitrothiobenzoates* may be easily prepared by reaction of the thiol and 3,5-dinitrobenzoyl chloride in the presence of pyridine as catalyst. However, many have low melting points which fall within a narrow range.

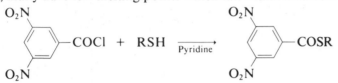

12.19.1c Procedure: 3,5-Dinitrothiobenzoates

In a dry test tube, place 200 mg of 3,5-dinitrobenzoyl chloride, 100 mg (five or six drops) of thiol, and three drops of pyridine. Heat the mixture in a beaker of boiling water for 10 min. Add 2 ml of water and a few drops of pyridine to destroy any remaining reagent. Stir with a glass rod to induce crystallization. Filter, wash with water, suspend the crude powder in several milliliters of sodium hydrogen carbonate solution, and warm if necessary to remove any 3,5-dinitrobenzoic acid. Collect, wash again with water, and recrystallize from dilute ethanol.

In those cases where the *disulfide* is a crystalline solid, derivation of thiols may be readily made by oxidation with sulfoxides.[67]

$$2\ RSH + CH_3\overset{\displaystyle O}{\overset{\displaystyle \|}{S}}CH_3 \longrightarrow RSSR + CH_3SCH_3 + H_2O$$

With thiophenols, the reaction may be carried out at room temperature for 24 hr. With aliphatic thiols, heating up to 160° must be applied for 24 hr. Disulfides may be also obtained by oxidation of thiols with hypoiodite, ferricyanide, and a variety of other mild oxidants.

12.19.2 Sulfides and Disulfides

Infrared spectroscopy alone cannot usually provide sufficient information to adequately classify a compound as belonging to the sulfide or disulfide class

---

[67]T. J. Wallace, and J. J. Mahon, *J. Org. Chem.*, **30**, 1502 (1965).

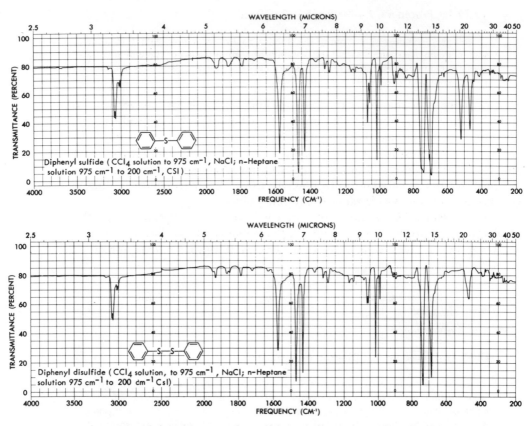

Fig. 12.30. Infrared spectra of a sulfide and the corresponding disulfide.   Note the striking similarity in the two spectra.

(Fig. 12.30). Both these classes of compounds exhibit a carbon-sulfur stretching frequency in the 800 to 600 cm^{-1} (12.5 μ to 16.7) region which is of little utility because it is easily confused with other bands in that region and, furthermore, is frequently obscured by the solvent. The sulfur-sulfur stretching frequency is a weak band in the 550 to 450 cm^{-1} (18.2 μ to 22.2) region, and it is thus inaccessible on most instruments.

Aliphatic and most aromatic sulfides and disulfides have unpleasant odors somewhat resembling that of hydrogen sulfide. Compounds with such odors which are not mercaptans or sulfoxides (as determined by the absence of characteristic infrared bands) should be strongly suspect as sulfides or disulfides. Most aliphatic sulfides and disulfides are liquids. The aromatic sulfides are liquids or low melting solids; aromatic disulfides are low melting solids. A number of polysulfides of general formula R—S$_n$—R are also known.

Most sulfides give *crystalline complexes with mercuric chloride.*

$$R_2S + nHgCl_2 \longrightarrow R_2S(HgCl_2)_n \xrightarrow[\text{NaOH}]{} R_2S$$

12.19.2a Procedure: Mercuric chloride complexes

The sulfide, neat or in alcohol solution, is added dropwise to a saturated solution of mercuric chloride in alcohol. The complex, which usually precipitates immediately, varies in ratio of mercuric chloride to sulfide, depending on the structure of the sulfide and the exact conditions of preparation. The sulfides may be regenerated from the complex by aqueous sodium cyanide or dilute hydroxide. The procedure may be used as a test for sulfides, or as a method for obtaining a solid derivative.

The most useful solid derivatives of the sulfides are the *sulfones* produced by oxidation with permanganate or peroxides.

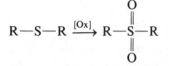

12.19.2b Procedure: Oxidation to sulfones

(a) Dissolve the sulfide in a small volume of warm glacial acetic acid. Shake and add 3% potassium permanganate solution in 2- or 3-ml portions as fast as the color disappears. Continue the addition until the color persists after shaking for several minutes. Decolorize the solution by addition of sodium hydrogen sulfite. Add 2 to 3 volumes of crushed ice. Filter off the sulfone and recrystallize from ethanol.

(b) Dissolve 1 mmole of the sulfide in 3 ml of benzene and 3 ml of methanol. Add 0.5 to 1 mmole of vanadium pentoxide and heat to 50° with stirring. Slowly add a slight excess of *t*-butyl hydroperoxide until the excess can be detected by acidified starch iodide paper. Add solid sodium hydrogen sulfite in small portions, and stir for 5 to 10 min. Filter and concentrate to a heavy oil. Add a small amount of water, extract with methylene chloride or other suitable solvent, and dry over magnesium sulfate.

(c) Dissolve the sulfide in glacial acetic acid. Add excess hydrogen peroxide and allow to stand overnight. Warm the solution to 50 to 60° for 5 to 10 min. Pour into ice water and collect the crystalline sulfone.

Other derivatives of sulfides which may prove useful include the sulfoxides, sulfilimines, and sulfonium salts.

Sulfoxides are cleanly produced by oxidation of sulfides with aqueous or aqueous-alcoholic sodium periodate at 0° for aliphatic sulfides and at room temperature for aromatic sulfides.[68]

$$\text{>}S + IO_4^- \longrightarrow \text{>}S{=}O + IO_3^-$$

The *N-p*-toluenesulfonyl sulfilimines are readily prepared by shaking a suspension of the sulfide in an aqueous solution of Chloramine-*T*.

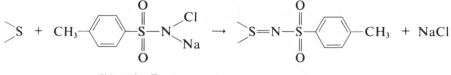

Chloramine-*T*

[68] N. J. Leonard and C. R. Johnson, *J. Org. Chem.*, **27**, 282 (1962).

Methiodides of sulfides may be prepared by warming the sulfide neat or in benzene solution with methyl iodide.

$$\text{>}S + CH_3I \longrightarrow \text{>}S^+ {-} CH_3 \ I^-$$

Sulfides and disulfides (as well as thiols, sulfones, sulfoxides, and many other sulfur compounds) undergo hydrogenolysis when treated with Raney nickel catalyst.[69]

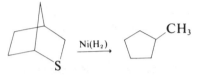

Dithio- and hemithioacetals and ketals form a special class of sulfides. The most common are the ethylene thioketals and ethylenehemithioketals. The following equations illustrate reactions which may be useful in their structure determination.

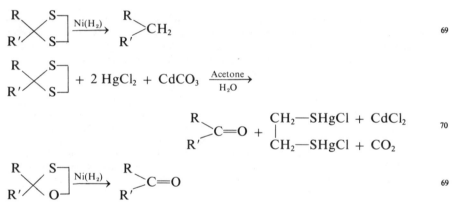

69

70

69

Disulfides may be reduced to thiols by the action of zinc and dilute acid, or they may be oxidized to sulfonic acids by potassium permanganate.

$$RSSR \quad \xrightarrow[H^+]{Zn} \quad 2\ RSH$$

$$RSSR \quad \xrightarrow{KMnO_4} \quad 2\ RSO_3H$$

12.19.3 Sulfoxides

Sulfoxides are usually low melting, highly hygroscopic solids with a noticeable odor similar to sulfides—which, in fact, is often due to trace amounts of

[69]G. R. Pettit and E. E. van Tamelen, Org. Reactions, **12**, 356 (1962).

[70]M. L. Wolfrom, *J. Am. Chem. Soc.*, **51**, 2188 (1929); H. W. Arnold and W. L. Evans, *Ibid*, **58**, 1950 (1936); J. English, Jr., and P. H. Griswold, *Ibid.*, **67**, 2040 (1945).

contaminating sulfide. The sulfoxide grouping is pyramidal and configura-
tionally stable; hence, sulfoxides may exist as geometrical or optical isomers.
Sulfoxides are weak bases; isolable salts are formed with some strong acids
including nitric acid. CAUTION: *Sulfoxides react explosively on contact
with perchloric acid.*

The sulfoxide grouping is readily noted by the characteristic strong
infrared band at 1050 to 1000 cm^{-1} (9.52 to 10.0 μ) (Fig. 12.31). This band is
quite sensitive to hydrogen bonding; the addition of methanol to the solution
causes a significant shift to lower frequency. Likewise, a shift to lower
frequency is observed in going from carbon tetrachloride to chloroform.

In the nmr, hydrogens alpha to sulfoxide appear in the $\delta 2.0$ to 2.5 region.
Methylene protons adjacent to sulfoxide often appear as an *AB* system.

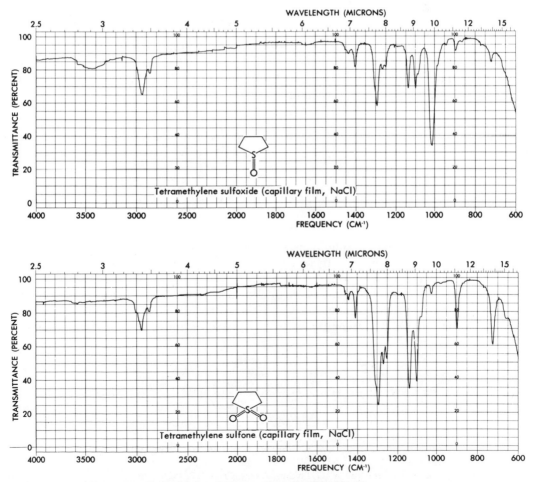

Fig. 12.31. Typical infrared spectra of a sulfoxide and sulfone.

Sulfoxides may be oxidized to sulfones by the methods outlined under sulfides.

$$R_2SO \xrightarrow{[Ox]} R_2SO_2$$

Conversion to sulfilimines may be brought about by reaction with sulfonyl isocyanates[71] or N-sulfinylsulfonamines.[72]

$$\begin{matrix} R \\ \diagdown \\ \diagup \\ R' \end{matrix} S{=}O \ + \ CH_3{-}\!\!\!\bigcirc\!\!\!{-}SO_2N{=}G{=}O \ \longrightarrow$$

$$CH_3{-}\!\!\!\bigcirc\!\!\!{-}SO_2N{=}S\!\!\begin{matrix} \diagup R \\ \diagdown \\ R' \end{matrix} \ + \ GO_2$$

$$G = C \text{ or } S$$

Most sulfoxides form crystalline alkoxysulfonium salts when treated with triethyloxonium fluoroborate. These salts may be characterized by their melting points and neutralization equivalents (titrate with aqueous sodium hydroxide to phenolphthalin endpoint.)[73]

$$\begin{matrix} R \\ \diagdown \\ \diagup \\ R' \end{matrix} S{=}O \ + \ (C_2H_5)_3O^+BF_4^- \ \longrightarrow \ \begin{matrix} R \\ \diagdown \\ \diagup \\ R' \end{matrix} S^+{-}OC_2H_5 \ \ BF_4^- \ + \ (C_2H_5)_2O$$

$$\downarrow NaOH$$

$$O{=}S\!\!\begin{matrix} \diagup R \\ \diagdown \\ R' \end{matrix} \ + \ NaCl \qquad + \ C_2H_5OH$$

Like sulfides, sulfoxides also form crystalline complexes with mercuric chloride. However, these complexes are sometimes difficult to purify to constant composition.

Like amine oxides, sulfoxides which contain β-hydrogen may be pyrolyzed to yield olefins.[74] The primary product is an unstable sulfinic acid which usually decomposes to a variety of unisolated sulfur containing materials.

$$R{-}\overset{\overset{\displaystyle O}{\|}}{S}{-}CH_2{-}CH_2{-}R' \xrightarrow{\Delta} [RSOH] + CH_2{=}CH{-}R$$

---

[71]C. King, *J. Org. Chem.*, **25**, 352 (1960).

[72]G. Schultz and G. Kresze, *Angew. Chem.*, **75**, 1022 (1963); J. Day and D. J. Cram, *J. Am. Chem. Soc.*, **87**, 4398 (1965).

[73]C. R. Johnson and W. G. Phillips, *J. Org. Chem.*, **32**, 1926 (1967).

[74]L. Bateman, M. Cain, T. Colclough, and J. I. Cunneen, *J. Chem. Soc.*, 3570 (1962).

12.19.4 Sulfones

Sulfones are colorless, odorless solids; however, a number of the dialkyl sulfones have low melting points.

When the sulfone is the primary functionality in the molecule, the two strong sulfone bands at 1325 and 1175 cm^{-1} (7.55 and 8.51 μ) are the most prominent in the infrared spectrum (Fig. 12.31).

Sulfones are relatively inert substances; the group is not readily subject to chemical alteration. Some sulfones may be reduced to sulfide by lithium aluminum hydride.[75] The sulfone group markedly increases the acidity of α-hydrogens. Sulfone anions, readily formed by action of strong bases (e.g., butyllithium), undergo the expected reactions of halogenation, alkylation, and acylation.

Infrared, nuclear magnetic resonance, and mass spectral data (the SO_2 grouping is transparent down to 185 mμ in the ultraviolet), along with physical constants, constitute the best method of identification.

Aromatic sulfones yield phenols when fused with solid potassium hydroxide.

$$ArSO_2R \xrightarrow[\text{(2)} \ H_3O^+]{\text{(1)} \ KOH, \Delta} ArOH$$

12.19.5 Sulfenic Acids and Derivatives.

RSOH   RSCl

Alkane- or arenesulfenic acid   Sulfenyl chloride

RSOR'   R—S—NR$_2'$

Sulfenate   Sulfenamide

The free sulfenic acids are highly unstable substances; only a few have been characterized. They are produced by the pyrolysis of sulfoxides.

The sulfenyl chlorides (prepared by reaction of chlorine and disulfides) are somewhat more stable. The most commonly encountered ones are 2,4-dinitrobenzene-, and benzene-, and trichloromethane sulfenyl chloride. Derivatives may be made by reaction with amines to produce amides, reaction with alkoxides in a nonpolar solvent (ether) to produce sulfenates, or by radical addition to olefins to produce β-chlorosulfides.

12.19.6 Sulfinic Acids and Derivatives

Alkane- or arenesulfinic acid   Sulfinate

---

[75]F. G. Bordwell and W. H. McKellin, *J. Am. Chem. Soc.*, **73**, 2251 (1951).

The free sulfinic acids are relatively unstable, but their sodium and potassium salts are quite stable. The salts are readily oxidized (permanganate) to sulfonic acid salts.

The sodium or potassium sulfinates may be alkylated to produce sulfones.

$$ArSO_2Na + CH_3I \longrightarrow Ar\underset{\underset{O}{\overset{\|}{}}}{\overset{\overset{O}{\overset{\|}{}}}{S}}CH_3 + NaI$$

The sulfinyl chlorides are readily produced by reaction of the sodium or potassium salts with thionyl chloride. The sulfinyl chlorides undergo the expected reactions with alcohols, etc. The esters (sulfinates), which are capable of existing in optically active forms, readily undergo transesterification with alcohols in the presence of hydrogen chloride, and they react with Grignard reagents to produce sulfoxides.

12.19.7 Sulfonic Acids; RSO₃H
Alkane- or Arenesulfonic Acids

Sulfonic acids are strong acids, comparable to sulfuric acid. The free acids and their alkali metal salts are soluble in water and insoluble in nonpolar organic solvents (ether, etc.).

The infrared spectrum of sulfonic acids exhibits three SO bands—1250 to 1160 cm^{-1}, 1080 to 1000 cm^{-1}, and 700 to 610 cm^{-1} (8.00 to 8.62 μ, 9.26 to 10.0 μ, and 14.28 to 16.39 μ). The first of these is the most intense.

The free acids are very hygroscopic and do not give sharp melting points. The acids are listed in *Tables of Identification of Organic Compounds* in order of increasing melting point of the sulfonamide.

The preparation of *S-benzylthiuronium sulfonates* from free sulfonic acids and their salts is highly recommended because of the ease with which they are made and the extensive data available.

$$RSO_3Na + C_6H_5CH_2SC(NH_2)_2{}^+ Cl^- \longrightarrow$$
$$C_6H_5CH_2SC(NH_2)_2{}^+ RSO_3{}^- + NaCl$$

12.19.7a Procedure: S-Benzylthiuronium sulfonates

To a solution of 0.5 g of the salt in 5 ml of water, add several drops of 0.1 N hydrochloric acid. Add a cold solution of 1 g of *S*-benzylthiuronium chloride in 5 ml of water. Cool in an ice bath, collect and recrystallize the derivative from aqueous ethanol.

If the free acid is available, dissolve 0.5 g in 5 ml of water and neutralize to phenolphthalein endpoint with sodium hydroxide. Then add 2 to 3 drops of 0.1 N hydrochloric acid and proceed as before.

Probably the most commonly employed method for the derivation of sulfonic acids and salts is the *conversion to sulfonamides* via the sulfonyl chloride.

$$RSO_3Na + PCl_5 \longrightarrow RSO_2Cl + POCl_3 + NaCl$$

$$RSO_2Cl + 2\,NH_3 \longrightarrow RSO_2NH_2 + NH_4Cl$$

12.19.7b Procedure: Sulfonamides

Place 1.0 g of the free acid or anhydrous salt and 2.5 g of phosphorous pentachloride in a small, dry flask equipped with a reflux condenser. Heat in an oil bath at 150° for 30 minutes. Cool and add 20 ml of benzene; then warm and stir to effect complete extraction. [If desired, the sulfonyl chloride may be isolated by removal of the solvent (benzene) or chloroform). Recrystallize from petroleum ether.] Add the benzene solution slowly, with stirring, to 10 ml of concentrated ammonia solution. If the sulfonamide precipitates, isolate it by filtration. Otherwise, it may be isolated by evaporation of the benzene layer. Recrystallization may be accomplished from water or ethanol.

Substituted sulfonamides may be obtained by treatment of the foregoing sulfonyl chloride with amines in the usual manner.

The preparation of *arylamine salts* of sulfonic acids is represented by the following equation.

$$RSO_3{}^-Na^+ + CH_3-\!\!\!\left\langle\!\!\bigcirc\!\!\right\rangle\!\!-NH_3{}^+Cl^- \longrightarrow$$

$$CH_3-\!\!\!\left\langle\!\!\bigcirc\!\!\right\rangle\!\!-NH_3{}^+ RSO_3{}^- + NaCl$$

12.19.7c Procedure: Arylamine salts

Dissolve 0.5 g of the sulfonic acid or its salt in a minimum volume of boiling water, and add a saturated aqueous solution of 1 g of *p*-toluidine hydrochloride. Cool, and recrystallize from hot water containing a drop of concentrated hydrochloric acid or from dilute ethanol.

Salts of other aromatic amines may be prepared in a similar manner; those of aniline, *o*-toluidine, pyridine, and phenylhydrazine have been employed.

12.19.8 Sulfonyl Chlorides; RSO₂Cl

The simple aliphatic and a number of the aromatic sulfonyl chlorides are liquids. High molecular weight aliphatic and substituted aromatics are solids. They are insoluble in water and soluble in most organic solvents. Their vapors have a characteristic irritating odor and are lacrymatory. In the infrared, strong bands are observed at 1360 cm^{-1} and 1170 cm^{-1} (7.35 μ and 8.55 μ).

Sulfonyl chlorides are easily converted to amides or substituted amides by reaction with concentrated ammonia solution or with amines in the presence of aqueous alkali (see Sec. 12.14.2 on amines).

12.19.9 Alkyl Sulfonates; RSO₂R'

The sulfonate group is an excellent leaving group from carbon; consequently, these esters, which may be liquids or crystalline solids, are highly reactive in nucleophilic displacement reactions.   Probably the most commonly encountered ones are the *p*-toluenesulfonates (*tosylates*) (Fig. 12.32).

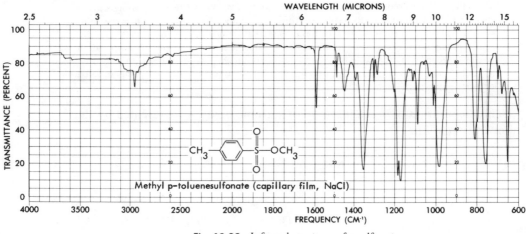

Fig. 12.32. Infrared spectrum of a sulfonate.

Most sulfonates are solvolyzed by heating in water, alcohols, or acetic acid to yield the sulfonic acid and alcohols, ethers, or acetates, respectively. Primary and some secondary alkyl sulfonates (especially tosylates), can be reduced to corresponding hydrocarbons with lithium aluminum hydride.

$$RO_3S-\!\!\left\langle\!\!\bigcirc\!\!\right\rangle\!\!-CH_3 \;+\; LiAlH_4 \;\longrightarrow\; RH \;+\; CH_3-\!\!\left\langle\!\!\bigcirc\!\!\right\rangle\!\!-SO_3^-$$

The primary and secondary sulfonates may be used as general alkylating agents to form quaternary ammonium salts, sulfonium salts, etc.

12.19.10 Sulfonamides; RSO₂NH₂

Most sulfonamides are colorless crystalline solids of high melting point and insoluble in water. The primary and secondary sulfonamides are soluble in aqueous alkali (Hinsberg test).   In the infrared sulfonamides exhibit the

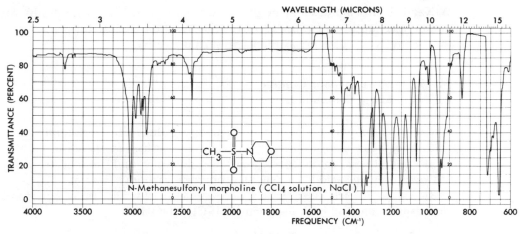

Fig. 12.33. Infrared spectrum of a sulfonamide.

appropriate —N—H stretching and deformation bands, as well as asymmetric [1370 to 1330 cm^{-1} (7.30 to 7.52 μ)] and symmetric [1180 to 1160 cm^{-1} (8.47 to 8.62 μ)] stretching of the sulfonyl grouping (Fig. 12.33).

A very sensitive color test for sulfonamides has been developed employing N,N-dimethyl-α-naphthylamine and nitrous acid.[76]

The only general method for characterization of sulfonamides is *hydrolysis to the sulfonic acid and amine.* The hydrolysis may be effected in a number of ways. Refluxing with 25% hydrochloric acid requires 12 to 36 hours. Eighty percent sulfuric acid or a mixture of 85% phosphoric and 80% sulfuric may also be used. The later method is particularly efficient.

$$CH_3\!-\!\bigcirc\!-\!SO_2\!-\!NHR \xrightarrow{H^+} CH_3\!-\!\bigcirc\!-\!SO_2H + RNH_3^+$$

12.19.10a Procedure: Hydrolysis of sulfonamides

In a test tube mix 1 ml of 80% sulfuric acid with 1 ml of 85% phosphoric acid. Add 0.5 g of the sulfonamide and heat the mixture to 160° for 10 min or until the sulfonamide has dissolved. A dark, viscous mixture will result. Cool and add 6 ml of water. Continue cooling and, while stirring, render the solution alkaline with 25% sodium hydroxide solution. Separate the amine by extraction or distillation. Prepare an appropriate derivative of the amine (benzoyl) and of the sodium sulfonate (S-benzylthiuronium salt).

The sodium salts of primary and secondary sulfonamides may be alkylated by refluxing with an appropriate alkyl halide or sulfate in ethanol.

$$RSO_2NHNa + R'X \longrightarrow RSO_2NHR' + NaX$$

---

[76]C. Hackmann, *Deut. med. Wechschr.*, **72**, 71 (1947); *Chem. Abstracts*, **41**, 4824 (1947).

N-Xanthylsulfonamides may be prepared from primary sulfonamides by reaction with xanthydrol (Sec. 12.15.1c); the reaction is slow (up to 1.5 hr), and relatively few data are available on such derivatives.

12.19.11 Miscellaneous Organic Sulfur Compounds

The most commonly encountered *thioacids* are thioacetic and thiobenzoic acid. Such acids could, in theory, exist in two forms. In the case of thioacetic acid, these would correspond to thiolacetic and thionacetic acid. However only one acid is known; its properties [carbonyl band at 1720 cm^{-1} (5.81 μ)] and reactions are best accounted for by the thiol structure. (Mercaptoacetic acid has the structure $HSCH_2COOH$).

$$CH_3C\overset{\displaystyle O}{\underset{\displaystyle SH}{<}}\qquad CH_3C\overset{\displaystyle S}{\underset{\displaystyle OH}{<}}$$

Thiolacetic acid   Thionacetic acid

The following equations illustrate characteristic reactions which are of value in the preparation of derivatives of thioacids.

$$RCOSH + C_6H_5NH_2 \xrightarrow{25°} RCONHC_6H_5 + H_2S$$

$$RCOSH + H_2O_2 \longrightarrow RCOOH$$

$$RCOSH + CH_2{=}CHC_6H_5 \longrightarrow RCOSCH_2CH_2C_6H_5$$

$$RCOS^- + R'X \longrightarrow RCOSR' + X^-$$

Thiolesters are readily hydrolyzed by acid or base to give the carboxylic acid and the thiol.

$$RC\overset{\displaystyle O}{\underset{\displaystyle SR'}{<}} \xrightarrow{H_2O} RCOOH + R'SH$$

The carbonyl band of aliphatic thiolesters appears near 1690 cm^{-1} (5.29 μ). Thiolesters are oxidized to sulfonic acids by hydrogen peroxide.[77]

$$CH_3C\overset{\displaystyle O}{\underset{\displaystyle SR}{<}} \xrightarrow{H_2O_2} CH_3COOH + RSO_3H$$

Thionesters hydrolyze to give an acid, an alcohol, and hydrogen sulfide. Relatively few of these esters are known.

$$RC\overset{\displaystyle S}{\underset{\displaystyle OR'}{<}} \xrightarrow{H_2O} RC\overset{\displaystyle O}{\underset{\displaystyle OH}{<}} + R'OH + H_2S$$

---

[77]J. S. Showell, J. R. Russell, and D. Swern, *J. Org. Chem.*, **27**, 2853 (1962).

Because of their instability, *dithioacids* are not well known. The *dithio-esters* are reddish-to-yellow oils, many of which are thermochromic. They are very sensitive to atmospheric oxygen, and they are readily desulfurized by Raney nickel.

$$C_6H_5CH_2\overset{\displaystyle S}{\overset{\|}{-C}}-SCH_3 \xrightarrow{H_2(Ni)} C_6H_5CH_2CH_3$$

Thioamides are among the more common and more stable of the compounds containing a thiocarbonyl group. They form addition compounds with mercuric chloride, are alkylated on sulfur to produce salts (analogous to the *S*-alkylthiuronium salts), and react with many oxidants to produce the corresponding amides.

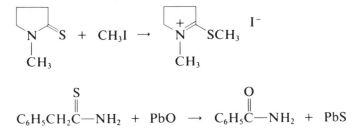

$$C_6H_5CH_2\overset{\displaystyle S}{\overset{\|}{C}}-NH_2 + PbO \longrightarrow C_6H_5\overset{\displaystyle O}{\overset{\|}{C}}-NH_2 + PbS$$

Relatively few *thials* and *thiones* are known. Most are very unstable with respect to dimerization, trimerization, or polymerization.

$$3CH_2S \longrightarrow \text{[cyclic structure with S]}$$

Thiobenzophenone is a fairly stable blue compound with a thiocarbonyl band at 1220 cm^{-1} (8.20 μ).

The nitrile band in aliphatic *thiocyanates* appears as a very strong absorption near 2140 cm^{-1} (4.67 μ); the aromatic derivatives appear at slightly higher frequencies.

$$RSCN + R'OH + HCl \longrightarrow RSC\underset{\displaystyle OR'}{\overset{\displaystyle +NH_2\ Cl^-}{<}} \longrightarrow RS\overset{\displaystyle O}{\overset{\|}{C}}NH_2$$

$$RSCN + [Ox] \longrightarrow RSO_3H + HCN$$

Isothiocyanates have an extremely intense absorption in the infrared in the 2140 to 1990 cm^{-1} (4.67 to 5.26 μ) region. The majority appear at the higher-frequency end of the range.

$$R-N=C=S + R'NH_2 \longrightarrow R-NH-\overset{\displaystyle S}{\overset{\|}{C}}-NH-R'$$

13

Searching the Literature

The problem of locating specific data in the original literature is becoming an ever greater challenge to the chemist, particularly in view of the recent tremendous proliferation and expansion of primary literature. An example of the rate of expansion of the current literature is indicated by the fact that, in 1942, *Chemical Abstracts* covered about 4000 periodicals, whereas in 1967, this number was approximately 10,000. Not only has the number of periodicals increased, but the total number of pages has increased even more dramatically. For example, in this same 25-year period, *The Journal of the American Chemical Society* has increased from 3066 to 7242 pages, *The Journal of Organic Chemistry* from 590 to 4220 pages, and the *Journal of the Chemical Society* (London) from 758 to 5480 pages. It is obvious then, that the retrieval of desired information from the original literature potentially poses a very time-consuming and complex task.

The difficulty of completing a thorough literature search depends on the amount and kind of data available to the searcher, and on the kind of information desired. Specific information on the physical and chemical properties of a specific compound can usually be found with relatively little difficulty. Knowledge of only the molecular formula and functional class requires considerable culling of the literature. Literature reviews concerning chemical reactions are quite involved, entailing a search of the functional group(s) involved, reagents used, general chemical process involved (i.e., hydrolysis, esterification, etc.), and the type of products formed.

Fortunately, several abstract journals are available which provide coverage of the current literature. Figure 13.1 illustrates the various abstract journals and the years of publication. Generally, it is necessary to use only *Chemical Abstracts* and *Chemisches Zentralblatt*, since they provide a very extensive coverage of the literature.

Several extensive compilations of physical and chemical data on com-

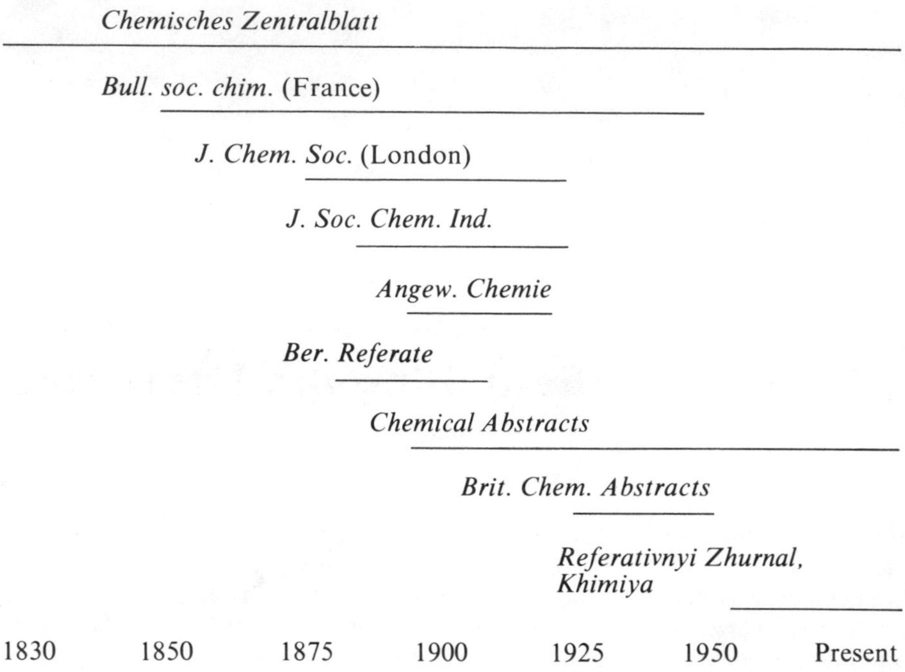

Chemisches Zentralblatt

Bull. soc. chim. (France)

J. Chem. Soc. (London)

J. Soc. Chem. Ind.

Angew. Chemie

Ber. Referate

Chemical Abstracts

Brit. Chem. Abstracts

*Referativnyi Zhurnal,
Khimiya*

| 1830 | 1850 | 1875 | 1900 | 1925 | 1950 | Present |

Fig. 13.1. Abstract journals.

pounds have appeared (see Fig. 13-2). However, none of these compilations
are complete to the present time due to the rapid increase in the rate at which
new compounds are produced.

The ability to readily locate a specific compound in the abstract journals,
and to find certain of the compilations of data on individual compounds,
requires determining the preferred name of the compound. It is advisable to
search under several alternate names of the compound in that changes in

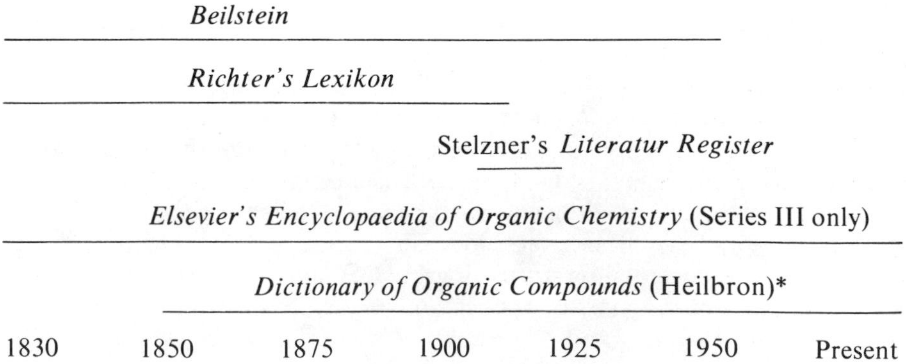

Beilstein

Richter's Lexikon

Stelzner's *Literatur Register*

Elsevier's Encyclopaedia of Organic Chemistry (Series III only)

Dictionary of Organic Compounds (Heilbron)*

| 1830 | 1850 | 1875 | 1900 | 1925 | 1950 | Present |

*Contains only selected compounds

Fig. 13.2. Compilations of data on individual compounds.

nomenclature may have occurred during the span of time covered by the abstract journal. When dealing with particularly complex heterocyclic compounds, the chemist is advised to refer to *The Ring Index* by A. M. Patterson, L. T. Capell, and D. F. Walker [published by the Chemical Abstracts Service of the American Chemical Society (1960)] for the preferred name and numbering system for substituted derivatives. The formulas of molecules in the indices of the abstract journals, and some of the data compilations, are arranged following two basic systems: the Hill and the Richtor systems, which will be outlined in the following sections.

When undertaking a literature search, it is usually advisable to begin with the most recent and comprehensive indices. An exception to this generalization might be in attempting to find the physical properties of a specific compound; here, Beilstein might be consulted first. In general, the more recent literature is more likely to contain the most reliable data and provide more extensive coverage of the literature.

The various abstract journals and compilations of data on specific compounds will be outlined, and methods for their use will be discussed. The extent of the discussions are, of necessity, limited, and the student is referred to the more comprehensive literature guides listed at the end of this chapter.

13.1   ABSTRACT JOURNALS

The two most important abstract journals are *Chemical Abstracts* and *Chemisches Zentralblatt*. These two journals provide nearly the same data, the only minor differences appearing in the methods of indexing. Specific compound information is most readily located by use of the subject and formula indices. Author and patent indices are also provided, the former being particularly useful in locating or tracing a particular scientist's research.

13.1.1 Chemical Abstracts

Chemical Abstracts first appeared in 1907. Yearly subject and author indices are continuous since 1907, the formula and patent indices being introduced in 1920 and 1935, respectively. Decennial indices have appeared in 1916, 1926, 1936, 1946, and 1956, with five-year indices appearing in 1961 and 1966.

Specific information on a given compound is most readily located by use of the subject and formula indices. The subject index indicates briefly in what context the compound was used, for example physical characterization, derivatization, or participation in chemical reactions, with reference numbers locating the abstract in *Chemical Abstracts*. Use of the subject index requires the determination of the preferred name for the compound. This can be most readily accomplished by locating the compound in the formula index in which the preferred name for each compound listed is given. When formula indices are not available, all suitable names for the compound must be employed. The methods of nomenclature for certain classes of compounds have been altered several times since the inception of *Chemical Abstracts*, and the searcher must be aware of such changes.

The formula indices are arranged according to the Hill system. Molecular formulas are arranged alphabetically, except when the compound contains carbon; in this case, the carbon is listed first, followed by hydrogen if present. In compounds containing carbon and hydrogen, the remaining elements are listed in alphabetical order. For example, ammonium silver chloroaurate appears as $Ag_2Au_3Cl_{17}H_{24}N_6$ and 5-bromo-4-methyl-2-thiazolecarboxylic acid is listed as $C_5H_5BrNO_2S$. Under each formula, the preferred names of the compounds included are arranged in alphabetical order. Following the names are reference numbers locating the abstract.

Author indices are used when a specific work, or series of articles in an area, by a given scientist is to be located. The patent indices list only patent numbers and reference numbers to the abstract.

The abstracts provided by *Chemical Abstracts* generally contain only a portion of the essential material contained in the original paper. If at all possible, the original literature should be consulted for the desired information.

13.1.2 *Chemisches Zentralblatt*

Chemisches Zentralblatt is the oldest abstract journal, first appearing in 1830. A number of minor changes in the title and in the numbering of yearly volumes have occurred.

The information provided by *Chemisches Zentralblatt* is nearly identical with that provided by *Chemical Abstracts*. The main difference is in the organization of the formula indices which employ the Richter system of classification of carbon-containing compounds, beginning with the sixth collective index (1922 to 1924). Compounds are tabulated according to their molecular formula, beginning with those compounds containing one carbon atom. The sequence of elements in combination with carbon are given in the sequence H, O, N, Cl, Br, I, F, S, and P, the remaining elements of the periodic table following in alphabetical order. A "guide number" is assigned each formula. This guide number is composed of an Arabic numeral referring to the number of carbon atoms present and a Roman numeral referring to the number of other elements. For example, 5-bromo-4-methyl-2-thiazolecarboxylic acid would be assigned a guide number 5V. One then simply looks at the top of the pages in the formula index until the guide number is located. The compound may then be located on that page.

13.1.3 *Miscellaneous Abstract Journals*

Several other abstract journals are available. Many of these cover special, and thus more limited, areas. However, owing to their more limited nature, they may well cover highly specialized journals, which in some cases may be rather obscure. These abstract journals include *Journal of the Chemical Society*, 1871–1925; *Journal of the Society of Chemical Industry*, 1882–1925; *British Chemical Abstracts*, 1926–1953; *Bulletin de da Societe chimique de France*, 1863–1946; *Berichte der deutschen chemischen Gesellschaft (Referate)*, 1880–95; *Nuclear Science Abstracts*, 1948–　, and *Science Abstracts*, 1898–

13.2 CURRENT LITERATURE COVERAGE SERVICES

The yearly indices of the various abstract journals usually lag significantly behind the publication of the abstracts. To browse the abstract journals requires considerable time. In order to help keep the scientist abreast of the current literature, current literature surveying journals have been introduced recently.

Chemical Titles, published by the Chemical Abstracts Service of the American Chemical Society, publishes a keyword index of titles of papers from selected chemical journals. Key words, words of greatest importance and meaning, are selected from a title, there may be several such words in a single title, and they are arranged in alphabetical order. This allows the literature reviewer to review the up-to-date literature in a rapid fashion.

Current Contents, Chemical Sciences, a weekly publication of the Institute for Scientific Information, provides a reproduction of the tables of contents of chemical journals appearing that week. Journal article titles can be scanned and checked against a listing of representative structural diagrams which are displayed adjacent to the contents pages. A key-word index and author index listing authors' addresses are also featured.

Current Chemical Papers, published by the Chemical Society (London) compiles only the titles and authors. The list of titles is divided into areas of interest, thus saving considerable time involved in reviewing all the titles.

13.3 COMPILATIONS OF CHEMICAL AND/OR PHYSICAL DATA

Numerous attempts have been made to tabulate the physical data, and, in many cases, the chemical data of specific compounds. Many of these compilations are included in Fig. 13.2 illustrating their time-period coverage.

Several of the most important of these compilations will be discussed individually.

13.3.1 Beilstein

Perhaps the most important compilation of chemical and physical data on individual compounds, and hence frequently referred to as the "organic chemists' bible," is Beilstein's *Handbuch der organischen Chemie.* Beilstein is composed of a main series (29 volumes), which covers all organic compounds through 1909, and three supplementary series covering the years 1910–1919, 1920–1929, and 1930–1949 (the last was not complete at the time of publication of this text).

The classification of compounds in Beilstein is complex and will not be discussed in detail in this text. For a detailed description of the classification system, the student is referred to pages 1 through 46 of volume 1 of the main series, and to "A Brief Introduction of Beilstein's *Handbuch der organischen Chemie,*" by E. H. Huntress and published by John Wiley & Sons, Inc., 1938.

Locating a desired compound in Beilstein is greatly facilitated by the availability of a compound formula index covering the main series and the first two supplementary series. The formula index gives the names of various isomers possessing a given formula, and it gives a reference number. The reference number is composed of a Roman numeral indicating the volume (band) number and an Arabic numeral which indicates on what page of the main series the compound reference will be found. Occasionally, a compound will not be found in the main series; the Beilstein reference number then indicates the volume and page on which the compound would appear if contained in the main series. The supplementary series contain the Beilstein reference number at the top center of each page; the student then simply searches for the reference number to locate a given compound in the supplementary series.

The citation for each compound includes bibliographical citations, mode of preparation or occurrence, physical properties, chemical reactions, physiological properties, uses, methods of analysis for, and derivatives of.

13.3.2 Richter's Lexikon

Richter's *Lexikon* contains a compilation of all known organic compounds up to the date of its publication in 1910. The compounds are classified according to the Richter system. The four volumes of Richter's *Lexikon* are classified as follows: Volume I, C_1 to C_9 (1I to 9III); Volume II, C_9 to C_{13} (9III to 13IV); Volume III, C_{13} to C_{20} (13IV to 20II); and Volume IV, C_{20} and higher (20II to 1039IV). Richter's *Lexikon* contains melting and boiling point data and a few references to the original literature.

13.3.3 Stelzner's Literatur-Register

Stelzner's *Literatur-Register* is a compilation of physical data on compounds covering the period of 1910–1921. The accompanying references are more extensive than those contained in Richter's *Lexikon*. The entries are classified according to the Richter system. The five volumes of this work are divided on the basis of time covered and not compound classification, as is Richter's *Lexikon*; Volume I covers the period 1910–1911; Volume II, 1912–1913; Volume III, 1914–1915; Volume IV, 1916–1918; and Volume V, 1919–1921.

13.3.4 Compilations by Mulliken and Huntress

A series of volumes prepared by Mulliken and/or Huntress have appeared tabulating physical properties for a variety of compounds. The original volume in this series by Mulliken, entitled *A Method for the Identification of Pure Organic Compounds*, Vol. 1 (1904), also describes certain techniques of structure determination.

The compounds appearing in these volumes are classified according to the types of atoms present, possession of color and functional groups, and their physical properties. Compounds are divided into *orders* on the basis of the types of atoms present and into two *suborders*, depending on whether the compound is colored or colorless. Further subdivision into *genus* is based on the types of functional groups present. The final classifications by *divisions* and *sections* are dependent on the physical state (solid or liquid) and solubility or density data, respectively. In the respective sections, compounds are listed in order of increasing melting or boiling points. In the last two titles listed below, location numbers are assigned to each section (see the introduction section of these two volumes), and the compounds are then located in the text by use of these location numbers. These volumes provide data on physical properties, derivatives, literature references, and frequency on the chemical properties.

1. Mulliken, S. P., *Identification of Pure Organic Compounds*, Vol. II. New York, John Wiley & Sons, Inc., 1916.
2. Mulliken, S. P., *Identification of Pure Organic Compounds*, Vol. IV. New York, John Wiley & Sons, Inc., 1922.
3. Huntress, E. H., and Mulliken, S. P., *Identification of Organic Compounds of Order I*, (containing C, H or C, H, and O). New York, John Wiley & Sons, Inc., 1941.
4. Huntress, E. H., *Organic Chlorine Compounds*, (order III containing chlorine in addition to C, H, and C, H, and O). New York, John Wiley & Sons, Inc., 1948.

13.3.5 Melting Point Tables
of Organic Compounds

Melting Point Tables of Organic Compounds [second supplemented edition by
W. Utermark and W. Schicke, and published by Interscience Publishers,
Inc., New York, N. Y. (1963)] is a tabulation of physical and chemical
properties of pure organic compounds. The tables present the melting point,
boiling point, crystalline form, other physical properties, chemical reactions,
and the Beilstein reference number. Although the compounds are listed in
order of increasing melting point, a formula index facilitates locating a
specific compound.

Another compilation essentially identical to the one just described is
Schmelzpunktstabellen by R. Kempf and F. Kutter (Druck und Verlag von
Friedr. Vieweg & Sohn Akt.-Ges., Braunschweig, Germany, 1928).

13.3.6 Elsevier's Encyclopaedia
of Organic Compounds

Elsevier's *Encyclopaedia of Organic Compounds* (Elsevier Publishing Company, Amsterdam) was originally designed "to give complete information on
all chemical and physical properties and on the most important physiological
properties of organic compounds." The accomplishment of such a proposed
design is truly a tremendous task and has proved, apparently, to be too
great. Although the area of organic chemistry was originally divided into
four major series (Series I, acyclic compounds; Series II, cyclic noncondensed
compounds; Series III, cyclic condensed compounds; and Series IV, heterocyclic compounds) only Volumes 12, 13, and 14 of Series III have appeared.
Volume 12 covers bicyclic compounds, Volume 13 covers tricyclic compounds, and Volume 14 covers tetracyclic and higher polycyclic compounds.
Several supplements have appeared in recent years, making a total of 19
individual covers.

The compounds in each category are divided according to the total
number of carbon atoms and, if cyclic, to the smallest cycle first. Further
classification is according to the types of functional groups present in the
order hydrocarbons, halides, nitrogen-containing, hydroxy compounds,
carbonyl compounds, carboxylic acids, and followed by compounds containing S, Se, Te, P, As, and other metals. The information provided for
each compound includes the physical properties, mode of formation or
source, references, chemical reactions, and derivatives. Each greater system
(particular carbon structure system) is introduced by a detailed systematic
survey covering the various modes of formation of the carbon structure.

13.3.7 Dictionary of Organic Compounds

The *Dictionary of Organic Compounds* (Oxford University Press), commonly
referred to as *Heilbron*, is a compilation of physical and chemical proper-

ties, with references to the original literature, of selected compounds. The first edition published in 1934, 1936, and 1937 was composed of three volumes. The Dictionary is currently composed of five volumes plus supplements (fourth edition published in 1965), covering approximately 12,000 compounds. The compounds are listed alphabetically according to the I.U.P.A.C. 1957 nomenclature rules.

13.3.8 Merck Index

Merck and Company, Rahway, N. J., publishes periodically a compilation of the physical and chemical properties of drugs and chemicals related to or used in the drug industry. This compilation is particularly useful in this rather limited area.

13.3.9 Other Sources of Physical Properties of Compounds

The following references provide additional, but quite limited, sources of data on the physical properties of compounds. No attempt to describe the use of these references will be made, since they are all quite self-explanatory.

1. Cheronis, N. D., Entrikin, J. B., and E. M. Hodnett, *Semimicro Qualitative Organic Analysis*, 3rd ed. New York, Interscience Publishers, Inc., 1965.
2. *Chemical Rubber Handbook*. Chemical Rubber Publishing Company, Cleveland, Ohio.
3. *Tables for Identification of Organic Compounds*. Chemical Rubber Publishing Company, Cleveland, Ohio, 1960.
4. *Lange's Handbook of Chemistry*. Handbook Publishers, Inc., Sandusky, Ohio.
5. Shriner, R. L., Fuson, R. C., and Curtin, D. Y., *The Systematic Identification of Organic Compounds*, 5th ed. New York, John Wiley & Sons, Inc., 1964.

13.3.10 Spectra Compilations

Extensive compilations of spectra in the various regions of the electromagnetic spectrum have been published. The references for such compilations appear at the end of the respective spectroscopy sections (Secs. 3.10, 4.11, 5.18, 6.6, 8.9).

13.4 REFERENCE TEXTS ON THE USE OF THE LITERATURE

Several more extensive texts are available describing the use of library resources. These are now listed.

1. Crane, E. J., Patterson, A. M., and Marr, E. B., *A Guide to the Literature of Chemistry*, 2nd ed. New York, John Wiley & Sons, Inc., 1957.
2. Mellon, M. G., *Chemical Publications—Their Nature and Their Use*. New York, McGraw-Hill Book Company, 1958.
3. Soule, B. A., *Library Guide for the Chemist*. New York, McGraw-Hill Book Company, 1938.

A series of articles by Hancock [*J. Chem. Ed.*, **45,** 193, 260, 336 (1968)] are highly recommended for further reading.

14

Structural Problems

14.1 INTRODUCTION

The successful solution of a structure problem requires recognition of the various partial structures indicated to be present in the molecule by the various bits of chemical and physical (primarily spectral) data. It is the assembling of these bits of data which is the most difficult part of the problem. It is necessary to be able to recognize certain common features (atoms or groups of atoms) which may appear in two or more of the partial structures which may be used as cement in putting the various fragments together. There are certain techniques which are very useful in solving structure problems which will be outlined by the use of specific examples.

Sample Problem 1. The ir and nmr spectra of Compound **A** are given in Figs. 14.1 and 14.2. Compound **A**, $C_{10}H_{14}O$, reacts with the Lucas reagent to produce an insoluble oil.

Oxidation of **A** with chromic acid produces **B**, $C_{10}H_{12}O$, which produces an orange-red precipitate when treated with 2,4-dinitrophenylhydrazine. The ir and nmr spectra of **B** are provided in Figs. 14.3 and 14.4.

Deduce the structure of **A** and **B**.

The first objective is to determine the number of sites of unsaturation in **A** by use of the equation

$$N = \frac{\sum_i n_i(v_i - 2) + 2}{2}$$

We find

$$N = \frac{10(4\text{-}2) + 14(1\text{-}2) + 1(2\text{-}2) + 2}{2} = \frac{8}{2} = 4$$

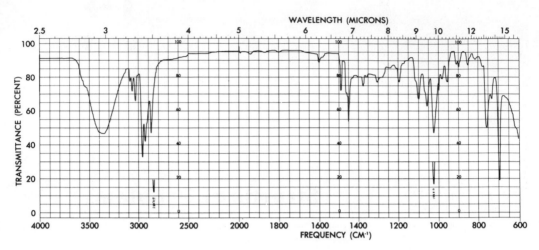

Fig. 14.1. Ir spectrum (film) of unknown **A**, sample problem 1.

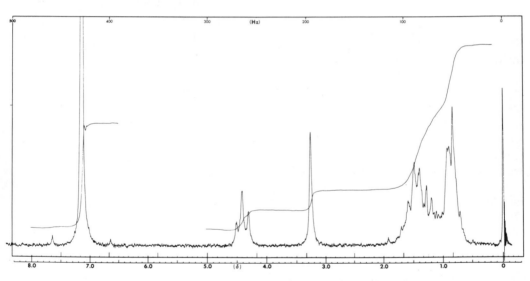

Fig. 14.2. Nmr spectrum of unknown **A**, sample problem 1.

The infrared spectrum of **A** indicates the presence of a hydroxyl group (3360 cm^{-1}). The nmr spectrum shows a singlet at $\delta 7.13$ with a relative intensity of 5 consistent with a monoalkyl substituted benzene. The slightly distorted triplet at $\delta 4.41$ (relative intensity 1) is substantially deshielded and is adjacent to a methylene (CH_2) group. This system probably represents the partial structure

$$
\begin{array}{c}
\text{H} \\
| \\
\text{C—OH} \\
| \\
\text{CH}_2
\end{array}
$$

WAVELENGTH (MICRONS)

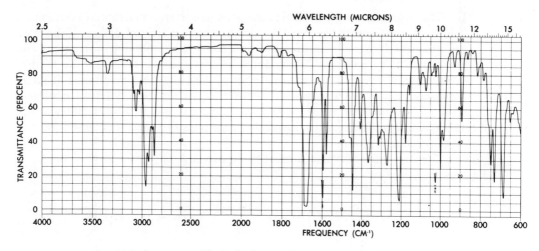

Fig. 14.3. Ir spectrum (film) of unknown **B**, sample problem 1.

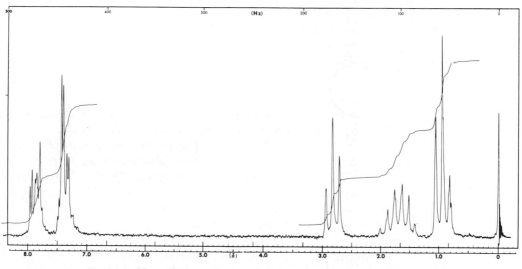

Fig. 14.4. Nmr spectrum of unknown **B**, sample problem 1.

The positive Lucas test indicates the presence of an alcohol (or other functional group capable of reacting with the Lucas reagent such as an epoxide) which reacts via a $S_{N}1$ mechanism (tertiary, allylic, or benzylic). This chemical test suggests that a benzene ring is attached to the carbinol carbon (this will be confirmed by the ir data of **B**). The nmr spectrum displays a highly distorted system at $\delta 0.83$ (relative intensity of 3) which probably represents one methyl bonded to a methylene group (this pattern is typical of an $A_2 B_3$ spin system).

The ir spectrum of **B** indicates the presence of an aryl ketone (1680 cm^{-1}) and is consistent with the partial structure of **A** indicated above. The nmr spectrum of **B** readily clarifies the nature of the remaining portion of the molecule. The triplet at $\delta 2.81$ (relative intensity of 2) represents the methylene group adjacent to the carbonyl group. The high field resonance at $\delta 0.98$ represents a methyl attached to methylene of not too different chemical shift. The multiplet at $\delta 1.72$ (relative intensity of 2) must represent the methylene group bonded to the methyl group and to the methylene attached to the carbonyl group.

The structures of **A** and **B** must therefore be

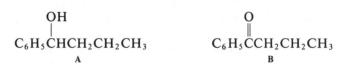

Sample Problem 2. Sarkomycin, $C_7H_8O_3$, is a very unstable acid isolated from a certain type of bacteria. Sarkomycin displays characteristic peaks in the infrared region at 3450 (broad), 1710 (intense and broad), 1639, and 886 cm^{-1}. Sarkomycin adsorbs one mole of hydrogen giving a new acid which displays peaks in the infrared region at 3450 (broad), 1739, and 1710 cm.$^{-1}$ Ozonolysis of sarkomycin gives formaldehyde and an acid (not isolated) which on oxidation with potassium permanganate gives succinic acid. Devise suitable structures for sarkomycin and all degradation products.

Sarkomycin contains four sites of unsaturation of which one site is present in the carboxyl group which must be present in the molecule.

The infrared spectrum of sarkomycin indicates the presence of a bonded —OH which, in conjunction with the carbonyl absorption at 1710 cm^{-1}, represents the —COOH group. The bands at 1639 and 886 cm^{-1} indicate the presence of a carbon-carbon double bond which is probably a terminal methylene group which we will represent as =CH_2. This is confirmed by the formation of formaldehyde on ozonolysis.

Hydrogenation of sarkomycin produces a dihydroderivative which still displays —COOH absorption at 3450 and 1710 cm^{-1}. However, a new band has been generated in the carbonyl region at 1739 cm^{-1}, which is characteristic of either a five-membered cyclic ketone or a six-membered or acyclic ester. The latter possibility is ruled out on the basis that an insufficient number of oxygen atoms are present to accommodate both a —COOH group and a —C$\overset{\displaystyle O}{\diagup}$O— group of an ester. The production of the 1739 cm^{-1} band on hydrogenation requires the conversion of an α,β-unsaturated five-membered ring ketone (1710 cm^{-1}) to a saturated system. Therefore the partial structure **1** must be present in sarkomycin. It is therefore only necessary to locate the position of the —COOH group.

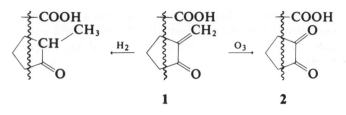

1   **2**

Ozonolysis of sarkomycin must give partial structure **2**, which on further oxidation gives succinic acid. This oxidation involves the loss of two carbon atoms, probably as carbon dioxide. Only two possibilities exist for the place-ment of the —COOH group in **2**, either adjacent to the carbonyl group (**3**), or the 4-position (**4**). Permanganate oxidation of **4**, in the enol form in which it probably exists, would give **5**, which on decarboxylation would give

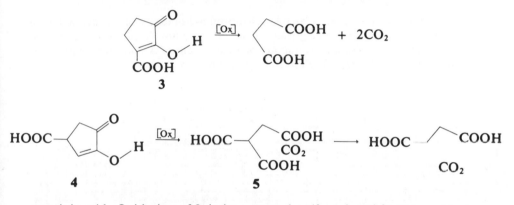

3

4   **5**

succinic acid. Oxidation of **3**, in its expected preferred enol form, would give succinic acid directly.

It is obvious that no unique structure can be proposed for sarkomycin strictly on the basis of the information provided. Three structures are possi-ble and are given below (**6, 7,** and **8**). The indicated instability of sarko-

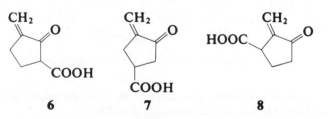

6   **7**   **8**

mycin would tend to favor **6**, in which a β-ketoacid is present which may undergo decarboxylation. However, no pertinent data concerning this point are given. The availability of the nuclear magnetic resonance spectrum would readily indicate the correct structure. The reader is encouraged to consider the expected differences in the resonance spectra of **6, 7,** and **8.**

In cases in which no unique structure can be derived, the student should consider what additional information may be needed and how to derive that information (either chemical or physical). An important part of any structure problem is recognizing the limitations of the data, and what additional data are needed.

14.2   PROBLEMS

In the following problems infrared data will be given only in cm^{-1}. The nmr data will be given as δ. Immediately following the δ value the multiplicity[1], coupling constant, and relative integrated intensity will be given in parentheses, e.g., $\delta 2.09$ $(d, J = 6.0$ Hz; 3).

1. Reaction of commercially available 1,4-dibromo-2-butene with diethyl malonate in ethanol containing two equivalents of sodium ethoxide produces compound **A** in high yield. The high resolution mass spectrum of **A** indicates a molecular formula of $C_{11}H_{16}O_4$. The infrared spectrum of **A** is shown in Fig. 14.5. At

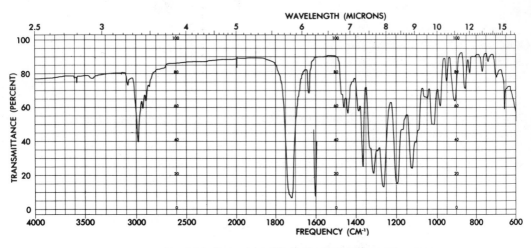

Fig. 14.5. Ir spectrum (film) of unknown **A**, problem 1.

450° compound **A** is cleanly isomerized to **B**. The nmr spectrum of **B** is reproduced in Fig. 14.6. Suggest structures for **A** and **B**.

2. Reaction of cyclobutanone with isopropenyl magnesium chloride in tetrahydrofuran, followed by treatment with aqueous ammonium chloride provides compound **X**. When **X** is reacted with t-butyl hypochlorite in chloroform solution in the dark, compound **Y** is produced. The ir spectrum of **Y** exhibits no significant absorption above 3000 cm^{-1} but does contain a very strong band at 1743

---

[1] Multiplicity is indicated by s (singlet), d (doublet), t (triplet), and m (complex multiplet).

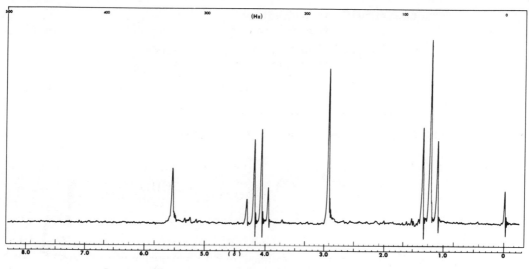

Fig. 14.6. Nmr spectrum of unknown **B**, problem 1.

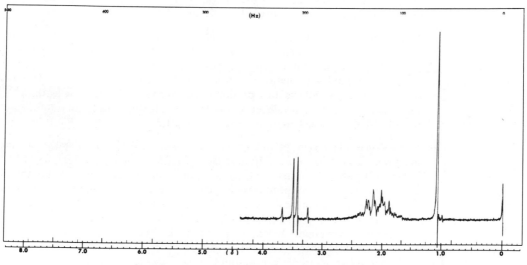

Fig. 14.7. Complete nmr spectrum of unknown **Y**, problem 2.

cm^{-1}. The complete nmr spectrum of **Y** is shown in Fig. 14.7. Suggest structures for **X** and **Y**.

3. Beginning with enamine **I** the following reaction sequence was achieved:

$$C_6H_5\overset{\overset{\displaystyle N(CH_3)_2}{|}}{C}=CH_2 + CH_3SO_2Cl \xrightarrow[\text{Benzene}]{\text{Triethylamine}} C_{11}H_{15}NO_2S$$

I   **II**

$$\downarrow \begin{array}{l} \text{(1) } H_2O_2 \\ \text{(2) } \Delta \end{array}$$

$$CH_3SO_2CH_3 + V \xleftarrow[\Delta]{OH^-} IV \xleftarrow{OH^-} C_9H_8O_2S$$

                                                   III

Compound **III** Ir: 1600, 1310 (s), 1123 (s) cm^{-1}.
 Nmr: $\delta 4.77$ (s, 2), 6.92 (s, 1), 7.42 (s, 5).

Compound **IV** Ir: 1675 (s), 1325 (s), 1151 (s) cm^{-1}.
 Nmr: $\delta 3.12$ (s, 3), 4.16 (s, 2), 7.8 (m, 5).

Compound **V** Mp 121°; neut. equiv. 121.

Suggest structures for **II** through **V**.

4. Compound **A** was isomerized to **B** ($\nu_{C=O}^{max}$ 1712 cm^{-1}) by boiling in water under

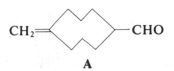

$$CH_2 =\!\!\!\!<\!\!\!\!>\!\!\!\!-\!CHO$$

A

a nitrogen atmosphere. Treatment of **B** with hydrogen peroxide gives **C**. Titration of **C** with 0.01 N sodium hydroxide indicated a neutralization equivalent of 167. The nmr of **C** included a singlet (3H) at $\delta 1.1$ and no absorptions due to olefinic hydrogens. Suggest structures of **B** and **C**.

5. Compound **I**, $C_{11}H_{16}O$, displays absorption in the infrared region at 1683 and 1600 cm^{-1}. The ultraviolet spectrum displays maxima at 239 (6500) and 320 mμ (75). Kuhn-Roth determination indicates the presence of 0.8 C—CH$_3$.

Treatment of **I** with methyl magnesium iodide produces an alcohol **II** which on dehydration yields as the predominant product **III**, $C_{12}H_{18}$ with λ_{max}^{EtOH} 283 mμ (16,000). Dehydrogenation of **III** with selenium produces 1,3-dimethylnaphthalene. Propose structures for **I**, **II**, and **III**.

6. Extraction of black Hemiptera bugs produces a colorless oil **I** possessing a characteristic nauseating odor. Microanalysis (*Anal*. Found: C, 75.02; H, 10.62) indicates the absence of nitrogen, sulfur, and halogen. Treatment of **I** with 2,4-dinitrophenylhydrazine produces a yellow precipitate. Compound **I** gives positive tests with both Tollen's and Fehling's solutions. The infrared spectrum of **I** displays characteristic peaks at 1725, 1380, and 975 cm^{-1}.

Synthetic compound **I** can be produced by thermal rearrangement of 3-(1-pentenyl) vinyl ether.

Propose a structure for compound **I**.

7. The reaction of benzylidine aniline ($C_6H_5CH=NC_6H_5$) with acetic anhydride produces compound **A**, $C_{17}H_{17}NO_3$. The infrared spectrum of **A** is devoid of absorption above 3050 cm^{-1}, but displays strong bands at 1750 and 1676 cm^{-1}. Treatment of **A** with sodium ethoxide in ethanol for two hours produces ethyl acetate, benzaldehyde, and acetanilide. Extended treatment of **A** with ethanol produces compound **B**, $C_{17}H_{19}NO_2$. The infrared spectrum of **B** displays no absorption above 3050 cm^{-1}, but shows an intense band at 1652 cm^{-1}. Propose structures for **A** and **B**.

8. Oxidation of pyrrole with hydrogen peroxide in acetic acid gives compound **I** (*Anal*. Found: C, 63.99; H, 6.54; N, 18.61). The infrared spectrum of **I** displays

characteristic bands at 3460 to 3490 and 1700 cm^{-1}, and λ_{max}^{EtOH} 215 mμ. Compound **I** does not react with carbonyl reagents. Chromic acid oxidation of **I** yields succinic acid. Basic permanganate oxidation followed by esterification with diazomethane yields methyl 2-pyrrolecarboxylate. Further oxidation of **I** with hydrogen peroxide in acetic acid yields succinimide. Ozonolysis of **I** gives pyroglutamic acid (2-pyrrolidone-5-carboxylic acid). Propose a structure for **I**.

9. Treatment of 2-benzyl-3-phenylpropyltrimethylammonium iodide with sodium amide in liquid ammonia produces several $C_{16}H_{16}$ compounds which are separated by preparative gas-liquid chromatography. The nmr spectra of three of these compounds are given below. Identify each compound.

   Compound **I**   Nmr: δ3.2 (broad s, 2), 4.8 (broad s, 1), and 7.1 (broad s, 5).

   Compound **II**   Nmr: δ1.80 (d, J = 1.0 Hz, 3), 3.62 (d, J = 1.2 Hz, 2), 6.58 (m, 1), and 7.2 to 7.3 (m, 10).

   Compound **III** Nmr: δ0.96 (m, 2), 1.28 (m, 1), 1.73 (m, 1), 2.65 (AB portion of an ABX system, J's = 12.0, 6.5, and 6.0 Hz, 2), and 7.2 (m, 10).

10. Pyrolysis of 6,7-dimethylbicyclo [3.2.0] hept-6-en-2-ol at 405° produces two compounds **A** and **B** with molecular formula $C_7H_{14}O$. Compounds **A** and **B** display the following spectral properties:

    Compound **A**   Ir: 1670 and 1650 cm^{-1}.
                     Uv: λ_{max}^{EtOH} 238 mμ (11,300).
                     Nmr: δ1.18 (d, J = 6.8 Hz, 3), 1.85 (m, 4), 1.91 (d, J = 1.5 Hz, 3), 2.50 (m, 3), and 5.72 (m, 1).

    Compound **B**   Ir: 1705 and 1661 cm^{-1}.
                     Uv: λ_{max}^{EtOH} 295 mμ (450).
                     Nmr: δ1.80 (s, 6), 2.00 (m, 2), 2.40 (m, 4), and 3.02 (s, 2).

11. The addition of hydrogen bromide to allene in the dark produces four compounds which are separable by gas-liquid chromatography. Identify each of the four compounds from the information provided.

    Compound **I**   Mass spectrum: 120 (100) and 122 (97.7).
                      Nmr: δ2.3 (s, 3) and 5.3 (broadened AB doublets, J = 1.3 Hz, 2).

    Compound **II**   Mass spectrum: 200 (100), 202 (195), and 204 (95.5)
                      Nmr: δ2.2.

    Compound **III** Mass spectrum: 240 (100), 242 (195), and 244 (95.5).
                      Nmr: δ1.95 (s, 3) and 3.4 (s, 2).

    Compound **IV**   Mass spectrum: 240 (100), 242 (195), and 244 (95.5).
                      Nmr: δ1.8 (s, 3) and 3.2 (AB doublets, J = 11.5 Hz, 2).

12. Compound **I** reacts with 2,4-dinitrophenylhydrazine and possesses the following spectral properties:

                     Ir: 1687 and 1595 cm^{-1}.

Uv: 222 mμ.
Nmr: δ1.2 (t, J = 6.5 Hz, 3), 3.4 (m, 2), 5.40 (double d, J = 10 and 2.5 Hz, 1), 6.30 (double t, J = 10 and 5 Hz, 1) and 11.2 (d, J = 2.5 Hz, 1).

Mild oxidation of **I** gives **II**. The spectral properties of **II** are as follows:

Ir: 3200, 1678, and 1595 cm^{-1}.
Uv: 209 mμ.
Nmr: δ1.20 (t, J = 6.5 Hz, 3), 2.40 (m, 2), 5.5 (d, J = 10.2 Hz, 1), 6.55 (double t, J = 10.2 and 5.5 Hz, 1), and 12.2 (s, 1).

Deduce structures for **I** and **II**.

13. Extraction of the leaves of *Trewia nudiflora* Linn. produces compound **A**, $C_7H_6N_2O$, which contains one *N*-methyl group. The spectral properties of **A** are as follows:

Ir: 2200, 1670, and 1605 cm^{-1}.
Uv: 206 (log ϵ 4.27), 254 (4.32), and 306 mμ (3.89).
Nmr: δ3.57 (s, 3), 6.55 (d, J = 9.5 Hz, 1), 7.34 (double d, J = 9.5 and 3.0 Hz, 1), and 7.90 (d, J = 3.0 Hz, 1).

Hydrolysis of **A** in 57% sulfuric acid yields an acid **B**, $C_7H_7NO_3$, which possesses the following spectral properties:

Ir: 3020, 1695, 1660, and 1605 cm^{-1}.
Uv: 206 (4.14), 255 (4.20), and 300 mμ (3.72).

Esterification of **B** with diazomethane produces the methyl ester $C_8H_9NO_3$ (Ir: 1720, 1662, and 1610 cm^{-1}).
Propose structures for compounds **A** and **B**.

14. Compound **A**, $C_9H_{16}O$, displays characteristic bands in the infrared region at 3570, 1620, and 800 cm^{-1}. Mild oxidation with chromic acid produces a ketone **B**, ν_{max} 1680 cm^{-1} and λ_{max}^{EtOH} 239 mμ. Ozonolysis of **B** followed by oxidative workup produces a ketoacid **C**, $C_8H_{14}O_3$. Treatment of **C** with hypobromite produces a diacid **D**, $C_7H_{12}O_4$. The nmr spectrum of **D** displays peaks at δ1.1 (s, 3), 2.2 (s, 2), and 10.8 (s, 1).
Propose structures for compounds **A**, **B**, **C**, and **D**.

15. Treatment of acetophenone with hydrogen chloride and hydrogen sulfide in absolute ethanol produces "anhydro acetophenone disulfide," $C_{24}H_{22}S_2$. The nmr spectrum of "anhydro acetophenone disulfide" displays the following peaks: δ1.45 (s, 3), 1.79 (s, 3), 6.24 (s, J < 0.1 Hz, 1), and an unsymmetrical multiplet with centers at 7.22 and 7.55 (15). Ozonolysis of the compound in methylene chloride followed by oxidation with formic acid and hydrogen peroxide produces benzoic acid as the only product. Ozonolysis in methylene chloride-pyridine (nonoxidative workup) did not produce acetophenone.
Propose a suitable structure for "anhydro acetophenone disulfide."

16. Compound **I**, $C_{15}H_{26}O_2$, possesses the following spectral properties:

Ir: 3300, 1380, 1120, and 825 cm^{-1}.

Nmr: $\delta 0.76$ (*d*, *J* = 7.2 Hz, 3), 0.94 (*d*, *J* = 7.2 Hz, 3), 1.28 (*s*, 3), 1.67 (broadened *s*, 3), 3.85 (apparent *t*, 1), 4.95 and 5.19 (singlets which are easily exchanged in deuterium oxide, 1 each), and 5.34 (broadened *s*, 1) in addition to other unresolved systems.

Treatment of **I** with selenium produces 1,6-dimethyl-4-isopropyl-naphthalene. Chromic acid oxidation of **I** gave **II**, $C_{15}H_{24}O_2$, possessing the following spectral properties:

Ir: 3500, 1680, 1129, and 810 cm^{-1}.
Uv: 236 mμ (14,640).

Compound **I** forms a monoacetate **III**, $C_{17}H_{28}O_3$ (Ir: 3500 and 1740 cm^{-1}), which on treatment with phosphorous oxychloride gives as the major product **IV**, $C_{17}H_{26}O_2$. The spectral properties of **IV** are as follows:

Ir: 1740, 1650, 893, and 825 cm^{-1}.
Nmr: $\delta 0.79$ (*d*, *J* = 7.2 Hz, 3), 0.94 (*d*, *J* = 7.2 Hz, 3), 1.72 (broadened *s*, 3), 2.03 (*s*, 3), 4.95 (apparent *t*, 1), 5.05 (*m*, 2) and 5.44 (*d*, 1), in addition to other unresolved systems.

Treatment of **I** with *p*-toluenesulfonyl chloride produced a monotosylate which on elimination of *p*-toluenesulfonic acid gave compound **V**, $C_{15}H_{24}O$ [Uv: λ_{max} 260 mμ (2606)].

Propose structures for **I** through **V**. What aspects of the total structure are unknown? How would you derive the information necessary to complete the structure assignments?

17. Irradiation of α-tropolone methyl ether (**I**) with ultraviolet light in the presence

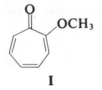

I

of oxygen and a sensitizer in methanol produces a very unstable compound **II**. The spectral properties of **II** are as follows:

Ir: 3500, 1725, and 1625 cm^{-1}.
Uv: λ_{max}^{EtOH} 220 mμ (note: this is at slightly longer wavelength than predicted).
Nmr: $\delta 3.8$ (*s*) (as the only recognizable system due to apparent decomposition of **II**).

Chromic acid oxidation of **II** gives **III** (*Anal.* Found: C, 54.38; H, 5.06), which possesses the following spectral properties:

Ir: 1728, 1672, and 1622 cm^{-1}.
Uv: λ_{max}^{EtOH} 240 mμ.

Nmr: $\delta 3.9$ (s, 3), 6.9 and 7.4 (AB doublets, J = 16 Hz, 1 each).

Compound **II** slowly rearranges on standing to give **IV**, isomeric with **II**, which possesses the following spectral properties:

Ir: 1740, 1713, 1680, and 1626 cm^{-1}.
Nmr: $\delta 2.2$ ($A_2 B_2$ system, 4), 3.4 (s, 3), 3.8 (s, 3), and 6.87 and 7.30 (AB doublets, J = 16 Hz, 1 each).

Attempted chromatography of **II** results in the formation of **V** (*Anal.* Found: C, 57.07; H, 4.83) which possesses the following spectral properties:

Ir: 1784, 1744, 1615, and 1590 (weak) cm^{-1}.
Nmr: $\delta 3.6$ (d, J = 8 Hz, 2), 3.8 (s, 3), 5.6 (t, J = 8 Hz, 1) and 6.3 and 7.5 (AB doublets, J = 6 Hz, 1 each).
Mass spectrum: 168 (parent peak).

Deduce structures for **II**, **III**, **IV**, and **V**.

18. Treatment of 1,1-di-*t*-butylprop-2-yn-1-ol with methyl lithium followed by acetyl chloride produces compound **I** (Ir: 3320 (sharp), 2270 and 1743 cm^{-1}). Treatment of **I** with potassium *t*-butoxide in *t*-butanol produces products **II** ($C_{15}H_{28}O$) and **III** ($C_{11}H_{18}$). The spectral properties of these products are as follows:

Compound **II**   Ir: 1935 and 1955 cm^{-1}.
                   Nmr: $\delta 1.20$ (s, 18), 1.26 (s, 9), and 3.75 (s, 1).

Compound **III**   Ir: 2020 and 1990 cm^{-1}.
                   Uv: 273 (185,000), 308 (28,900), and 316 mμ (33,000).
                   Nmr: $\delta 1.28$ (s) (only peak).

Compound **III** on melting was converted to **IV**. The spectral properties of **IV** are as follows:

Ir: 1950 and 1925 cm^{-1}.
Uv: 250 (18,500), 258 (16,300), 268 (13,200), 298 (1720), and 316 mμ (2040).
Nmr: $\delta 1.20$ (s) (only peak).
Mass spectrum: 300 (molecular ion).

19. The reaction of bis(trifluoromethyl)diazomethane [$(CF_3)_2 CN_2$] with *cis*-2-butene produced three isomeric $C_7 H_8 F_6$ compounds **A**, **B**, and **C**. The spectral properties of these three compounds are given below.

Compound **A**   Ir: No characteristic absorption in the 1500 to 1700 cm^{-1} region.
                  Nmr: ^{19}F resonance spectrum: $\delta 5.50$ (s).
                  ^{1}H resonance spectrum: $\delta 0.8$ to 1.5 (m).

Compound **B**   Ir: No characteristic absorption in the 1500 to 1700 cm^{-1} region.
                  Nmr: ^{19}F resonance spectrum: $\delta 0.84$ (q, 1) and -9.83 (q, 1).
                  ^{1}H resonance spectrum: $\delta 0.8$ to 1.5 (m).

Compound **C**   Ir: 1647 cm^{-1}.
Nmr: ^{19}F resonance spectrum: δ10.1 (m)
^{1}H resonance spectrum: δ1.19 (t, J = 7.0 Hz, 3), 2.11 (broadened s, 3), and 2.46 (q, J = 7.0 Hz with further unresolved fine splitting, 2).

Propose structures for **A**, **B**, and **C**.

20. Compound **A**, $C_{10}H_{14}O$, gives positive ketone tests. The spectral properties of **A** are as follows:

Ir: 1780 and 801 cm^{-1}.
Uv: 310 mμ (260).
Nmr: δ1.12 (s, 3), 1.19 (s, 3), 1.75 (broadened s, 3), 2.55 (m, 3), 4.03 (m, 1), and 5.46 (m, 1).

Hydrogenation of **A** produces **B**, $C_{10}H_{16}O$, possessing the following spectral properties:

Ir: 1778 cm^{-1}.
Uv: 309 mμ (53).
Nmr: δ0.96 (s, 3), 1.03 (d, J = 6 Hz, 3), 1.22 (s, 3), 1.5 to 2.7 (m, 5), 3.54 (t, actually overlapping doublets with both J's = 7 Hz, 2).

Treatment of **A** with methanolic potassium hydroxide produces an acidic compound **C**, $C_{10}H_{16}O_2$. Treatment of **C** with diazomethane gives the corresponding methyl ester **D**, identical with a product obtained from **A** on treatment with methanolic hydrogen chloride. The spectral properties of **D** are as follows:

Ir: 1734 and 802 cm^{-1}.
Nmr: δ1.08 (s, 3), 1.13 (s, 3), 1.72 (broadened s, 3), 1.85 to 2.40 (m, 5), 3.67 (s, 3), and 5.20 (m, 1).

Treatment of **A** with 30% sulfuric acid, or **C** with p-toluenesulfonic acid in refluxing toluene, or **B** with trifluoroperacetic acid, produces, among other products, compound **E**, $C_{10}H_{16}O_2$. The spectral properties of **E** are as follows:

Ir: 1776 cm^{-1}
Nmr: δ1.15 (d, J = 6 Hz, 3), 1.20 (s, 3), 1.30 (s, 3), 1.50 to 2.70 (m, 6) and 4.80 (overlapping doublets with J's = 6 Hz, 1).

Propose structures for **A** through **E**.

21. The reaction of allyl alcohol with a catalytic amount of palladium chloride at approximately 40° followed by vacuum distillation yields a mixture of two isomeric $C_6H_{10}O_2$ compounds **A** and **B**. Both compounds readily decolorize bromine water and react with acetyl chloride.
Compound **A** displays the following spectral properties:

Ir: 3410 (broad), 3073, 2985, 1661, and 895 cm^{-1}.
Nmr: δ2.47 (m, 2), 3.55 (d, 2), 4.00 (m, 1), 4.30 (s, 1) overlapping 4.30 (m, 2), and 4.95 (m, 2).

Compound **B** displays the following spectral properties:

Ir: 3410 (broad), 3010, 1661, and 840 cm $^{-1}$.
Nmr: δ1.74 (finely split s, 3), 3.55 (d, 2), 4.30 (s, 1) overlapping 4.46 (m, 2), 4.72 (m, 1), and 5.44 (m, 1).

Esterification of **A** with acetic anhydride in pyridine results in the formation of a monoacetate whose nmr spectrum differed from that of **A** only by displacement of the δ3.55 doublet to δ4.05 and the disappearance of the δ4.30 singlet. Similar treatment of **B** results in the displacement of the δ3.55 doublet of **B** to δ4.05 and the disappearance of the δ4.30 singlet.

Hydrogenation of **A** and **B** results in the uptake of one mole of hydrogen and produces identical products **C**, $C_6H_{12}O_2$. The spectral properties of **C** are as follows:

Ir: 3400 and 1030 cm $^{-1}$ (in addition to numerous other weak peaks in the 1200 to 900 cm $^{-1}$ region).
Nmr: δ0.96 (d, 3), 1.24 (m, 1), 1.93 (m, 1), 2.22 (m, 1), 3.26 (m, 1), 3.39 (d, 2), 3.75 (m, 1), 3.81 (m, 1), and 4.15 (s, 1).

Propose structures for **A**, **B**, and **C**.

22. The condensation of pulegone (**I**) with ethyl acetoacetate in alcoholic sodium ethoxide yields "pulegone acetone." Pulegone acetone has been formulated by various authors as **IIa, IIb, IIc,** and **IId**.

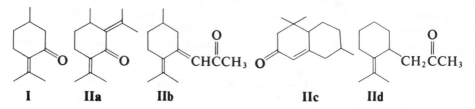

I   **IIa**   **IIb**   **IIc** **IId**

Pulegone acetone does not give a positive iodoform test. Pulegone acetone absorbs two equivalents of hydrogen to produce a saturated alcohol. The spectral properties of the pulegone acetone are as follows:

Ir: 1675 and 1640 cm $^{-1}$.
Uv: λ_{max}^{EtOH} 241 mμ (15,200).
Nmr: δ0.97 (d, 3), 1.11 (s, 6), 2.0 to 2.2 (m, 8), 2.30 (s, 2), and 6.1 (s, 1).

Ozonation of pulegone acetone yields an acid **III** ($C_{13}H_{20}O_4$) which on heating yields **IV** ($C_{12}H_{18}O_2$). Compound **IV** shows no olefinic hydrogens in the nmr. The infrared spectrum of **IV** displays a strong band at 1775 cm $^{-1}$ and a weak band at 1686 cm $^{-1}$.

Choose the correct structure for pulegone acetone and derive suitable structures for compounds **III** and **IV**.

23. Treatment of γ,γ-dimethylallyl phenyl sulfide with ethyl trichloroacetate and sodium methoxide in petroleum ether produces as the predominant product compound **I**, $C_{12}H_{14}SCl_2$. The spectral properties of **I** are as follows:

> Ir: 1640, 995,and 925 cm^{-1}.
> Nmr: $\delta 1.50$ (s, 6), 5.17 (AB portion of an ABX system, 2), 6.34 (X portion of the ABX system, 1), and 7.20 (m, 5).

Chromatography of **I** on silica gel produces **II**, $C_{12}H_{14}OS$, along with hydrogen chloride. The spectral properties of **II** are as follows:

> Ir: 1690, 1630, 960, and 910 cm^{-1}.
> Nmr: $\delta 1.30$ (s, 6), 5.20 (AB portion of an ABX system, 2), 6.06 (X portion of the ABX system, 1), and 7.23 (m, 5).

Treatment of **II** with refluxing aqueous-ethanolic sodium hydroxide produces, among others, compound **III**, $C_6H_{10}O_2$, possessing the following spectral properties:

> Ir: 3000 to 2500 (broad), 1710, 1645, 960, and 910 cm^{-1}.
> Nmr: $\delta 1.30$ (s, 6), 5.17 (AB portion of an ABX system, 2), 6.16 (X portion of the ABX system, 1), and 12.50 (s, 1).

Propose structures for **I, II**, and **III**.

24. Compound **A**, $C_{11}H_{16}O$, is a natural product found in various mint oils including American spearmint. Compound **A** possesses the following spectral properties:

> Ir: 1717, 1640, 965 cm^{-1}.
> Uv: λ_{max}^{EtOH} 239 mμ.

Catalytic hydrogenation of **A** gives **B**, $C_{11}H_{20}O$. Compound **B** displays a band in the infrared region at 1746 cm^{-1}.

Ozonolysis of **A** followed by oxidative workup gives an acid **C**, $C_9H_{12}O_5$, and a volatile acid **D**. Acid **D** gives a neutralization constant of 59 $\pm$ 1. Acid **C** produces a color with ferric chloride, and reacts with sodium hypoiodite to yield, after acidification, a dibasic acid **E**, $C_8H_{10}O_6$. Acid **E** reacts with alkaline hydrogen peroxide giving as the sole product succinic acid. Acid **C**, after heating with hydrazine in base, gives an acid identified as *n*-nonanoic acid.

Deduce structures for compounds **A** through **E**.

25. Adrenaline (**I**) on treatment with acetic anhydride in aqueous bicarbonate solu-

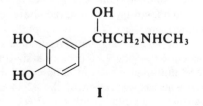

I

tion gives compound **II**. Compound **II** is converted to **III** on treatment with thionyl chloride in benzene followed by refluxing with anhydrous methanolic hydrogen chloride in benzene. Compound **III** is not soluble in organic solvents but is reasonably soluble in water. Compound **III** gives an immediate precipitate with aqueous silver nitrate and produces a color with ferric chloride. Dissolution of III in pH 2-5 aqueous buffer followed by recovery of the organic material gives predominantly compound IV, whereas dissolution of II in aqueous bicarbonate produces V. Compound IV is water soluble (organic insoluble) and gives positive tests with ferric chloride and silver nitrate. IV is converted into V on refluxing in pyridine. Compound IV is converted to VI on treatment with methanolic hydrogen chloride. Compound VI is soluble in water, is not appreciably soluble in organic solvents, and gives positive tests with ferric chloride and silver nitrate.

On the basis of the above information, and the spectral information provided below, propose structures for compounds I through V.

Compound II $C_{15}H_{19}NO_6$
Ir: 3220 (broad), 1770, and 1627 cm^{-1}.
Nmr (CDCl$_3$): δ2.07 (s, 3), 2.27 (s, 6), 2.82 (s, 3), 3.58 (AB portion of an ABX system with an overlapping singlet, 3 total), 4.83 (X portion of the ABX system, 1), and 7.23 (m, 3).

Compound III $C_{11}H_{14}ClNO_3$
Ir: 3110 (broad) and 1667 cm^{-1}.
Nmr (D$_2$O): δ2.82 (s, 3), 3.24 (s, 3), 3.22 (m, 2), 4.43 (X portion of an ABX system, 1), and 6.92 (m, 3).

Compound IV $C_{11}H_{16}ClNO_4$
Ir: 3090 (broad) and 1740 cm^{-1}.
Nmr (D$_2$O): δ2.19 (s, 3), 2.82 (s, 3), 3.38 (AB portion of an ABX system, 2), 5.85 (X portion of an ABX system, 1), and 6.92 (m, 3).

Compound V $C_{11}H_{15}NO_4$
Ir (nujol mull): 3120 (broad) and 1624 cm^{-1}.
Nmr: Unavailable due to lack of solubility in all solvents tried.

Compound VI $C_{10}H_{16}ClNO_3$
Ir: 3225 (broad) cm^{-1}.
Nmr (D$_2$O): δ2.70 (s, 3), 3.20 (s, 3) overlapping a multiplet at 3.20 (2), 4.35 (X portion of an ABX system, 1), and 8.1 (m, 3).

26. Reaction of 1,5-cyclooctadiene with sulfur dichloride in methylene chloride produces a 96.6% crude yield of compound A, $C_8H_{12}Cl_2S$. Compound A possesses reactive chlorines, producing an immediate precipitate on treatment with silver nitrate, as well as undergoing very facile nucleophilic displacements (positive sodium iodide test). The nuclear magnetic resonance spectrum of A displays overlapping multiplets at δ2.34 and 2.80 (5) and a six-line pattern at δ4.68 (1). Saturation of the δ2.80 multiplet results in collapse of the δ4.68 multiplet to a triplet with $J = 8$ Hz, whereas saturation of the δ2.34 multiplet results in collapse of the δ4.68 multiplet to a doublet with $J = 3$ Hz.

Treatment of A with lithium aluminum hydride produces compound B, $C_8H_{14}S$. Oxidation of B with sodium metaperiodate in methanol produces a single, homogeneous sulfoxide C, $C_8H_{14}SO$. Raney nickel desulfurization of B produces cyclooctane.

Deduce the structure of A with appropriate stereochemistry.

27. A naturally occurring ketone **I**, $C_{10}H_{14}O$, possesses the following spectral properties:

Ir: 1676 and 1600 cm^{-1}.
Uv: λ_{max}^{EtOH} 253 mμ (does not correlate with Woodward's rules.)
Nmr: $\delta 0.7$ (s, 3), 0.9 (s, 3), 1.6 (m, 2), 2.0 (double d, J = 8 and 3.5 Hz, 1), 2.6 (double d, J = 7 and 3 Hz, 2), 2.3 (d, J = 1.9 Hz, 3), and 5.97 (q, J = 1.0 Hz, 1).

Hydrogenation of **I** produces **II** which displays carbonyl absorption in the infrared region at 1705 cm^{-1} with no intense absorption in the ultraviolet region. The only resolvable diagnostic peaks in the nmr spectrum were singlets at $\delta 0.9$ and 1.0 (3 each) and a doublet at $\delta 1.1$ (3). Compound **II** displays a positive anomalous Cotton effect curve.

Ozonolysis of **I** followed by oxidation produces an optically active acid **III**, $C_9H_{14}O_3$, which on treatment with hypobromite gives an optically inactive diacid **IV**, $C_8H_{12}O_4$.

Propose structures for **I**, **II**, and **III** including the absolute stereochemistry.

28. Compound **X**, $C_7H_{14}O$, is converted to **Y** on treatment with trifluoroperacetic acid. Compound **Y** on basic hydrolysis gives (+)-2-butanol and propionic acid. Compound **X** is converted to **Z**, C_7H_{16}, on treatment with hydrazine and base. Compound **Z** is optically inactive.

What is the structure of **X**, **Y**, and **Z**? Derive the absolute stereochemistry of **X**, **Y**, and the 2-butanol employing Brewster's rules.

29. A naturally occurring sesquiterpene **A**, $C_{15}H_{24}$, possesses the following spectral properties.

Ir: 3080, 2975, 1655, 1383, and 900 cm^{-1}.
Uv: Weak low end absorption.
Nmr: $\delta 0.85$ (s, 3), 1.77 (m, J ≈ 1.5 Hz, 3), 4.85 (m, J ≈ 1.5 Hz, 2), 4.70 (m, J ≈ 1.5 Hz, 2), plus other nonresolved systems.

Dehydrogenation of **A** over selenium produces a hydrocarbon identified as 1-methyl-7-isopropylnaphthalene. (Alkyl groups situated on fully substituted carbon atoms are lost in this process.)

Ozonolysis of **A** produces a diketone **B**, $C_{13}H_{20}O_2$, with ν_{max} 1740 cm^{-1}. The nmr spectrum of **B** displays a characteristic absorption at $\delta 0.8$ (s, 3) and 2.2 (s, 3). **B** is transformed into a ketoacid by careful treatment with sodium hypobromite. **B** is esterified with diazomethane to give a ketoester **C**, $C_{13}H_{20}O_3$.

Compound **C** is not isomerized by treatment with sodium methoxide in refluxing methanol (indicating that all epimerizable groups are in the most stable configuration). Several treatments of **C** with sodium methoxide in methanol-O-D produce **C** containing four deuterium atoms. **C** gives a monobenzylidene derivative on treatment with benzaldehyde and base.

The ketoester displays a negative optical rotatory dispersion curve.

Propose structures for **A**, **B**, and **C** indicating the absolute stereochemistry of each of the structures.

30. Ether extraction of a culture of the Basidiomycete fungus *Fistulina hepatica* produces a mixture of compounds which is separated into nonpolar and polar fractions. Separation of the polar fraction by countercurrent distribution gives a
weakly optically active, colorless, crystalline compound **A** with molecular formula $C_{13}H_{12}O_4$. Kuhn-Roth analysis indicates the presence of one $C-CH_3$.
The spectral properties of **A** are as follows:

Ir: 3300 (broad), 2220, 2175, 2150 cm^{-1}.
Uv: Multitude of sharp, intense bands between 207 and 358 mμ.
Nmr: $\delta 2.1$ (*s*, 3), 3.82 (*d*, 2), and complex absorption in the
3.5 to 5.0 region.

Compound **A** reacts with slightly over three moles of $NaIO_4$ to produce
formaldehyde (identified as its dimedone derivative), an aldehyde ($C_{10}H_4O$) and
other unidentified fragments. Reduction of the aldehyde with sodium borohydride followed by hydrogenation produces 1-decanol.

Treatment of **A** with dilute base produces compound **B** ($C_{13}H_{12}O_4$). The
infrared spectrum of compound **B** displays bands at 3300 (broad), 2200, and
1655 cm^{-1}. The ultraviolet spectrum displays a number of intense bands between 246 and 343 mμ. Compound **B** consumes one mole of periodate.

Ozonolysis of compound **B** produces a neutral compound **C** ($C_5H_8O_5$,
ν_{max} 1742 cm^{-1}) identical by paper chromatography with a compound obtained
by oxidizing D-lyxose with dilute iodine and base followed by acidification.
Compound **C** derived from **B** displays a sign of rotation opposite that of the
derivative derived from D-lyxose and consumes one mole of periodate.

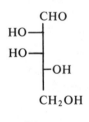

D-lyxose

Derive structures for **A**, **B**, and **C**, including absolute stereochemistry.

31. A naturally occurring optically active compound **A**, $C_{10}H_{16}O$, displays bands in
the infrared region at 1682 and 1624 cm^{-1} with λ_{max}^{EtOH} 237 mμ. Ozonolysis of **A**
in pyridine solution gives **B**. **B** is optically active, does not strongly absorb in the
ultraviolet, displays bands in the infrared at 1710 and 1724 cm^{-1}, and gives
positive aldehyde tests.

Treatment of **B** with potassium permanganate gives an optically inactive
acid **C** with neutralization equivalent of 73 ± 1, and an optically inactive acid **D**
with neutralization equivalent of 88 ± 1. The acid **D** displays a doublet at high
field and a septet at lower field in the nmr spectrum. **C** is converted to a new
substance **E** by heating. Compound **E** displays bands in the infrared region at
1855 and 1780 cm^{-1}.

Treatment of **A** with sodium amalgam and ethanol gives **F** which displays a band in the infrared at 1711 cm^{-1}, does not strongly absorb in the ultraviolet, and gives a positive optical rotatory dispersion curve.

Derive structures for **A** through **F** including the absolute configuration.

32. The condensation of diethyl malonate with methyl vinyl ketone produces a neutral compound **A** and an acid **B**.

Compound **A** is hydrolyzed with aqueous sulfuric acid to give acid **C** with melting point 76.5 to 77.0°. Compound **C** possesses a neutralization equivalent of 182 and gives a positive iodoform test. The spectral properties of **C** are as follows:

> Ir: 3330, 1710, 1685, and 1618 (w) cm^{-1}.
> Uv: λ_{max}^{EtOH} 248 mμ (7000).
> Nmr: δ1.81 (s, 3), 2.15 (s, 3), 11.29 (s, 1) in addition to other unresolved systems.

Compound **B**, $C_{11}H_{12}O_4$, which melts at 187 to 188° with evolution of a gas, is soluble in sodium bicarbonate, produces a deep red color with ferric chloride, and gives a negative iodoform test. Treatment of **B** with diazomethane produces **D**, $C_{13}H_{16}O_4$, whose spectral properties are as follows:

> Ir: 1752, 1667, and 1622 cm^{-1} (all intense).
> Nmr: δ3.7 (2 overlapping singlets, 3 each), 5.3 (m, 1), 5.1 (s, 1), 1.7 (broadened s, 3), and other unresolved systems.

Heating **B** at 210° for 5 minutes produces **E** and carbon dioxide. **E** is soluble in dilute sodium hydroxide and gives an intense color with ferric chloride. **E** displays bands in the infrared region at 3230, 1707, and 1608 cm^{-1}. The nmr spectrum of **E** retained the δ5.3 multiplet and δ1.7 broadened singlet observed in the spectrum of **D**.

Oxidation of **E** gives a diacid **F**, which on isomerization in base gives a diacid identical with the diacid formed from **C** on treatment with hypoiodite.

Derive structures for compounds **A** through **F**.

33. A hydrocarbon mixture isolated from a plant source is fractionated giving a portion with boiling point 80 to 110° at 1 mm. This hydrocarbon fraction was oxidized with chromic acid producing compound **A**, $C_{15}H_{20}O$. The spectral properties of **A** are as follows:

> Ir: 1685, 1618, 1602, 1573, 1515, 1450, 1370 (broad), and 827 cm^{-1}
> Uv: λ_{max}^{EtOH} 239 mμ (11,500).
> Nmr: δ1.28 (d, J = 7 Hz, 3), 2.06 (s, 3), 2.10 (s, 3), 2.25 (s, 3), 2.43 (AB portion of an ABX system, 2), 3.20 (m, 1), 6.00 (s, 1), and 7.0 (broad s, 4).
> Optical rotation: $[\alpha]_D$ + 68°.

A 2.0-gram portion of **A** is ozonized for 8 hours in ethyl acetate (65 mg ozone per hour). The reaction mixture was oxidized with hydrogen peroxide in formic acid to give a crude acid **B**. Compound **B** is purified via the S-benzylthiouronium salt and is esterified with ethanol and sulfuric acid to give the corresponding ethyl ester **C**, $C_{13}H_{18}O_2$. The spectral properties of **C** are as follows:

> Ir: 1742, 1604, 1574, 1515, 1450, 1410, 1370, and 819 cm^{-1}.
>
> Uv: λ_{max}^{EtOH} 251 (shoulder, 290), 257 (370), 263 (425), and 272 mμ (350).
>
> Nmr: δ1.13 (t, $J = 7$ Hz, 3), 1.23 (d, $J = 7$ Hz, 3), 2.25 (s, 3), 2.40 (AB portion of an ABX system, 2), 3.16 (m, 1), 3.97 (q, $J = 7$ Hz, 2), and 7.0 (broad s, 4).

Compound **C** is reduced to **D** with lithium aluminum hydride. **D** on treatment with *p*-toluenesulfonyl chloride in pyridine gives **E**. **E** in turn is treated with lithium aluminum hydride to give **F**, $C_{11}H_{16}$. The spectral properties of **F** are as follows:

> Ir: 1579, 1515, 1375, and 794 cm^{-1}.
>
> Uv: λ_{max}^{EtOH} 252 (shoulder, 130), 259 (185), 264 (230), and 272 mμ (180).
>
> Nmr: δ0.78 (t, $J = 7$ Hz, 3), 1.18 (d, $J = 7$ Hz, 3), 1.55 (m, 2), 2.29 (s, 3), 2.7 (m, 1), and 7.0 (broad s, 4).
>
> Optical Rotatory Dispersion Curve: $[\phi]_{589} + 25°$, $[\phi]_{578} + 26°$, $[\phi]_{546} + 30°$, $[\phi]_{436} + 54°$, $[\phi]_{364} + 91°$, and $[\phi]_{313} + 127°$.

Deduce structures for **A** through **F** including the absolute stereochemistry utilizing Brewster's rules.

APPENDIX I

Vapor Pressure-Temperature Nomograph

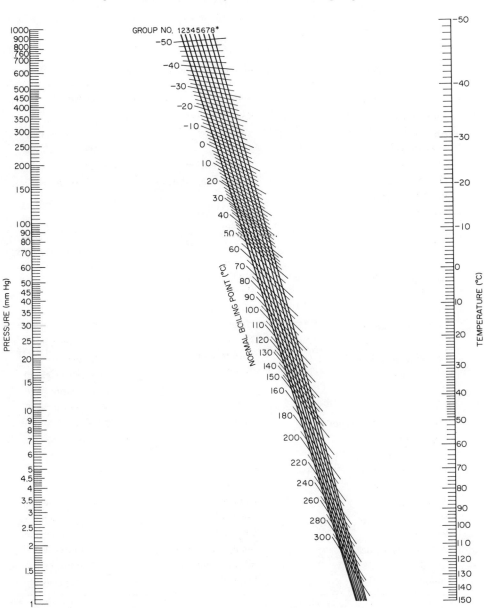

From *Industrial and Engineering Chemistry*, Vol. 38, p. 320, 1946. Copyright 1946 by the American Chemical Society and reproduced by permission of the copyright owner.

*See Table A.1.

Table A.1 Groups of Compounds Represented in Nomographs.

Group 1	Group 3	Group 5
Anthracene	Acetaldehyde	Ammonia
Anthraquinone	Acetone	Benzyl alcohol
Butylethylene	Amines	Methylamine
Carbon disulfide	Chloroanilines	Phenol
Phenanthrene	Cyanogen chloride	Propionic acid
Sulfur monochloride	Esters	
Trichloroethylene	Ethylene oxide	
	Formic acid	**Group 6**
Group 2	Hydrogen cyanide	Acetic anhydride
	Mercuric chloride	Isobutyric acid
Benzaldehyde	Methyl benzoate	Water
Benzonitrile	Methyl ether	
Benzophenone	Methyl ethyl ether	
Camphor	Naphthols	**Group 7**
Carbon suboxide	Nitrobenzene	
Carbon sulfoselenide	Nitromethane	Benzoic acid
Chlorohydrocarbons	Tetranitromethane	Butyric acid
Dibenzyl ketone		Ethylene glycol
Dimethylsilicane		Heptanoic acid
Ethers	**Group 4**	Isocaproic acid
Halogenated hydrocarbons	Acetic acid	Methyl alcohol
Hydrocarbons	Acetophenone	Valeric acid
Hydrogen fluoride	Cresols	
Methyl ethyl ketone	Cyanogen	
Methyl salicylate	Dimethylamine	**Group 8**
Nitrotoluenes	Dimethyl oxalate	*n*-Amyl alcohol
Nitrotoluidines	Ethylamine	Ethyl alcohol
Phosgene	Glycol diacetate	Isoamyl alcohol
Phthalic anhydride	Methyl formate	Isobutyl alcohol
Quinoline	Nitrosyl chloride	Mercurous chloride
Sulfides	Sulfur dioxide	*n*-Propyl alcohol

Vapor Pressure-Temperature Nomograph

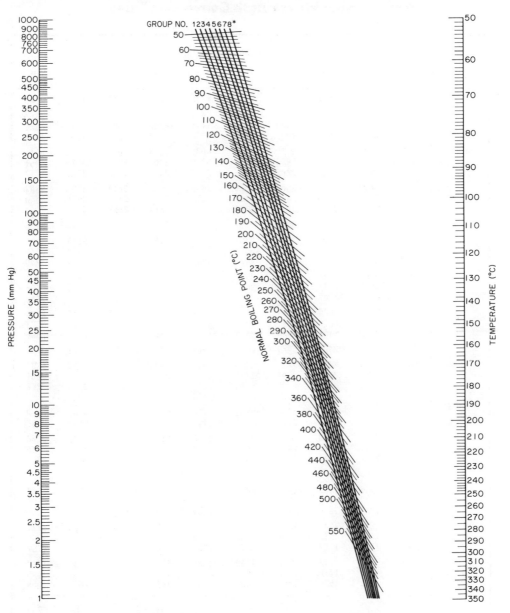

From *Industrial and Engineering Chemistry*, Vol. 38, p. 320, 1946. Copyright 1946 by the American Chemical Society and reproduced by permission of the copyright owner.

*See Table A.1.

APPENDIX II

Wave-number Wavelength Conversion Table

Wavelength (μ)	Wave number (cm⁻¹)									
	0	1	2	3	4	5	6	7	8	9
2.0	5000	4975	4950	4926	4902	4878	4854	4831	4808	4785
2.1	4762	4739	4717	4695	4673	4651	4630	4608	4587	4566
2.2	4545	4525	4505	4484	4464	4444	4425	4405	4386	4367
2.3	4348	4329	4310	4292	4274	4255	4237	4219	4202	4184
2.4	4167	4149	4132	4115	4098	4082	4065	4049	4032	4016
2.5	4000	3984	3968	3953	3937	3922	3906	3891	3876	3861
2.6	3846	3831	3817	3802	3788	3774	3759	3745	3731	3717
2.7	3704	3690	3676	3663	3650	3636	3623	3610	3597	3584
2.8	3571	3559	3546	3534	3521	3509	3497	3484	3472	3460
2.9	3448	3436	3425	3413	3401	3390	3378	3367	3356	3344
3.0	3333	3322	3311	3300	3289	3279	3268	3257	3247	3236
3.1	3226	3215	3205	3195	3185	3175	3165	3155	3145	3135
3.2	3125	3115	3106	3096	3086	3077	3067	3058	3049	3040
3.3	3030	3021	3012	3003	2994	2985	2976	2967	2959	2950
3.4	2941	2933	2924	2915	2907	2899	2890	2882	2874	2865
3.5	2857	2849	2841	2833	2825	2817	2809	2801	2793	2786
3.6	2778	2770	2762	2755	2747	2740	2732	2725	2717	2710
3.7	2703	2695	2688	2681	2674	2667	2660	2653	2646	2639
3.8	2632	2625	2618	2611	2604	2597	2591	2584	2577	2571
3.9	2564	2558	2551	2545	2538	2532	2525	2519	2513	2506
4.0	2500	2494	2488	2481	2475	2469	2463	2457	2451	2445
4.1	2439	2433	2427	2421	2415	2410	2404	2398	2392	2387
4.2	2381	2375	2370	2364	2358	2353	2347	2342	2336	2331
4.3	2326	2320	2315	2309	2304	2299	2294	2288	2283	2278
4.4	2273	2268	2262	2257	2252	2247	2242	2237	2232	2227
4.5	2222	2217	2212	2208	2203	2198	2193	2188	2183	2179
4.6	2174	2169	2165	2160	2155	2151	2146	2141	2137	2132
4.7	2128	2123	2119	2114	2110	2105	2101	2096	2092	2088
4.8	2083	2079	2075	2070	2066	2062	2058	2053	2049	2045
4.9	2041	2037	2033	2028	2024	2020	2016	2012	2008	2004
5.0	2000	1996	1992	1988	1984	1980	1976	1972	1969	1965
5.1	1961	1957	1953	1949	1946	1942	1938	1934	1931	1927
5.2	1923	1919	1916	1912	1908	1905	1901	1898	1894	1890
5.3	1887	1883	1880	1876	1873	1869	1866	1862	1859	1855
5.4	1852	1848	1845	1842	1838	1835	1832	1828	1825	1821
5.5	1818	1815	1812	1808	1805	1802	1799	1795	1792	1789
5.6	1786	1783	1779	1776	1773	1770	1767	1764	1761	1757
5.7	1754	1751	1748	1745	1742	1739	1736	1733	1730	1727
5.8	1724	1721	1718	1715	1712	1709	1706	1704	1701	1698
5.9	1695	1692	1689	1686	1684	1681	1678	1675	1672	1669
	0	1	2	3	4	5	6	7	8	9

(Reprinted from K. Nakanishi, *Infrared Absorption Spectroscopy—Practical*, Holden-Day, Inc., San Francisco, 1962.)

Wavelength(μ)	Wave number(cm^{-1})									
	0	1	2	3	4	5	6	7	8	9
6.0	1667	1664	1661	1658	1656	1653	1650	1647	1645	1642
6.1	1639	1637	1634	1631	1629	1626	1623	1621	1618	1616
6.2	1613	1610	1608	1605	1603	1600	1597	1595	1592	1590
6.3	1587	1585	1582	1580	1577	1575	1572	1570	1567	1565
6.4	1563	1560	1558	1555	1553	1550	1548	1546	1543	1541
6.5	1538	1536	1534	1531	1529	1527	1524	1522	1520	1517
6.6	1515	1513	1511	1508	1506	1504	1502	1499	1497	1495
6.7	1493	1490	1488	1486	1484	1481	1479	1477	1475	1473
6.8	1471	1468	1466	1464	1462	1460	1458	1456	1453	1451
6.9	1449	1447	1445	1443	1441	1439	1437	1435	1433	1431
7.0	1429	1427	1425	1422	1420	1418	1416	1414	1412	1410
7.1	1408	1406	1404	1403	1401	1399	1397	1395	1393	1391
7.2	1389	1387	1385	1383	1381	1379	1377	1376	1374	1372
7.3	1370	1368	1366	1364	1362	1361	1359	1357	1355	1353
7.4	1351	1350	1348	1346	1344	1342	1340	1339	1337	1335
7.5	1333	1332	1330	1328	1326	1325	1323	1321	1319	1318
7.6	1316	1314	1312	1311	1309	1307	1305	1304	1302	1300
7.7	1299	1297	1295	1294	1292	1290	1289	1287	1285	1284
7.8	1282	1280	1279	1277	1276	1274	1272	1271	1269	1267
7.9	1266	1264	1263	1261	1259	1258	1256	1255	1253	1252
8.0	1250	1248	1247	1245	1244	1242	1241	1239	1238	1236
8.1	1235	1233	1232	1230	1229	1227	1225	1224	1222	1221
8.2	1220	1218	1217	1215	1214	1212	1211	1209	1208	1206
8.3	1205	1203	1202	1200	1199	1198	1196	1195	1193	1192
8.4	1190	1189	1188	1186	1185	1183	1182	1181	1179	1178
8.5	1176	1175	1174	1172	1171	1170	1168	1167	1166	1164
8.6	1163	1161	1160	1159	1157	1156	1155	1153	1152	1151
8.7	1149	1148	1147	1145	1144	1143	1142	1140	1139	1138
8.8	1136	1135	1134	1133	1131	1130	1129	1127	1126	1125
8.9	1124	1122	1121	1120	1119	1117	1116	1115	1114	1112
9.0	1111	1110	1109	1107	1106	1105	1104	1103	1101	1100
9.1	1099	1098	1096	1095	1094	1093	1092	1091	1089	1088
9.2	1087	1086	1085	1083	1082	1081	1080	1079	1078	1076
9.3	1075	1074	1073	1072	1071	1070	1068	1067	1066	1065
9.4	1064	1063	1062	1060	1059	1058	1057	1056	1055	1054
9.5	1053	1052	1050	1049	1048	1047	1046	1045	1044	1043
9.6	1042	1041	1040	1038	1037	1036	1035	1034	1033	1032
9.7	1031	1030	1029	1028	1027	1026	1025	1024	1022	1021
9.8	1020	1019	1018	1017	1016	1015	1014	1013	1012	1011
9.9	1010	1009	1008	1007	1006	1005	1004	1003	1002	1001
	0	1	2	3	4	5	6	7	8	9

Wavelength(μ)	Wave number(cm^{-1})									
	0	1	2	3	4	5	6	7	8	9
10.0	1000.0	999.0	998.0	997.0	996.0	995.0	994.0	993.0	992.1	991.1
10.1	990.1	989.1	988.1	987.2	986.2	985.2	984.3	983.3	982.3	981.4
10.2	980.4	979.4	978.5	977.5	976.6	975.6	974.7	973.7	972.8	971.8
10.3	970.9	969.9	969.0	968.1	967.1	966.2	965.3	964.3	963.4	962.5
10.4	961.5	960.6	959.7	958.8	957.9	956.9	956.0	955.1	954.2	953.3
10.5	952.4	951.5	950.6	949.7	948.8	947.9	947.0	946.1	945.2	944.3
10.6	943.4	942.5	941.6	940.7	939.8	939.0	938.1	937.2	936.3	935.5
10.7	934.6	933.7	932.8	932.0	931.1	930.2	929.4	928.5	927.6	926.8
10.8	925.9	925.1	924.2	923.4	922.5	921.7	920.8	920.0	919.1	918.3
10.9	917.4	916.6	915.8	914.9	914.1	913.2	912.4	911.6	910.7	909.9
11.0	909.1	908.3	907.4	906.6	905.8	905.0	904.2	903.3	902.5	901.7
11.1	900.9	900.1	899.3	898.5	897.7	896.9	896.1	895.3	894.5	893.7
11.2	892.9	892.1	891.3	890.5	889.7	888.9	888.1	887.3	886.5	885.7
11.3	885.0	884.2	883.4	882.6	881.8	881.1	880.3	879.5	878.7	878.0
11.4	877.2	876.4	875.7	874.9	874.1	873.4	872.6	871.8	871.1	870.3
11.5	869.6	868.8	868.1	867.3	866.6	865.8	865.1	864.3	863.6	862.8
11.6	862.1	861.3	860.6	859.8	859.1	858.4	857.6	856.9	856.2	855.4
11.7	854.7	854.0	853.2	852.5	851.8	851.1	850.3	849.6	848.9	848.2
11.8	847.5	846.7	846.0	845.3	844.6	843.9	843.2	842.5	841.8	841.0
11.9	840.3	839.6	838.9	838.2	837.5	836.8	836.1	835.4	834.7	834.0
12.0	833.3	832.6	831.9	831.3	830.6	829.9	829.2	828.5	827.8	827.1
12.1	826.4	825.8	825.1	824.4	823.7	823.0	822.4	821.7	821.0	820.3
12.2	819.7	819.0	818.3	817.7	817.0	816.3	815.7	815.0	814.3	813.7
12.3	813.0	812.3	811.7	811.0	810.4	809.7	809.1	808.4	807.8	807.1
12.4	806.5	805.8	805.2	804.5	803.9	803.2	802.6	801.9	801.3	800.6
12.5	800.0	799.4	798.7	798.1	797.4	796.8	796.2	795.5	794.9	794.3
12.6	793.7	793.0	792.4	791.8	791.1	790.5	789.9	789.3	788.6	788.0
12.7	787.4	786.8	786.2	785.5	784.9	784.3	783.7	783.1	782.5	781.9
12.8	781.3	780.6	780.0	779.4	778.8	778.2	777.6	777.0	776.4	775.8
12.9	775.2	774.6	774.0	773.4	772.8	772.2	771.6	771.0	770.4	769.8
13.0	769.2	768.6	768.0	767.5	766.9	766.3	765.7	765.1	764.5	763.9
13.1	763.4	762.8	762.2	761.6	761.0	760.5	759.9	759.3	758.7	758.2
13.2	757.6	757.0	756.4	755.9	755.3	754.7	754.1	753.6	753.0	752.4
13.3	751.9	751.3	750.8	750.2	749.6	749.1	748.5	747.9	747.4	746.8
13.4	746.3	745.7	745.2	744.6	744.0	743.5	742.9	742.4	741.8	741.3
13.5	740.7	740.2	739.6	739.1	738.6	738.0	737.5	736.9	736.4	735.8
13.6	735.3	734.8	734.2	733.7	733.1	732.6	732.1	731.5	731.0	730.5
13.7	729.9	729.4	728.9	728.3	727.8	727.3	726.7	726.2	725.7	725.2
13.8	724.6	724.1	723.6	723.1	722.5	722.0	721.5	721.0	720.5	719.9
13.9	719.4	718.9	718.4	717.9	717.4	716.8	716.3	715.8	715.3	714.8
14.0	714.3	713.8	713.3	712.8	712.3	711.7	711.2	710.7	710.2	709.7
14.1	709.2	708.7	708.2	707.7	707.2	706.7	706.2	705.7	705.2	704.7
14.2	704.2	703.7	703.2	702.7	702.2	701.8	701.3	700.8	700.3	699.8
14.3	699.3	698.8	698.3	697.8	697.4	696.9	696.4	695.9	695.4	694.9
14.4	694.4	694.0	693.5	693.0	692.5	692.0	691.6	691.1	690.6	690.1
14.5	689.7	689.2	688.7	688.2	687.8	687.3	686.8	686.3	685.9	685.4
14.6	684.9	684.5	684.0	683.5	683.1	682.6	682.1	681.7	681.2	680.7
14.7	680.3	679.8	679.3	678.9	678.4	678.0	677.5	677.0	676.6	676.1
14.8	675.7	675.2	674.8	674.3	673.9	673.4	672.9	672.5	672.0	671.6
14.9	671.1	670.7	670.2	669.8	669.3	668.9	668.4	668.0	667.6	667.1
	0	1	2	3	4	5	6	7	8	9

Wavelength(μ)	Wave number(cm⁻¹)									
	0	**1**	**2**	**3**	**4**	**5**	**6**	**7**	**8**	**9**
15.0	666.7	666.2	665.8	665.3	664.9	664.5	664.0	663.6	663.1	662.7
15.1	662.3	661.8	661.4	660.9	660.5	660.1	659.6	659.2	658.8	658.3
15.2	657.9	657.5	657.0	656.6	656.2	655.7	655.3	654.9	654.5	654.0
15.3	653.6	653.2	652.7	652.3	651.9	651.5	651.0	650.6	650.2	649.8
15.4	649.4	648.9	648.5	648.1	647.7	647.2	646.8	646.4	646.0	645.6
15.5	645.2	644.7	644.3	643.9	643.5	643.1	642.7	642.3	641.8	641.4
15.6	641.0	640.6	640.2	639.8	639.4	639.0	638.6	638.2	637.8	637.3
15.7	636.9	636.5	636.1	635.7	635.3	634.9	634.5	634.1	633.7	633.3
15.8	632.9	632.5	632.1	631.7	631.3	630.9	630.5	630.1	629.7	629.3
15.9	628.9	628.5	628.1	627.7	627.4	627.0	626.6	626.2	625.8	625.4
16.0	625.0	624.6	624.2	623.8	623.4	623.1	622.7	622.3	621.9	621.5
16.1	621.1	620.7	620.3	620.0	619.6	619.2	618.8	618.4	618.0	617.7
16.2	617.3	616.9	616.5	616.1	615.8	615.4	615.0	614.6	614.3	613.9
16.3	613.5	613.1	612.7	612.4	612.0	611.6	611.2	610.9	610.5	610.1
16.4	609.8	609.4	609.0	608.6	608.3	607.9	607.5	607.2	606.8	606.4
16.5	606.1	605.7	605.3	605.0	604.6	604.2	603.9	603.5	603.1	602.8
16.6	602.4	602.0	601.7	601.3	601.0	600.6	600.2	599.9	599.5	599.2
16.7	598.8	598.4	598.1	597.7	597.4	597.0	596.7	596.3	595.9	595.6
16.8	595.2	594.9	594.5	594.2	593.8	593.5	593.1	592.8	592.4	592.1
16.9	591.7	591.4	591.0	590.7	590.3	590.0	589.6	589.3	588.9	588.6
17.0	588.2	587.9	587.5	587.2	586.9	586.5	586.2	585.8	585.5	585.1
17.1	584.8	584.5	584.1	583.8	583.4	583.1	582.8	582.4	582.1	581.7
17.2	581.4	581.1	580.7	580.4	580.0	579.7	579.4	579.0	578.7	578.4
17.3	578.0	577.7	577.4	577.0	576.7	576.4	576.0	575.7	575.4	575.0
17.4	574.7	574.4	574.1	573.7	573.4	573.1	572.7	572.4	572.1	571.8
17.5	571.4	571.1	570.8	570.5	570.1	569.8	569.5	569.2	568.8	568.5
17.6	568.2	567.9	567.5	567.2	566.9	566.6	566.3	565.9	565.6	565.3
17.7	565.0	564.7	564.3	564.0	563.7	563.4	563.1	562.7	562.4	562.1
17.8	561.8	561.5	561.2	560.9	560.5	560.2	559.9	559.6	559.3	559.0
17.9	558.7	558.3	558.0	557.7	557.4	557.1	556.8	556.5	556.2	555.9
18.0	555.6	555.2	554.9	554.6	554.3	554.0	553.7	553.4	553.1	552.8
18.1	552.5	552.2	551.9	551.6	551.3	551.0	550.7	550.4	550.1	549.8
18.2	549.5	549.1	548.8	548.5	548.2	547.9	547.6	547.3	547.0	546.7
18.3	546.4	546.1	545.9	545.6	545.3	545.0	544.7	544.4	544.1	543.8
18.4	543.5	543.2	542.9	542.6	542.3	542.0	541.7	541.4	541.1	540.8
18.5	540.5	540.2	540.0	539.7	539.4	539.1	538.8	538.5	538.2	537.9
18.6	537.6	537.3	537.1	536.8	536.5	536.2	535.9	535.6	535.3	535.0
18.7	534.8	534.5	534.2	533.9	533.6	533.3	533.0	532.8	532.5	532.2
18.8	531.9	531.6	531.3	531.1	530.8	530.5	530.2	529.9	529.7	529.4
18.9	529.1	528.8	528.5	528.3	528.0	527.7	527.4	527.1	526.9	526.6
19.0	526.3	526.0	525.8	525.5	525.2	524.9	524.7	524.4	524.1	523.8
19.1	523.6	523.3	523.0	522.7	522.5	522.2	521.9	521.6	521.4	521.1
19.2	520.8	520.6	520.3	520.0	519.8	519.5	519.2	518.9	518.7	518.4
19.3	518.1	517.9	517.6	517.3	517.1	516.8	516.5	516.3	516.0	515.7
19.4	515.5	515.2	514.9	514.7	514.4	514.1	513.9	513.6	513.3	513.1
19.5	512.8	512.6	512.3	512.0	511.8	511.5	511.2	511.0	510.7	510.5
19.6	510.2	509.9	509.7	509.4	509.2	508.9	508.6	508.4	508.1	507.9
19.7	507.6	507.4	507.1	506.8	506.6	506.3	506.1	505.8	505.6	505.3
19.8	505.1	504.8	504.5	504.3	504.0	503.8	503.5	503.3	503.0	502.8
19.9	502.5	502.3	502.0	501.8	501.5	501.3	501.0	500.8	500.5	500.3
	0	**1**	**2**	**3**	**4**	**5**	**6**	**7**	**8**	**9**

Wavelength(μ)	Wave number(cm^{-1})									
	0	1	2	3	4	5	6	7	8	9
20.0	500.0	499.8	499.5	499.3	499.0	498.8	498.5	498.3	498.0	497.8
20.1	497.5	497.3	497.0	496.8	496.5	496.3	496.0	495.8	495.5	495.3
20.2	495.0	494.8	494.6	494.3	494.1	493.8	493.6	493.3	493.1	492.9
20.3	492.6	492.4	492.1	491.9	491.6	491.4	491.2	490.9	490.7	490.4
20.4	490.2	490.0	489.7	489.5	489.2	489.0	488.8	488.5	488.3	488.0
20.5	487.8	487.6	487.3	487.1	486.9	486.6	486.4	486.1	485.9	485.7
20.6	485.4	485.2	485.0	484.7	484.5	484.3	484.0	483.8	483.6	483.3
20.7	483.1	482.9	482.6	482.4	482.2	481.9	481.7	481.5	481.2	481.0
20.8	480.8	480.5	480.3	480.1	479.8	479.6	479.4	479.2	478.9	478.7
20.9	478.5	478.2	478.0	477.8	477.6	477.3	477.1	476.9	476.6	476.4
21.0	476.2	476.0	475.7	475.5	475.3	475.1	474.8	474.6	474.4	474.2
21.1	473.9	473.7	473.5	473.3	473.0	472.8	472.6	472.4	472.1	471.9
21.2	471.7	471.5	471.3	471.0	470.8	470.6	470.4	470.1	469.9	469.7
21.3	469.5	469.3	469.0	468.8	468.6	468.4	468.2	467.9	467.7	467.5
21.4	467.3	467.1	466.9	466.6	466.4	466.2	466.0	465.8	465.5	465.3
21.5	465.1	464.9	464.7	464.5	464.3	464.0	463.8	463.6	463.4	463.2
21.6	463.0	462.7	462.5	462.3	462.1	461.9	461.7	461.5	461.3	461.0
21.7	460.8	460.6	460.4	460.2	460.0	459.8	459.6	459.3	459.1	458.9
21.8	458.7	458.5	458.3	458.1	457.9	457.7	457.5	457.2	457.0	456.8
21.9	456.6	456.4	456.2	456.0	455.8	455.6	455.4	455.2	455.0	454.8
22.0	454.5	454.3	454.1	453.9	453.7	453.5	453.3	453.1	452.9	452.7
22.1	452.5	452.3	452.1	451.9	451.7	451.5	451.3	451.1	450.9	450.7
22.2	450.5	450.2	450.0	449.8	449.6	449.4	449.2	449.0	448.8	448.6
22.3	448.4	448.2	448.0	447.8	447.6	447.4	447.2	447.0	446.8	446.6
22.4	446.4	446.2	446.0	445.8	445.6	445.4	445.2	445.0	444.8	444.6
22.5	444.4	444.2	444.0	443.9	443.7	443.5	443.3	443.1	442.9	442.7
22.6	442.5	442.3	442.1	441.9	441.7	441.5	441.3	441.1	440.9	440.7
22.7	440.5	440.3	440.1	439.9	439.8	439.6	439.4	439.2	439.0	438.8
22.8	438.6	438.4	438.2	438.0	437.8	437.6	437.4	437.3	437.1	436.9
22.9	436.7	436.5	436.3	436.1	435.9	435.7	435.5	435.4	435.2	435.0
23.0	434.8	434.6	434.4	434.2	434.0	433.8	433.7	433.5	433.3	433.1
23.1	432.9	432.7	432.5	432.3	432.2	432.0	431.8	431.6	431.4	431.2
23.2	431.0	430.8	430.7	430.5	430.3	430.1	429.9	429.7	429.6	429.4
23.3	429.2	429.0	428.8	428.6	428.4	428.3	428.1	427.9	427.7	427.5
23.4	427.4	427.2	427.0	426.8	426.6	426.4	426.3	426.1	425.9	425.7
23.5	425.5	425.4	425.2	425.0	424.8	424.6	424.4	424.3	424.1	423.9
23.6	423.7	423.5	423.4	423.2	423.0	422.8	422.7	422.5	422.3	422.1
23.7	421.9	421.8	421.6	421.4	421.2	421.1	420.9	420.7	420.5	420.3
23.8	420.2	420.0	419.8	419.6	419.5	419.3	419.1	418.9	418.8	418.6
23.9	418.4	418.2	418.1	417.9	417.7	417.5	417.4	417.2	417.0	416.8
24.0	416.7	416.5	416.3	416.1	416.0	415.8	415.6	415.5	415.3	415.1
24.1	414.9	414.8	414.6	414.4	414.3	414.1	413.9	413.7	413.6	413.4
24.2	413.2	413.1	412.9	412.7	412.5	412.4	412.2	412.0	411.9	411.7
24.3	411.5	411.4	411.2	411.0	410.8	410.7	410.5	410.3	410.2	410.0
24.4	409.8	409.7	409.5	409.3	409.2	409.0	408.8	408.7	408.5	408.3
24.5	408.2	408.0	407.8	407.7	407.5	407.3	407.2	407.0	406.8	406.7
24.6	406.5	406.3	406.2	406.0	405.8	405.7	405.5	405.4	405.2	405.0
24.7	404.9	404.7	404.5	404.4	404.2	404.0	403.9	403.7	403.6	403.4
24.8	403.2	403.1	402.9	402.7	402.6	402.4	402.3	402.1	401.9	401.8
24.9	401.6	401.4	401.3	401.1	401.0	400.8	400.6	400.5	400.3	400.2
	0	1	2	3	4	5	6	7	8	9

INDEX